HAZARDOUS AND INDUSTRIAL WASTE MANAGEMENT AND TESTING: THIRD SYMPOSIUM

A symposium
sponsored by
ASTM Committee D-34
on Waste Disposal
Philadelphia, Pa., 7–10 March 1983

ASTM SPECIAL TECHNICAL PUBLICATION 851
Larry P. Jackson, Western Research Institute,
Alan R. Rohlik, SOHIO, and
Richard A. Conway, Union Carbide Corp.,
editors

ASTM Publication Code Number (PCN)
04-851000-16

1916 Race Street, Philadelphia, Pa. 19103

Library of Congress Cataloging in Publication Data

Hazardous and industrial waste management and testing.

(ASTM special technical publication ; 851)
"Papers presented at the International Symposium on Industrial and Hazardous Solid Wastes"—Foreword.
"ASTM publication code number (PCN) 04-851000-16."
Includes bibliographies and index.
1. Hazardous wastes—Congresses. 2. Factory and trade waste—Congresses. 3. Hazardous wastes—Testing—Congresses. 4. Factory and trade waste—Testing—Congresses. I. Jackson, Larry P. II. Rohlik, Alan R. III. Conway, Richard A. IV. ASTM Committee D-34 on Waste Disposal. V. International Symposium on Industrial and Hazardous Solid Wastes (3rd : 1983 : Philadelphia, Pa.) VI. Series.
TD811.5.H382 1984 628.4'456'0287 84-70137
ISBN 0-8031-0405-7

Library of Congress Catalog Card Number: 84-70137

NOTE

The Society is not responsible, as a body, for the statements and opinions advanced in this publication.

Printed in Baltimore, Md. (b)
December 1984

Foreword

This publication, *Hazardous and Industrial Waste Management and Testing: Third Symposium*, contains papers presented at the International Symposium on Industrial and Hazardous Solid Wastes, which was held 7–10 March 1983 in Philadelphia, Pa. This symposium was the third in a series of annual symposia on hazardous waste sponsored by ASTM Committee D-34 on Waste Disposal and was also the first international conference in this series.

Chairing the event were Larry P. Jackson, Western Research Institute, who also served as one of the editors of this publication, and Carlo Merli, Institute of Applied and Industrial Chemistry, University of Rome. In addition, Alan R. Rohlik, Standard Oil Co. (Ohio), and Richard A. Conway, Union Carbide Corp., also served as editors of this publication.

Related ASTM Publications

Hazardous and Industrial Solid Waste Testing: Second Symposium, STP 805 (1983), 04-805000-16

Hazardous Solid Waste Testing: First Conference, STP 760 (1981), 04-760000-16

Aquatic Toxicology and Hazard Assessment: Sixth Symposium, STP 802 (1983), 04-802000-16

Aquatic Toxicology and Hazard Assessment (Fifth Conference), STP 766 (1982), 04-766000-16

Aquatic Toxicology and Hazard Assessment (Fourth Conference), STP 737 (1981), 04-737000-16

Permeability and Groundwater Contaminant Transport, STP 746 (1981), 04-746000-38

A Note of Appreciation to Reviewers

The quality of the papers that appear in this publication reflects not only the obvious efforts of the authors but also the unheralded, though essential, work of the reviewers. On behalf of ASTM we acknowledge with appreciation their dedication to high professional standards and their sacrifice of time and effort.

ASTM Committee on Publications

ASTM Editorial Staff

Contents

Introduction

This is the third ASTM Special Technical Publication (STP) resulting from the annual symposium held by ASTM Committee D-34 on Waste Disposal. The first two symposia have been published as *ASTM STP 760* and *STP 805.* The symposium on which this publication is based was something of a departure from the first two in that it contained many papers by members of the international scientific community interested in wastes and waste disposal practices. The international contributions are the result of an association begun in 1980 with members of the European community who sought the participation of Committee D-34 in an international conference held in Rome in 1981. The results of this conference were so fruitful in the interest generated among the attendees, that a decision was made to make this international meeting a biennial event, alternating between the United States and various international sites. The contributions making up this volume come from the second international conference (the third symposium in this series), which was titled the International Symposium on Industrial and Hazardous Solid Wastes. The third biennial conference (which will be the fifth symposium in this series) will be held in 1985 in Alexandria, Egypt.

As with any ASTM activity, the principal objective of the current symposium was to generate interest in the activities of the ASTM sponsoring committee and to provide a forum for information exchange that could lead to future standards. In addition, this symposium promoted the circulation of information across international boundaries and gave participants a wider view of waste management problems. The papers selected fell into four general subject areas:

(*a*) analysis of wastes and waste disposal sites,
(*b*) amelioration of wastes in the disposal environment,
(*c*) waste as resource, and
(*d*) national perspectives in waste management.

These topics present a broader view of waste disposal than has been presented in this symposium series in the past. The first two gatherings focused on the immediate need, generated by the regulatory climate in this country, for standards related to characterization and testing of wastes. While this need still exists and is as critical as ever, this symposium stepped away a little and tried to present a look at other issues relevant to safe, efficient waste disposal.

The changes that occur in a waste after disposal are as important as the character of the waste at the time of disposal. While the latter issue dictates the classification of the waste as hazardous or nonhazardous for the purpose of disposal, the issue of waste behavior is the heart of risk assessment over the life of the disposed material and, hence, defines its long-term impact on man and the environment. As the understanding of waste characteristics grows and our attitude towards waste management matures beyond the immediate question of hazardous versus nonhazardous designations, the need for standards relating to waste behavior in the disposal environment will become a dominant need. This area of research will expand into the broader activity of risk assessment, which is a very important area of interest to various ASTM committees.

The use of wastes for their resource potential can serve three purposes (1) alteration of the character of the waste, (2) reduction of its volume, and (3) conservation of natural resources. This is definitely a major direction for waste disposal in the future and, as such, is of interest to the members of ASTM Committee D-34 and the general public. The papers on the various national perspectives on waste management show different thinking on the subject: these papers point out new problems and thus, serve to keep our thinking fresh and encourage creative problem solving in the future.

Larry P. Jackson

Western Research Institute, Laramie, Wyo. 82071; symposium chairman and editor.

Alan R. Rohlik

Standard Oil Co. (Ohio), Cleveland, Ohio 44128; editor.

Richard A. Conway

Union Carbide Corp., South Charleston, W.Va. 25303; editor.

Analysis of Wastes and Waste Disposal Sites

Amelia J. Janisz[1] *and W. Scott Butterfield*[2]

Biological Sampling Methods and Effects of Exposure to Municipal and Chemical Landfill Leachate on Aquatic Organisms

REFERENCE: Janisz, A. J. and Butterfield, W. S., **"Biological Sampling Methods and Effects of Exposure to Municipal and Chemical Landfill Leachate on Aquatic Organisms,"** *Hazardous and Industrial Waste Management and Testing: Third Symposium, ASTM STP 851*, Larry P. Jackson, Alan R. Rohlik, and Richard A. Conway, Eds., American Society for Testing and Materials, Philadelphia, 1984, pp. 5–23.

ABSTRACT: Extensive biological sampling on five abandoned hazardous waste sites in New York, New Jersey, and Puerto Rico was undertaken during 1981 and 1982 to determine the impact of priority pollutants on aquatic fauna and, potentially, on human health. The selection criteria for sites, sampling equipment, problems in personnel protection, and sample handling procedures are presented.

The effects of the hazardous waste sites were assessed using a wide range of fish and invertebrate species. Tissue specimens from eleven vertebrate and eight invertebrate species were analyzed. Forty samples of these tissue specimens were analyzed for all inorganic priority pollutant parameters; an additional 35 samples were analyzed for organic priority pollutants or an appropriate subset of them. High concentrations of polychlorinated biphenyls (PCBs) were found in aquatic organisms exposed to chemical landfill leachate; the results of the tissue analyses at other sites were negative.

KEY WORDS: fish, landfill leachate, hazardous waste, polychlorinated biphenyls (PCBs), metal, toxic interactions, risk analysis, environmental surveys, aquatic biology, hazardous wastes, industrial wastes

Between 1980 and 1982, Fred C. Hart Associates, Inc., acting as a field investigation team (FIT) for the U.S. Environmental Protection Agency (EPA) Region II, investigated over 200 uncontrolled hazardous waste sites. Activities at these sites included extensive sampling programs to determine the extent of contamination. Standard approaches to characterizing the relative toxicity of

[1] Project manager, Fred C. Hart Associates, Inc., New York, N.Y. 10036.
[2] Senior aquatic ecologist, Roy F. Weston, Inc., Raritan Center, Edison, N.J. 08837.

a site included sampling of the soil and sediment, surface waters, groundwater, leachate, drums, oil lagoons, and methane gas vents; drilling monitor wells, and monitoring ambient air. Sampling the aquatic fauna was part of an integrated approach to site investigation when surface waters were suspected to be contaminated with inorganic or organic compounds with high bioconcentration potentials.

In this paper, the authors discuss the sampling procedures and the effects of chemical landfill and municipal landfill leachate on aquatic fauna and present recommendations for future studies. The names and specific locations and identifying characteristics of sites mentioned are not identified in this paper.

The primary objective of the abandoned hazardous waste site investigations was to determine hazards to human health. Biological sampling programs, therefore, focused on tissue analyses for inorganic and organic materials that may be concentrated in aquatic organisms to levels toxic to humans. Traditional methodologies for determining impacts on aquatic ecosystems, such as diversity index comparisons [*1–3*] and bioassays [*3*], were used to complement the results of tissue analyses. In some situations these procedures provided evidence of impacts when tissue analyses indicated negative results.

Three criteria were considered in determining whether a site was a candidate for a biological sampling program. The first criterion was location. The site had to be located in or adjacent to a body of water or sufficiently close to permit an influx of contaminants through drainage or subsurface discharge. The second criterion was a documented history of uncontrolled disposal of hazardous materials at a site, particularly materials with the potential for bioconcentration in aquatic organisms to toxic levels. The bioconcentration potentials were determined from the literature [*4–8*]. The decisions to include biological sampling programs at three of the sites were based on allegations by knowledgeable individuals within local environmental agencies.

The final criterion was the regular utilization of the adjacent bodies of water by sport or sustenance fisherman. Thus, the presence of contaminated aquatic organisms could be linked to potential human impact. This criterion was a major factor at sites in Puerto Rico, where potentially contaminated bodies of water are utilized by local residents as major sources of food.

Under ideal conditions, detailed water and sediment analyses are conducted before the decision is made to sample biological organisms. Samples from the surfaces of adjacent bodies of water and leachate streams, as well as results of sediment samples at critical locations, should be considered. The contaminants aquatic organisms tend to accumulate exhibit low water solubilities [*7*], and the organic fractions, if present, tend to accumulate in the sediments [*7,9–11*].

Biological sampling programs were conducted at five sites (Sites A through E) between 1981 and 1982. Site C was a freshwater stream adjacent to a municipal landfill; Site D consisted of two lagoons which communicated with an allegedly contaminated stream running between both lagoons; Site E was a concrete discharge canal running from the plant outfall to a saltwater bay. Sites A and B are described in greater detail in this paper.

Sites A and B both met the three criteria necessary for a decision to conduct biological sampling. Both sites were adjacent to bodies of water—one was close to a tidal stream and the other to a freshwater creek. Soil samples collected by the state at one site showed contamination; the other site was observed to have contaminated wastes flowing directly into a tidal creek. Both sites had a history of uncontrolled hazardous waste disposal; site contamination was documented in both cases by sediment samples and additionally at one site by oil samples. Finally, the waters adjacent to both sites were used by local fishermen for sustenance or sport fishing. Sites C, D, and E were adjacent to bodies of water that were used by local fishermen; however, hazardous waste disposal at these sites was only alleged.

Procedures

The biological sampling programs were designed to be site specific. Each program considered site accessibility, species abundance, sampling equipment, and potential health hazards to sampling personnel. Station location, personnel protection, sample equipment, sampling procedures, sample handling, specimen preparation, and analytical techniques are discussed in the following sections.

Station Location

The sampling stations were tentatively identified after preliminary site investigations. These stations were initially selected to permit statistical comparisons between aquatic fauna from contaminated and uncontaminated areas. Where possible, sampling stations were located upstream, adjacent to, and downstream of known or suspected discharges, permitting upstream-downstream comparisons. At most sites, the distance between contaminated and uncontaminated stations was greater than the home range of the species collected. This precluded the exposure of the uncontaminated or "clean" organisms to a potentially contaminated area. The station locations, however, had to be modified during sampling operations because of difficulty in obtaining access to the stations. The only incidence of possible cross-contamination occurred at Site B when only migratory species of fish were collected, so that all the fish sampled were potentially exposed to contaminated sediments.

At Sites A and B, the station location was selected as follows. A reconnaissance of Site A in late fall determined the accessibility of the creeks. Seven sampling stations were selected initially (Fig. 1). Stations 1 and 2 were located at the head and near the mouth of Creek B, respectively. Three stations were initially established on Creek A: 3A, 4A, and 5. Station 3A was thought to be located at the discharge point of the lagoon. During the actual sampling, Sta-

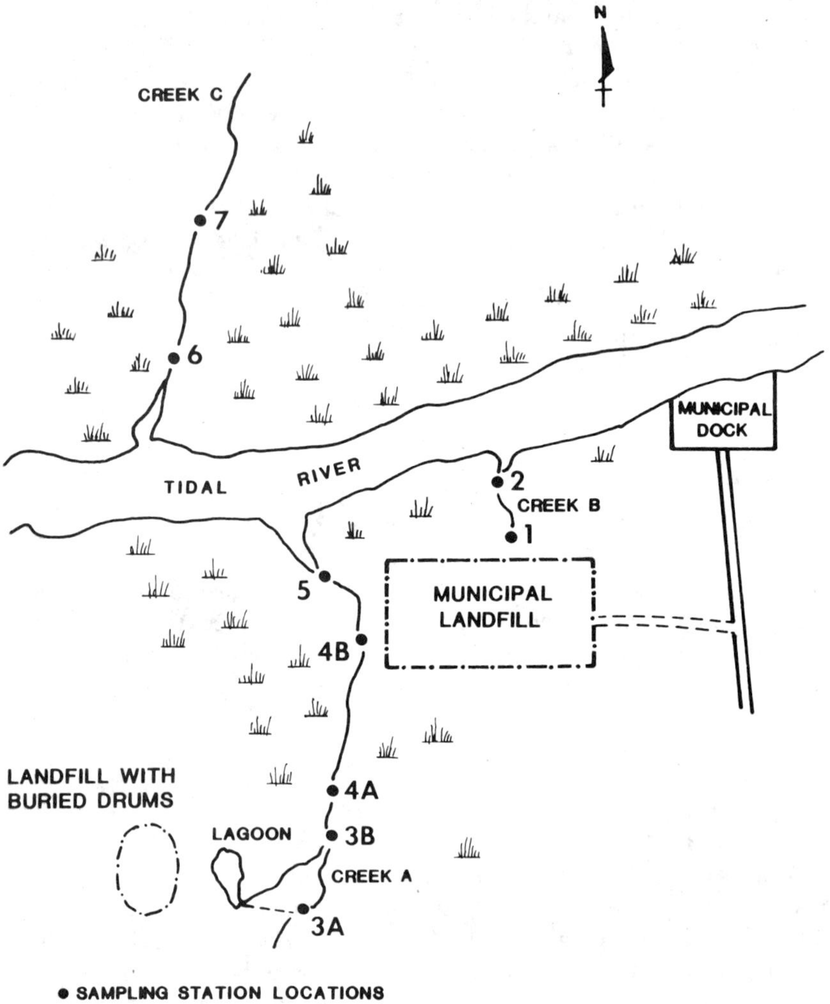

FIG. 1—*Site A—map and sampling station locations.*

tion 3A was determined to be a minor discharge point. The primary entry point of the lagoon leachate was at Station 3B. Station 4B was added during the invertebrate sampling program to provide a point approximately midway between the leachate discharge and the mouth of Creek A. Stations 6 and 7 were located on control Creek C, near its mouth and upstream, respectively. At Site B, three discrete sampling stations were selected within a 1500-m section of Big Creek (Fig. 2). The stations were located upstream, adjacent to, and downstream of the site.

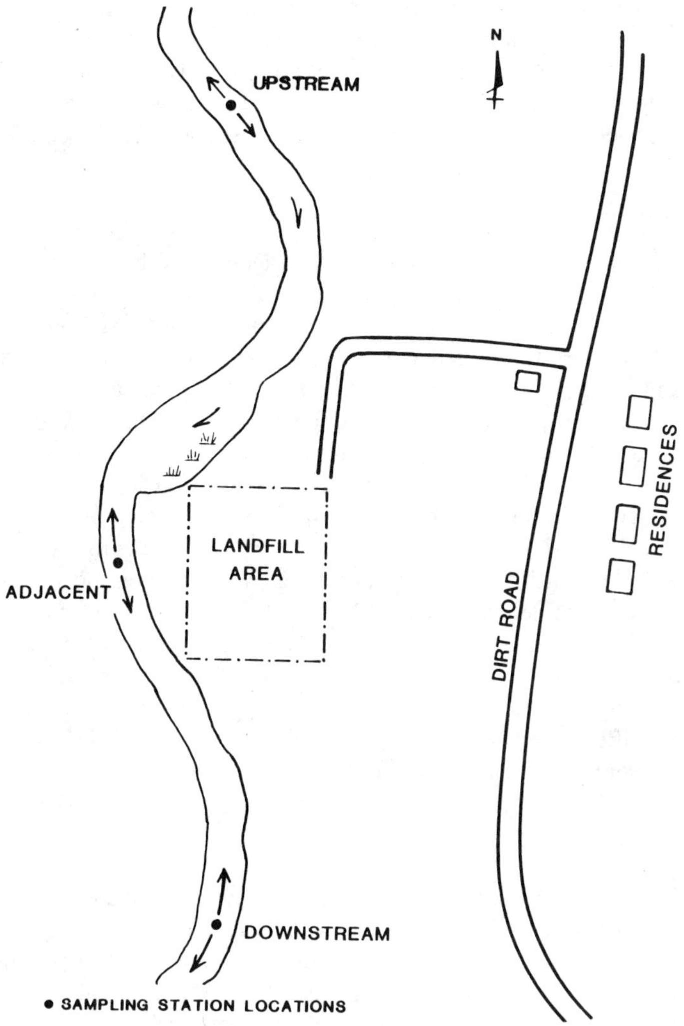

FIG. 2—*Site B—map and sampling station locations.*

Personnel Protection

Special consideration was given to protecting personnel from chemical hazards. Cartridge respirators, waders, and protective gloves were worn during the sampling operations in severely contaminated areas to minimize exposure. Two sites in Puerto Rico were located in freshwater areas of the island where a parasitic trematode, *Schistosoma masoni*, occurs. This parasite is transmitted to humans, its final host, through contact with water. Contact with the surface water was minimized. Waders were worn at all times as well as elbow-length waterproof gloves. Rain gear and hard hats with face shields were worn occasionally. When contact did occur, the areas of the body were flushed with rubbing alcohol and abraded with rough cloths. To kill any parasites, aquatic organisms collected at these sites were frozen before they were processed.

Equipment

The selection of sampling equipment was determined by the site characteristics, such as water depth, stream versus pond, water conductivity, stream bed substrate, and expected organisms. A portable backpack electrofishing unit was used in freshwater streams where the bottoms were firm and the water depths shallow enough to permit wading. Electrofishing at all freshwater sampling locations was performed on an equal "unit-effort" basis. Approximately 60 m of stream at each sampling location was shocked in order to compare the sampling locations. Gill, seine, or throw nets were used when large open-water habitats such as ponds or lagoons were sampled. Sampling narrow, soft-bottomed, brackish tidal creeks required setting minnow traps to collect residential small fish and using a Petite ponar dredge to collect benthic invertebrates. Burrowing and sessile organisms such as crabs and oysters were collected by hand and, in one instance, specimens were purchased from local fishermen. Sampling locations were chosen in order to minimize differences due to habitat. For example, in the freshwater streams, approximately equal numbers of pools and riffles were shocked at each location. Habitat differences were not a significant source of variation between samples.

Sampling Procedures

An attempt was made to collect a variety of aquatic organisms from bodies of water near each site. Factors considered in the selection of species included (1) the metabolic behavior of the organism with respect to the contaminant of concern [*12-14*], (2) whether the organism had limited mobility, and, thus, greater potential for exposure, for example, residential fish or benthic invertebrates, (3) whether the organism was sought by local fishermen, (4) the lipid content of the organism, (5) the abundance of the species, and (6) representation of the various trophic levels. Additional modifications to the sampling programs were made during implementation because all the desired species either were not caught or were not caught in sufficient numbers for laboratory analysis. Two or more invertebrate species from a single sampling station were combined to produce adequate specimen weights on several occasions. Only one sampling round was conducted at each site. At Sites A and B, the sampling procedures were as follows.

Samples of vertebrates and invertebrates were taken from the tidal creeks adjacent to Site A. Galvanized steel minnow traps were selected for fish collection based on the species captured during short hauls taken during the preliminary reconnaissance. The traps were baited with bread and raw chicken and set in the deepest area of the channel at Stations 1, 2, 3A, 4A, 5, 6, and 7 (Fig. 1). The traps were retrieved after approximately 24 h.

Fiddler crabs, *Uca minax*, were dug from the creek banks, where they were hibernating, at Stations 1, 2, 3B, 4B, 5, 6, and 7. Dip nets were used to capture shrimp, *Palaemonetes* sp., wherever sufficient weights for analysis of crabs were not collected. A few mud crabs, *Rhithropanepeus harrisi*, and amphipods, *Gammarus* sp., were also included in many of the invertebrate samples.

Only vertebrate samples were collected at Site B. Three discrete sampling stations were selected within a 1500-m section of Big Creek (Fig. 2). The stations were located upstream, adjacent to, and downstream of the site. A portable backpack electrofishing unit was used to stun the fish since this permitted efficient sampling of pooling areas adjacent to stumps and undercut banks. Sweeps of approximately 60-m stream sections were made at each station.

Sample Handling and Processing

An important aspect of our biological sampling program was handling and processing tissue specimens for analyses. Procedures used for handling and processing tissue specimens were developed by the EPA, New York Department of Environmental Conservation, and New Jersey Department of Environmental Protection [*15–17*].

These procedures specify techniques to (1) prevent cross-contamination of the tissue specimens to be analyzed, particularly polychlorinated biphenyls (PCBs), and (2) preserve tissues until analyses are conducted. These procedures are described in the following paragraphs.

Samples from each station and of each species collected were placed in separate laboratory-cleaned jars or unused plastic bags. The specimens were iced for a maximum of 8 h or frozen for a maximum of two weeks until they could be processed. Invertebrate specimens were thoroughly rinsed with ambient water before icing to remove adhering sediments. Dry ice was utilized for quick freezing when available. The samples taken in Puerto Rico presented unique problems because of high temperatures, difficulties in procuring ice and dry ice in remote areas, and parasite contamination. The collected organisms were packed in ice in coolers and shipped overnight to a laboratory facility when they were frozen at $-10°C$. The frozen organisms were later thawed for processing.

Tissue specimens for laboratory analyses were prepared under both field and laboratory conditions. All the equipment that came in contact with the specimens was detergent washed and solvent rinsed with methylene chloride. This cleaning procedure was repeated for samples from each station and for specimens of each species.

The size of fish collected determined the processing technique. Larger fishes were scaled and filleted while smaller fishes were scaled, deheaded, detailed, and eviserated. Muscle tissue was preferentially used since our study concentrated on potential human health hazards. The shells of oysters were removed while whole body samples of crabs and shrimp were used. The prepared tissue specimen from each sample was either minced or homogenized in a stainless steel blender and placed in detergent-washed, solvent-rinsed glass jars with Teflon-lined lids. The prepared specimens were frozen, packed in ice or dry ice in coolers, and shipped overnight to contract laboratories for analysis.

Field sample processing presented logistical problems and was conducted only at one site, where a limited sampling campaign was planned. Field processing required establishing a temporary work area with such equipment as scales, measuring boards, glass trays, and knives. The work area had to provide a clean environment to prevent contamination of specimens during processing. When this was conducted on site, consideration was given to possible adverse weather conditions, such as extremely high or low temperatures, precipitation, and winds which carried contaminants. Provisions were made for cleaning equipment between the preparation of specimens, discarding used solvent, and disposing of unused portions of the organisms.

Analytical Procedures

The specimens were analyzed according to the EPA Interim Methods for the Sampling and Analysis of Priority Pollutants in Sediments and Fish Tissue [*15*] for either a total priority pollutant scan or for PCBs, chlorinated pesticides, or metals. The lipid content was determined using the ether extraction method [*19*]. More limited analyses were done where the specific compounds, for example, PCBs or pesticides, were known or alleged to be discharging from the site. Recoveries of extractable priority pollutants in fish tissue were generally quite low, stemming from the use of hexane/aqueous acetonitrile partitioning. Low levels of benzene and toluene were detected in some specimens but not in the blanks. This has been observed in other kinds of fish or shellfish tissue.

Results and Discussion

Tissue specimens from 11 vertebrate and 8 invertebrate species were analyzed. Forty samples of these tissue specimens were analyzed for all inorganic priority pollutant parameters; an additional 35 samples were analyzed for organic priority pollutants or an appropriate subset of them. Table 1 shows the species name, sampling location, and total numbers of organisms collected at each site. A total of 588 vertebrates and over 137 invertebrates were collected from all the sites. The effects of chemical landfill leachate on aquatic organisms were most clearly seen at Site A.

The Site A landfill was located adjacent to a tidal wetlands near a heavily populated area. During the 1970s several thousand drums were buried at the site. A lagoon formed west of the landfill and was observed to be flowing into a tidal creek, Creek A, at a point north of the landfill (Fig. 1). The samples collected from 1978 to 1982 by the EPA from leachate seeps surrounding the landfill identified numerous hazardous organics and levels of PCBs ranging from 1 to over 6500 ppm.

The study design was based on "after-impact" data. Analyses of Site A's impact were complicated by the location of an adjacent large municipal sanitary landfill. Creek B, abutting only the municipal sanitary landfill, was selected to assess separately any effect of the municipal landfill on aquatic organisms.

The vertebrates trapped in Creeks A, B, and C were all the common mummichog with the exception of two eels. An equal catch per unit effort was expended at each sampling location (see the previous section on sampling procedures for more details). Table 2 shows the results of the tissue specimen analyses for PCBs. The PCB levels shown in Table 2 are total PCBs because an exact match of the fingerprints of the different Aroclors identified was not attempted by the laboratory. The Aroclors included in the total PCBs reported were Aroclor 1248, 1254, and 1260. All the aquatic organisms collected at Stations 3 through 5 in Creek A had levels of PCBs significantly above the values found in Creeks B and C (Table 3).

Elevated PCB levels were also present in the channel sediment of Creek A. Metal analyses on tissue specimens revealed elevated aluminum levels in both fish and invertebrates in Creek A. Cadmium was detected at higher levels in the invertebrate tissue specimens from Stations 4B and 5 (Creek A); mercury was found in only one specimen from Creek A. The PCB levels found in the aquatic organisms, 5.5 to 8.6 ppm, were above the Food and Drug Administration (FDA) action level of 5.0 ppm. No other priority pollutant organic compounds were detected.

Table 4 [*20*] shows an analysis of variance (ANOVA) used to compare the mean total lengths of specimens in the three creeks.

Statistically significant differences were noted in the length of Creek A mummichogs in comparison with those of the other two creeks; the mummichogs in Creek A were significantly smaller.

Parasites were found within the abdominal cavity of the fish during evisceration and appeared to be a single species, probably a monogenetic trematode, commonly found in mummichogs. No external parasitizations, fin rot, or other fungal diseases were recorded. The incidence of parasitization did not correspond to previously established baselines on the seasonal parasite burden (which for winter should be zero) or the relative size of the mummichogs [*21,22*].

A PCB index [*19*] developed by the New York State Department of Environmental Conservation (parts per million $\times$ percent lipid/100 = PCB index) was used to eliminate the effects of different percentages of lipids between the different species. The PCB index was found to be three to five times lower in Creek B and approximately ten times lower in Creek C than in Creek A. Creek A may be a "hot spot" in terms of PCB contamination.

As part of the National Pesticides Monitoring Program (NPMP) [*23*], fish

TABLE 1—*Species and numbers of fish and invertebrates collected by site and by station.*

Scientific Name	Common Name	Total Organisms Collected at Each Location, No.							
		1	2	3	4A	4B	5	6	7
		SITE A							
Anguilla rostrata	American eel[a]	...	...	1	...	...	...	...	1
Fundulus heteroclitus	mummichog[a]	58	46	62	72	...	59	44	45
Uca minax	fiddler crab[a]	...	...	...	...	...	...	...	...
Palaemonetes sp.	grass shrimp[a]	...	...	...	...	...	...	...	...
Hypaniolo grayi	polychaetes	6	9	...	...	1	6	...	...
Nereis succinea	polychaetes	9	11	...	...	...	1	...	...
Nereis sp.	polychaetes	...	...	...	...	...	...	1	...
Macoma balthica	bivalve mollusk	6	7	...	...	...	...	...	...
Hydrobia minuta	snails	...	1	...	...	...	...	...	...
Cyathura polita	isopods	...	...	...	...	1	...	5	10
Rhithropanopeus harrisi	decapods	...	...	...	...	...	...	...	...
Gammarus sp.	amphipods	...	...	...	...	...	...	...	...

SITE B

Scientific Name	Common Name	Upstream	Adjacent	Downstream
Salvelinus fontalis	brook trout[a]	2	...	...
Cyprinidae	minnows[a]	10	7	...
Cyprinus carpio	carp	1	...	...
Exoglossum maxillingua	cutlips minnow	...	2	2
Hybognathus nuchalis	silvery minnow	...	1	...
Notropis cornutus	common shiner	...	...	1
Notropis stramineus	sand shiner	1	1	2
Semotilus atromaculatus	creek chub	2	...	1
Semotilus corporalis	fallfish	...	...	1
Catostomus commersoni	white sucker[a]	7	6	9
Hypentelium nigricans	northern hog sucker	4	1	...
Moxostoma duquesnei	black redhorse	...	...	1

Micropterus dolomieui	smallmouth bass[a]	...	...	1
Etheostoma blennoides	greenside darter	...	1	...
Cottus cognatus	slimy sculpin	...	2	2

SITE C

		Lagoon 1	Lagoon 2
Lepomis auritus	redbreast sunfish[a]	17	3
Tilapia sp.	mouthbrooder[a]	39	31
Bathygobius soporator	frillfin goby[a]	1	2
Macrobrachium carcinus	shrimp[a]		
Macrobrachium sp.		100 g[b]	

SITE D

		Upstream	Adjacent	Downstream
Lepomis macrochirus	bluegill	...	1	2
Caranx latus	horse-eye jack[a]	...	...	1
Tilapia sp.	mouthbrooder[a]	9	9	17
Awaous tajasica	river goby	...	3	1
Atya scabra	freshwater shrimp[a]	...	...	2
Macrobrachium carcinus	freshwater shrimp	...	2	...
Macrobrachium heterochirus	freshwater shrimp	...	5	...
Xiphocaris elongata	freshwater shrimp	...	15	...

SITE E

		Outfall of Discharge Stream	Bay Area
Scomberomorus maculatus	Spanish mackerel[a]	...	4
Crassostrea virginica	American oyster[a]	21	...

[a] Indicates species used for tissue analysis.
[b] Indicates total weight sent for tissue analyses.

TABLE 2—*Mean PCB concentrations (ppm) in vertebrates, invertebrates, and channel sediments.*[a]

Location	Vertebrates	Invertebrates	Channel Sediments
Station 1	0.94	0.27	ND[b]
Station 2	2.0	0.36	1.42
Station 3A	5.5	...	0.08
	1.8 (eel)	...	...
Station 3B	...	6.30	67.80
Station 4A	6.0	...	9.78
Station 4B	...	2.92	...
Station 5	8.6	2.25	2.08
Station 6	0.64	0.24	ND
Station 7	0.71	0.61	ND
	1.6 (eel)	...	...

[a] The figures given identify the mean PCB concentration of all the specimens composited for a sample. The total number of specimens composited for each sample are presented in Table 1.
[b] Detection limit was 0.1 ppb.

TABLE 3—*Results of the* t-*test of the mean PCBs present in fish tissue in Creeks B and C versus the mean PCBs present in fish tissue in Creek A.*

Creek A	Creeks B and C
$X = 6.7$	$X = 1.07$
$S^2 = 2.77$	$s^2 = 0.40$
$n = 3$	$n = 4$

$ts = 8.4^*$, $df = 12$
$t_{0.05[12]} = 2.179$

TABLE 4—*Analysis of variance—total length of the common mummichog versus the location, taken from Ref* 20.[a]

	df	*ss*	*ms*	*F*-Value
Model	2	65.08	32.54	31.44**
Errors	383	396.38	1.03	$P \leq 0.01$
Corrected total	385	461.46		

[a] Rum Creek versus the control creek and Edmond's Creek.

have been trapped on a yearly basis at a gaging station upstream of Site A. Table 5 presents data regarding the PCB concentration, percent lipid, and PCB index for several species of finfish analyzed by the NPMP. High PCB concentrations in excess of the 5-ppm FDA action level were detected at this upstream location. The fish were collected under this program from one to five times a year in unidentified months. These fish were large (greater than 15 cm) with a home range of several miles and were potentially exposed to contami-

TABLE 5—*National Pesticide Monitoring Program comparison of PCB concentrations in upstream vertebrates, percent lipids, and PCB index (taken from Ref* 18*).*

Year	Species	PCB, ppm	Percent Lipid	PCB Index
1970	golden shiner	7.29	5.4	135
1970	rock bass	2.21	6.7	33
1970	white perch	4.04	7.3	55
1970	white perch	4.04	5.9	68
1971	golden shiner	0.89	4.5	20
1971	golden shiner	1.82	7.1	26
1971	white perch	3.65	9.8	37
1971	white perch	2.99	8.9	34
1971	white sucker	2.08	4.7	44
1971	white sucker	2.08	4.5	46
1972	golden shiner	1.50	3.5	43
1972	largemouth bass	4.90	4.3	114
1972	white perch	2.30	3.9	59
1972	white perch	3.20	6.0	53
1973	carp	2.10	3.1	67
1973	golden shiner	3.50	5.7	61
1973	golden shiner	1.50	4.5	33
1973	white perch	3.90	4.9	80
1976	carp	5.20	15.4	34
1978	largemouth bass	3.91	5.4	72
1978	white sucker	7.80	9.2	85
1978	white sucker	4.70	7.0	67

nants from Creek A as well as other sources. However, on a percent lipid basis the mean PCB index for the upstream location from 1970 through 1978 was 58, while for Creek A the PCB index in mummichogs averaged 308. The PCB index values were approximately 1.5 times higher for Creek B and 0.5 times lower in the Creek C than in the upstream Raritan River fish collection. The PCB index for the eel collected in Creek A was 4 times greater than for the eel collected in the control creek.

The PCB concentrations found in the mummichog tissue did not appear to be acutely toxic. However, the striking incidence of parasitization in the fish and statistically significant lower total organism lengths in Creek A suggest sublethal effects that may be related to the leachate discharging from the lagoon. Sublethal reproductive, physiological, and morphological effects of low concentrations of PCBs (1 to 5 ppb) as well as other pollutants have been documented for fish; furthermore, fish not adversely affected at the time of initial exposure later succumbed to fungal disease [*24*].

Site B was a freshwater stream located adjacent to a 13-acre farm. The farm was a suspected dumping site for large quantities of bulk and drummed liquid, solid, and semisolid wastes. The area was rural, and groundwater was the principal source of drinking water for nearby residents. Big Creek was used by local residents for trout and game fishing. During the actual sampling, only migratory species of fish were collected, so all the fish sampled were potentially exposed to areas of contaminated sediments.

Table 1 shows the species and number of fish collected at each station for Site B. Table 6 shows the metal concentrations for each species collected at Site B. Aluminum, chromium, iron, manganese, zinc, and boron were found at concentrations above the EPA detection limits. All the concentrations of these metals were within normal range for fish muscle tissue analyses. In addition, no metal contamination was found in the sediment samples taken from Big Creek, and no pesticides or PCBs were found during the tissue analyses.

At the remaining sites, Sites C, D, and E, no evidence of contamination was found in either sediment or fish tissue analyses. 4,4-Dichlorodiphenyldichloroethylene (4,4-DDE) was found in concentrations of less than 0.001 ppm in two samples from Site D and one sample from Site E. Extremely low levels of pesticides were found in occasional specimens.

The metal concentrations found in the fish tissue specimens at all five sites are presented in Table 6. All the metal concentrations were within the normal range for fish muscle tissue. Data analysis using multiple linear regression was not possible because of missing data values in Table 6. Product-moment correlation coefficients, which measure the strength of the relationship between two variables, were calculated between the average total length and the average metal concentrations for fish species only. Chromium ($r^2 = -0.63$), cadmium ($r^2 = -0.91$), and zinc ($r^2 = -0.61$) were statistically significant at $P \leq 0.01$; that is, as fish size increases, the concentration of these metals decreases. This finding is not unexpected. Copper, manganese, zinc, and iron are essential trace metals for fish metabolism; their concentrations are controlled by homeostatic regulation. The total body content of zinc has been found to remain relatively constant in spite of increasing body size. Product-moment correlations were also run between average sediment concentrations and the average tissue concentrations for metals; zinc in fish tissue appeared positively correlated to the amount of zinc in the sediment ($r^2 = 0.67$; $P \leq 0.001$). Evidence on metal accumulation in the food chain, however, is conflicting [*25*]. Correlation coefficients were computed among the metals. The metals that were positively correlated were chromium and iron ($r^2 = 0.68$), chromium and manganese ($r^2 = 0.88$), chromium and zinc ($r^2 = 0.69$), and iron and manganese ($r^2 = 0.89$). Other correlations which were positive but based on a smaller number of observations ($N \leq 8$) were barium and iron ($r^2 = 0.89$), barium and manganese ($r^2 = 0.92$) and barium and zinc ($r^2 = 0.87$) at $P \leq 0.01$. Concentrations of these metals may vary together through homeostatic regulation, although evidence for this is circumstantial. We present these metals data, in part, because of the lack of literature in this area and the difficulty in establishing background levels of metals in fish tissue.

The impacts of landfill leachate on aquatic organisms were mixed. Chemical landfill leachate from Site A had definite and specific effects on aquatic organisms that could be quantified by tissue analyses. Although the leachate draining Site B contained volatile organics, tissue analyses did not show any impact. Traditional biological community analyses, however, showed a reduction in the total number of species. The municipal landfill leachate effects were less clear. At Sites C, D, and E, where no evidence of contamination was found in the soil, tissue analyses did not show any positive results. At these three sites, allegations were made by knowledgeable individuals of hazardous waste dumping; the tissue analyses results showed no evidence of contamination.

Product-moment correlation coefficients indicated that some metals in fish tissue may vary together and are related to the total size of the fish. Bioconcentration factors (BCFs) following the octanol-water partitioning coefficients have been determined for a number of the priority pollutants and should be evaluated when biological sampling is being considered. Metals generally have low BCFs, with the notable exception of arsenic and mercury, and very little is likely to accumulate in tissue specimens. Priority pollutants with high BCFs such as PCBs (BCF = 45 600) [7] are more likely to be found in tissue analyses and to identify a site's impact on the ecosystem.

Procedures for tissue preparation may have caused some variation between the samples. The procedures using methylene chloride for cleaning equipment between individual sample preparation generally prevented cross-contamination between samples. (We note, though, that methylene chloride is no longer used because it is on the priority pollutant list.) The use of methylene chloride did not appear to affect the sample results. The freezing and thawing procedures, however, used to preserve the specimens before filleting, may have altered the results of the volatile organic analyses. Possible solutions to this problem could be the preparation of partially frozen instead of thawed specimens, or laboratory comparisons of fish tissue specimens prepared under different conditions and previously exposed to known concentrations of volatile organics, or both. The extent to which freezing and thawing affected the analyses was not quantifiable. We also felt that field processing of tissue specimens was generally difficult and should not be attempted under most circumstances, since the possibility of airborne contamination by dust particles exists at most hazardous waste sites. Different invertebrate species were combined to make up a sufficient weight for laboratory analyses. Although this is not a recommended practice, it would probably be necessary again under similar field conditions. In our analyses, combining several different species of invertebrates, all of which were exposed to the leachate containing PCBs, did not seem to affect the results; however, this is only one case. Further experiments in both the field and laboratory would be required to assess the effects of combining several species. Finally, our sampling was directed toward assessing the potential health effects on the human population, so we used fillets for tissue analyses rather than whole-body specimens. Whole-body analyses would probably have revealed higher levels of most pollutants.

The principal value of our studies was that they were applied studies done under natural conditions. The results are a summation of the multiple effects of a constantly changing aquatic ecosystem plus the effects of the hazardous waste sites. Both Hammons' methods [26] and the National Academy of Sci-

TABLE 6—*Comparison by site, species, and location of average total length to average metal concentrations*[a,b]

Site	Species	Location	Length, cm[c]	Metal Concentration, ppm[d]									
				Chromium	Barium	Cadmium	Iron	Copper	Manganese	Lead	Zinc	Boron	Aluminum
A	mummichog	1	7.19	...	...	...	20.0	...	3.2	0.54	28.0	...	9.0
A	mummichog	2	8.25	...	...	0.18	8.5	2.2	5.0	2.2	33.0	...	...
A	mummichog	3	6.83	...	...	0.21	31.0	2.5	4.4	1.6	33.0	...	42.0
A	mummichog	4	7.53	...	...	0.20	7.9	1.8	3.4	2.3	27.0	...	...
A	mummichog	5	7.60	...	...	0.18	6.7	1.8	5.6	2.4	30.0	...	...
A	mummichog	6	8.55	...	...	0.17	6.0	2.6	3.6	2.1	29.0	...	...
A	mummichog	7	8.17	0.1	...	0.18	5.5	...	4.8	2.7	32.0	...	...
B	trout	upstream	23.5	0.086	...	...	23.0	...	0.6	...	4.4	0.86	1.3
B	white sucker	upstream	25.8	0.069	...	...	8.3	...	0.6	...	9.9	0.93	3.4
B	carp	upstream	51.3	0.082	...	...	8.6	...	0.35	...	9.4	1.2	...
B	white sucker	adjacent	24.7	0.11	...	...	2.4	...	0.37	...	9.0	0.75	...
B	white sucker	downstream	25.4	...	...	...	3.2	...	0.24	...	9.3	0.8	...
B	minnows	adjacent	12.1	0.12	...	...	8.2	...	1.1	...	14.0	0.75	...
B	minnows	upstream	13.1	...	...	...	2.0	...	1.0	...	16.0	0.75	...
C	mouthbrooder	upstream	15.5	0.13	2.4	...	12.0	...	0.85	...	8.9	0.93	...

C	mouthbrooder	adjacent	14.4	0.093	...	...	6.7	...	0.85	...	3.6	0.79	6.6
C	mouthbrooder	downstream	22.8	...	...	...	4.6	...	...	...	3.2	1.2	...
C	horse-eye jack	downstream	11.5	0.25	0.75	...	9.7	...	4.0	...	10.0	0.97	12.0
C	shrimp composite	...	...	0.18	5.2	...	46.0	...	10.0	...	16.0	1.2	...
D	mouthbrooder	Lagoon 1	17.4	0.13	...	...	14.0	...	1.2	...	4.3	0.8	...
D	sunfish	Lagoon 1	6.9	0.40	3.1	...	21.0	...	6.7	...	18.0	1.1	...
D	mouthbrooder	Lagoon 1	6.4	0.44	7.2	...	260.0	...	21.0	...	22.0	2.4	...
D	mouthbrooder	Lagoon 1	9.9	0.37	4.0	...	70.0	...	12.0	...	12.0	0.14	...
D	goby	Lagoon 2	42.8	...	0.38	...	1.8	...	0.46	...	0.5	...	...
D	mouthbrooder	Lagoon 2	15.8	0.14	0.43	...	20.0	...	3.1	...	3.3	10.0	...
D	sunfish	Lagoon 2	13.5	0.23	2.1	...	3.6	...	2.3	...	7.1	0.9	...
D	shrimp	Lagoon 2	...	...	18.0	...	37.0	...	9.5	...	15.0	1.3	...
D	shrimp	Lagoon 1	...	0.13	1.8	...	90.0	...	14.0	...	47.0	1.4	...
D	shrimp	Lagoon 1	...	...	18.0	...	140.0	...	0.12	...	19.0	1.5	...
E	mackerel	bay	50.0	0.073	...	...	3.6	...	0.083	...	4.5	1.0	...
E	oyster	discharge canal	...	0.13	...	...	23.0	...	2.9	...	45.0	3.2	...

[a] Mercury, cobalt, and nickel are not included because they occur in only one sample.
[b] Blank spaces indicate metal concentrations below EPA detection limits.
[c] The average total length is based on the total number of specimens indicated in Table 1.
[d] All the samples were processed as wet weight.

ences review [27] consider the results of laboratory studies and speculate on the effects of priority pollutants on a natural system. Although development of baseline ecosystems [27] has been proposed, many ecotoxicologists may have to document the effects of hazardous wastes sites on the ecosystem in order to provide data for enforcement cases. Development of a baseline ecosystem will generally be impractical under these circumstances and most scientists will work with "after-impact" data. In addition, although laboratory studies such as bioassays may complement field studies, these lead only to predictions about the effects of pollutants on an ecosystem. Field studies, such as ours, validate or contradict these predictions and perhaps suggest new methodologies. In particular, although both references discuss the limited validity of single-species tests, and we were not concentrating on analyzing only a single species, we had clearly defined results from the effects of chemical leachate on just mummichogs from Site A. However, this was a single species naturally found in this habitat as opposed to species easily raised in the laboratory or commonly used as a testing animal. Single-species tests may, therefore, not necessarily be ruled out under natural conditions and may give an accurate indication of the effect of a hazardous waste site on the ecosystem.

Conclusions

Documented evidence of the discharge of priority pollutants to the aquatic environment is necessary before planning a biological sampling program. Biological sampling campaigns should not be planned at sites where there are only allegations of hazardous waste disposal.

Where evidence of hazardous waste disposal has been documented, both tissue analyses and traditional biological community analyses should be incorporated into analyses of aquatic ecosystem impact. Depending on the types of priority pollutants present, tissue analyses may show pronounced effects such as high levels of PCBs, or alternatively, the results may indicate community level impacts. Priority pollutants with high bioconcentration factors are good indicators for biological sampling. Residential species of fish or invertebrates, because of their prolonged exposure to the most contaminants, are most likely to demonstrate the impact of a particular hazardous waste site on the environment.

Acknowledgments

The work for this project was partially funded under EPA Contract No. 68-01-6056 while both authors were employed by Fred C. Hart Assoc., Inc., Newark, N.J. 07102.

References

[1] Gaufin, A. R. in *Biological Methods for the Assessment of Water Quality, ASTM STP 528*, American Society for Testing and Materials, Philadelphia, 1973, pp. 96–116.

[2] Cairns, J., Jr., Dickson, K. L., and Lanza, G. in *Biological Methods for the Assessment of Water Quality, ASTM STP 528*, American Society for Testing and Materials, Philadelphia, 1973, pp. 148–163.
[3] Sprague, J. J. in *Biological Methods for the Assessment of Water Quality, ASTM STP 528*, American Society for Testing and Materials, Philadelphia, 1973, pp. 6–30.
[4] Chiou, C. T., Freed, V. H., Schmedding, D. W., and Kohnerto, R. L., *Environmental Science and Technology*, Vol. 11, No. 5, 1977, pp. 475–478.
[5] Veith, G. D., Macek, K. L., Petrocelli, S. R., and Carroll, J. in *Aquatic Toxicology: Third Symposium, ASTM STP 707*, American Society for Testing and Materials, Philadelphia, 1980, pp. 116–129.
[6] Veith, G. D. and Morris, R. T., "A Rapid Method for Estimating Log P for Organic Chemicals," EPA 600/3-78-049, U.S. Environmental Protection Agency, Environmental Research Laboratory, Duluth, MN, 1978.
[7] Kenaga, E. E. and Goring, C. A. I. in *Aquatic Toxicology: Third Symposium, ASTM STP 707*, American Society for Testing and Materials, Philadelphia, 1980, pp. 78–115.
[8] Waldichuk, M. in *Pollution and Physiology of Marine Organisms*, Academic Press, New York, 1981, pp. 1–57.
[9] Zitko, V. in *Contaminants and Sediments: Volume 2—Analysis, Chemistry, Biology*, Ann Arbor Science, Ann Arbor, Mich. 1980, pp. 89–100.
[10] Halter, M. T. and Johnson, J. E. in *Aquatic Toxicology and Hazard Evaluation, ASTM STP 634*, American Society for Testing and Materials, Philadelphia, 1977, pp. 178–195.
[11] Prosi, F. in *Metal Pollution in the Aquatic Environment*, Springer-Verlag, New York, 1979, pp. 271–321.
[12] Skea, S. C., Simonin, A. S., Dean, H. J., Colquon, J. R., Spagnoli, J. J., and Veith, G. D., *Bulletin of Environmental Contamination and Toxicology*, Vol. 22, 1979, pp. 332–336.
[13] Spagnoli, J. J. and Skinner, J. C., *Pesticides Monitoring Journal*, Vol. 11, No. 2, 1977, pp. 69–87.
[14] "Review of PCB Levels in the Environment," EPA-560/7-76-001, U.S. Environmental Protection Agency, Washington, D.C., 1976.
[15] "Interim Methods for the Sampling and Analysis of Priority Pollutants in Sediments and Fish Tissue," U.S. Environmental Protection Agency, Environmental Monitoring and Support Laboratory, Cincinnati, 1977.
[16] "Fish Preparation Procedures for Contaminants Analysis," New York State Department of Environmental Conservations, Bureau of Environmental Protection, Albany.
[17] Jacangelo, D. J., "Technique for Processing Fish and Shellfish Specimens Collected for Polychlorinated Biphenyl (PCB) Analysis in New Jersey," State of New Jersey Department of Environmental Protection, Division of Fish, Game and Shell Fisheries, Bureau of Fisheries, Trenton.
[18] *Toxic Substances in Fish and Wildlife. November 1, 1981 to April 30, 1982*, Vol. 5, No. 2, New York State Department of Environmental Conservation, Division of Fish and Wildlife, Bureau of Environmental Protection, Albany, 1982.
[19] *Pesticide Analytical Manual*, Vol. I., U.S. Department of Health and Human Services, Washington, D.C., 1968.
[20] *SAS User's Guide*, MacMillan, New York, 1972.
[21] Barkman, L. L. and James, H. A., *Iowa State Journal of Research*, Vol. 54, No. 1, 1979, pp. 77–81.
[22] Valiela, I., Wright, J. E., Teal, J. M., and Volkmann, S. B., *Marine Biology*, Vol. 40, No. 2, 1977, pp. 135–144.
[23] Schmitt, C. J., Ludke, J. L., and Walsh, D. F., *Pesticides Monitoring Journal*, Vol. 4, No. 1, 1981, pp. 136–204.
[24] Linden, O., Sharp, J. R., Laughlin, R., Jr., and Neff, J. M., *Marine Biology*, Vol. 51, No. 2, 1979, pp. 101–110.
[25] Lewis, M. A. in *Aquatic Toxicology: Third Conference, ASTM STP 707*, American Society for Testing and Materials, Philadelphia, 1980, pp. 191–204.
[26] Hammons, A. S., *Methods for Ecological Toxicology: A Critical Review of Laboratory Multi-Species Tests*, EPA-560/11-80-26, Oak Ridge National Laboratories, Oak Ridge, TN, 1980.
[27] National Academy of Sciences, "Testing for Effects of Chemicals on Ecosystems," a report by the Committee to Review Methods for Ecotoxicology, National Academy of Sciences, Washington, D.C., 1981.

Peter Young[1] *and Alan Parker*[1]

Vapors, Odors, and Toxic Gases from Landfills

REFERENCE: Young, P. and Parker, A., **"Vapors, Odors, and Toxic Gases from Landfills,"** *Hazardous and Industrial Waste Management and Testing: Third Symposium, ASTM STP 851*, Larry P. Jackson, Alan R. Rohlik, and Richard A. Conway, Eds., American Society for Testing and Materials, Philadelphia, 1984, pp. 24-41.

ABSTRACT: Landfill gas from eight different sites in the United Kingdom was analyzed for trace constituents. Municipal, industrial, and liquid codisposal sites were included. A hazard assessment was made on the basis of the concentrations of potentially toxic components, and some of the causes of landfill odors were identified.

Three trapping methods were used to provide samples for analysis by gas chromatography/mass spectrometry. Supplementary techniques were employed to quantify the bulk gases, hydrogen sulfide, and metallic elements present.

Over 100 organic constituents were identified, but only 10 exceeded their individual recommended toxicity threshold limit values. Toxic compounds were more abundant at industrial sites. Odors were strongest from recently deposited waste, with methanethiol exceeding the odor threshold by a factor of more than 10^6. Hydrogen sulfide was not a significant cause of odor. Odors from the compounds identified and quantified depend more on the stage of decomposition than on the nature of the waste, and industrial wastes did not exacerbate the smell.

KEY WORDS: landfill gas, odors, analysis, municipal solid waste, industrial wastes, toxicity, hydrogen sulfide, hazardous wastes

The production of the major components of landfill gas, such as methane and carbon dioxide, is now well documented [*1,2*], but field data on the minor components are sparse. Landfill odors are undesirable, and concern has been expressed recently as to the possible long-term health risks associated with the gas. There is also a wide variation in the intensity of odor from different sites which is not easily related to the nature of the deposited waste.

The experimental program described in this paper was planned following a literature survey in July 1981. The literature demonstrated that although vapors

[1]Head, Waste Research Unit, and head, Landfill Research, respectively, Harwell Laboratory, Didcot, Oxfordshire, England, 0X11 ORA.

and odors are widely accepted as features of a landfill, their concentration has been largely unquantified and unspecified. With only scattered information on toxic gases available, it appears that hazardous releases are rare. Where releases have occurred, only site operators on the landfill itself have been involved [*3,4*]. The information available on the background concentrations of potentially hazardous components is not sufficient to allow any conclusive assessment of the long-term health risk landfill gas might represent. The short-term releases from a recently deposited waste are the only situations that have been considered [*5,6*]. One objective of this experimental work, therefore, was to identify the concentrations of potentially toxic components to allow an assessment of any health risk to be made.

Estimation of the releases of odorous chemicals from landfills is also hampered by the dearth of experimental data. As with toxic gases, mathematical models cannot be validated, because very little field data are available [*7*]. Attention has again been focused on a few particular components, with no attempt to estimate the cumulative effect. Odors are particularly renowned for their variability when interferences occur between different species. The second major objective was therefore to identify the causes of landfill odors and relate them to the nature and age of the waste present in each site.

A comprehensive survey of all domestic and industrial sites was not practical, so measurements were restricted to eight different landfills. These included sites filled with different forms of municipal waste as well as those containing industrial waste, the object being to represent a wide range of regimens. The principal analytical technique used was gas chromatography/mass spectrometry (GC/MS), and several trapping methods were developed to take samples. Atomic emission spectrometry was used to identify a wide range of metallic elements present as vapors or aerosols at four of the industrial waste sites.

After the majority of organic components in landfill gas had been identified and quantified, the sites were compared. However, the emphasis of this paper is on the application of these results to assessments of the hazardous and odorous gas releases.

Site Sampling Methods

Gas Sampling Probes

The objective of the gas sampling probes was to enable samples to be taken of the pure landfill gas as it was generated, without disturbing the conditions below the surface. With representative samples of landfill gas and additional measurements above ground, the results would have the potential to be used to estimate actual dilution factors. The attenuation of the gas could then be compared with the dilution required in air to remove any toxicity hazard or odor.

The method adopted made use of modified drive-in piezometers. These consist of a cadmium-plated mild steel point joined to a 19-mm internal-diameter galvanized mild steel pipe which is perforated for a length of 300 mm. Such a pipe can be connected to other lengths of galvanized iron pipe and hammered into the waste by a post hammer. All piezometer components were degreased, washed in acetone, dried at 100°C, cleaned with compressed air, and sealed in polyethylene until they were used on site. The use of these metal probes minimized both the risk of contamination, and the possible adsorption of any volatile components in the probe.

Insertion of the probes was largely successful and allowed sampling of gas at depths of down to 4.3 m with a minimum of disturbance and negligible air entrainment. At least three probes were inserted at each site at different locations and to varying depths.

Sampling Technique

The sampling technique devised was aimed at meeting three main criteria:

1. to facilitate the trapping of representative gas samples over a volume range from 25 cm^3 to about 100 dm^3,
2. to minimize the volume between the probe within the landfill and the trapping point, and
3. to eliminate surfaces likely to cause preferential absorption of some components or a release of organic materials into the gas stream.

The materials employed were therefore restricted to polytetrafluoroethylene (PTFE, teflon) and glass together with the metal sampling probes. Polyvinyl chloride (PVC) and rubber tubing were used only after the gas had passed through the trapping medium, since these polymers could influence the organic components in the landfill gas.

Figure 1 is a schematic diagram of the apparatus used to sample the gas. The aluminum sampling tube guard helped to ensure that no substantial particles were drawn into the tube and protected the sampling tube from the walls of the probe. The gas was then drawn through 5 m of 3-mm outside-diameter PTFE tubing to a three-way glass/PTFE tap by means of a small diaphragm pump. The T-tap allowed small representative samples to be drawn through trapping tubes by the syringe, providing a flow was maintained by the pump. The dead volume between the tube guard and the trapping medium was kept to between 3 and 5 mL. Sampling of bulk gas or condensate, or sampling by bubblers containing specific absorbing reagents, was carried out between the pump and the T-tap. A larger pump was installed when necessary to increase the sample volumes. The flow rate was adjusted using a needle valve attached to a rotameter with a soap bubble meter used for accurate measurement of the flow rate. The landfill sites were under various positive pressures during sampling of the gas, and at no time did the sampling method cause significant depletion of the gas entering the probe.

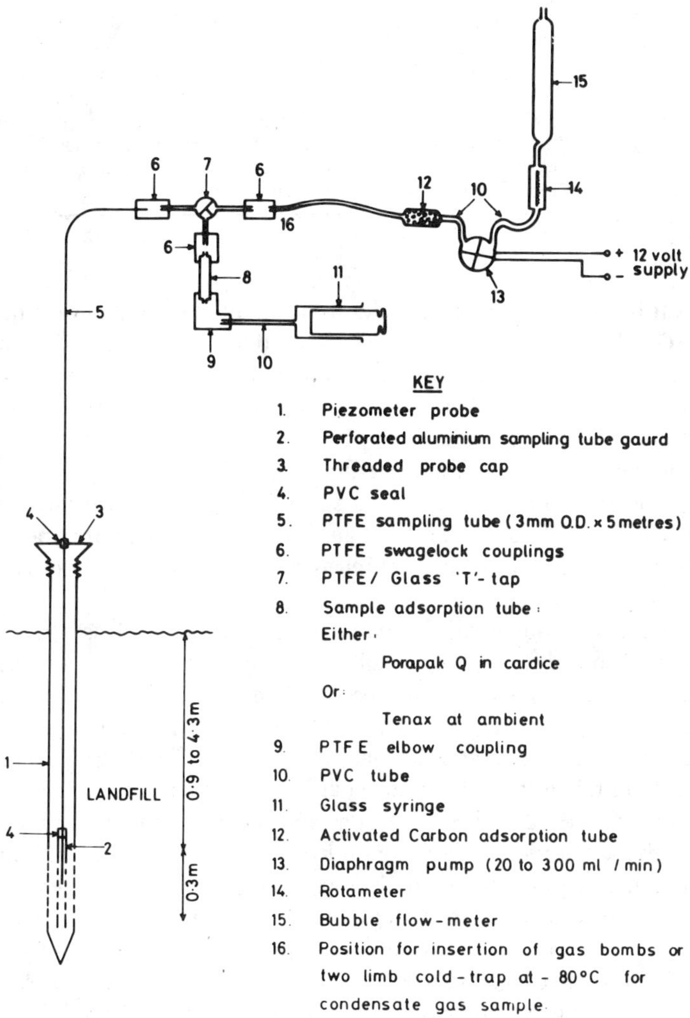

FIG. 1—*Schematic diagram of sampling apparatus (not to scale).*

Sampling Procedure

The full sampling procedure used three different trapping methods for subsequent GC/MS analysis. These were Tenax absorption tubes, Porapak Q absorption tubes, and a condensate sample taken at −80°C. The three trapping mediums allowed a comprehensive analysis of the gas, with each trap designed for a particular range of compounds. Separation of the components on the appropriate chromatographic packing allowed analysis as shown in Table 1.

TABLE 1—*Analysis of samples by GC/MS.*

Sample	Chromatograph Packing	Object of Analysis
Tenax	SE30	nonpolar species and general "fingerprinting"
Porapak Q	Chromosorb 101	polar and low-molecular-weight species
Condensate	Tenax GC	water-soluble bases and neutrals
Condensate	Chromosorb 101	water-soluble acids and neutrals

Tenax tubes were used at all the sites, but condensate samples were only analyzed by GC/MS from four sites, since no additional compounds were detected in these samples. The Porapak Q tubes were employed at most of the sites and provided quantitative data on a few key components such as the alcohols and low-molecular-weight organosulfurs.

The condensate trap used for collecting samples for analysis by emission spectroscopy consisted of a two-limbed glass trap immersed in a cardice-methanol mixture at −80°C. About 100 dm^3 of gas was passed through the trap, and no filter was used since the total metal releases were of interest, either as volatile species or in aerosols.

Measurements of the methane content were taken using a Gascope or Gascoseeker throughout the sampling as a check of the consistency of the gas being produced during sampling. Both these methane meters use the thermal conductivity of the gas to register methane and are very sensitive to hydrogen, which gives a threefold greater response than methane itself. Bulk gas samples were therefore taken in stainless steel gas bombs for analysis by GC for methane, carbon dioxide, hydrogen, oxygen, and nitrogen, and separately for hydrogen sulfide. Additional sampling for hydrogen sulfide was carried out using specific detection tubes marketed by Draeger and Gastec, but these were found to be grossly unreliable for landfill gas in which other impurities caused massive overestimates of the hydrogen sulfide present. A more sensitive method was employed for hydrogen sulfide at four sites. Two consecutive Dreschel bottles were used to bubble the gas through a solution which was 0.125 *M* in zinc acetate and 0.05 *M* in sodium acetate. The solutions were then analyzed by a spectrophotometric method developed by the U.S. Environmental Protection Agency (EPA) for car exhaust emissions [*8*].

At the last two sites investigated, additional sampling took place for carbonyl sulfide and total cyanide. The carbonyl sulfide was trapped successfully on absorbent tubes of Tenax at −80°C, and the cyanide was trapped by bubbling through 1 *M* potassium hydroxide at icebath temperature. The total cyanide method was adapted from another EPA car exhaust emission test [*8*], but the results did not indicate efficient trapping or quantitative recovery of cyanide from landfill gas. Carbonyl sulfide was present at concentrations of 20 to 30 $\mu g/m^3$.

Analytical Methods

GC/MS Analysis

The trapped samples were analyzed on a VG Micromass model MM16F mass spectrometer after concentration in a liquid-nitrogen-cooled loop and flash desorption onto the chromatographic column. The GC/MS analysis was terminated on detection of naphthalene in order to restrict the number of compounds under investigation. A few chromatograms were allowed to run on longer, and these confirmed that only a small proportion of the components were of sufficient molecular weight to be eluted after naphthalene and excluded from the results.

Since the object of the GC/MS analysis was to quantify the minor components present, a careful assessment of the validity of the data was made using four criteria:

1. The representativeness of the sample. Experiments were performed to establish that contamination of the various trapping methods did not occur during on-site sampling or the subsequent handling of the samples [*9*]. The sampling technique was also designed to minimize contamination and fractionation of the sample. Additional work is in hand to confirm that key components are, indeed, collected quantitatively, but the indications to date are that the analysis was representative of the gas present within the landfill.

2. The reproducibility of the results. A complete sampling procedure was carried out three times using one probe at one site (Site A). The complete analysis of these samples did not indicate any additional variation in the results because of difficulties in repeating the analytical procedures [*9*]. The different trapping methods were able to retain all of those compounds they were designed to trap, with condensate samples made largely redundant by the complete absence of soluble acids, bases, and neutrals.

Comparison of the three sets of data from Site A for Tenax showed that 66% of the compounds agreed within a factor of two of the mean, and only 8% had a discrepancy of greater than a factor of three. The Porapak Q tubes gave similar agreement, with 68% within a factor of two of the mean and only 6% varying by more than a factor of three. However, for compounds of less than seven carbon atoms, which the Porapak Q is specifically designed to trap, the agreement was always within a factor of two. At all the other sites investigated, a comparison of two samples from each trap gave similar levels of agreement.

3. The sensitivity in relation to the standards adopted. Anisole was adopted as the sole standard for all Tenax trap analyses. Additional standards were used for some of the more polar species detected using Porapak Q. These include the alcohols and methanethiol, but many other compounds such as dimethyl disulfide and the esters were found to give very similar responses to anisole. Both internal and external standards were used.

4. The errors associated with the analytical technique. The largest component of the uncertainty factor of about three associated with the GC/MS results quoted in this paper is undoubtedly derived from the interpretation of partially unresolved peaks. This interference also causes the detection limit to vary between 0.1 and 2.0 mg/m^3 according to the abundance of compounds at any given elution time for a particular sample.

A paper has been published which discusses the analytical technique and the validation experiments in more detail [*9*].

Analysis for Metals

The condensate samples for metals analysis were mixed with high-purity nitric acid and heated to 120°C to decompose any organic species present prior to analysis. The analytical technique adopted was somewhat unusual given the small sample available and the range of elements of interest. An initial atomic emission spectrographic survey was performed to select the elements for the second-stage analysis by a direct reading spectrometer and a programmable monochromator. In view of the possible volatility of species of interest the sample was analyzed directly by the inductively coupled plasma method and not by an evaporative procedure.

The first-stage survey uses a photographic plate as the detector. This allows simultaneous detection of the majority of metals and provides a permanent record,which is available for future reference. The method, however, does not readily provide quantitative data; its purpose is to indicate the different metals which are present and warrant measurement.

The direct reading spectrometer and programmable monochromator enable quantitative or semiquantitative data to be obtained simultaneously for the 39 elements for which the direct reader has been tuned, plus additional elements such as arsenic, selenium, and tellurium by sequential scanning of the monochromator.

Discussion of Results

The eight sites investigated are distinguished by a code letter. Table 2 lists the important features of each site. The original six sites investigated (A to F) were selected to be of widely different type and age, whereas the more recent investigations at Sites F, G, and H were aimed at providing more information or releases after about one year of degradation. Three locations were sampled at Site D, and these are coded D1, D2, and D3.

Toxic Components

Although 110 different components were identified, only 10 exceeded the time-weighted average threshold limit value (TLV) adopted by either the Na-

TABLE 2—*Key to sites investigated.*

Identification Letter	Type of Site	Type of Waste	Age of Waste at Sampling Point	Number of Locations on Site Sampled
A	stone quarry	crude municipal	5 to 6 years	2
B	stone quarry	baled municipal	7 months	1
C	stone quarry	pulverized municipal	3 weeks	1
D	clay pit	municipal, industrial, and liquids	>15 months	3
E	clay pit	municipal and liquids	6 months	1
F1	sand and gravel pit	industrial	5 years	2
F2			12 months	
G	old lime bed	industrial and liquids	9 months	1
H	sand and gravel pit	crude municipal with pulverized fuel ash cover	12 months	1

tional Institute of Occupational Safety and Health (NIOSH) or the American Conference of Governmental Industrial Hygienists (ACGIH) [*10,11*]. The components observed above their TLV in landfill gas are listed in Table 3. Dichlorofluoromethane is not included, although it was present in Site D at up to 93 mg/m^3 in Probe 3, because the ACGIH TLV of 40 mg/m^3 is not consistent with current toxicity information on halocarbons. The NIOSH TLV of 4200 mg/m^3 was therefore adopted in this case; in all other cases the lower of the TLVs was adopted.

The most widespread compound in Table 3 is benzene, which was particularly prevalent at industrial waste sites. At the municipal waste sites benzene concentrations were at least an order of magnitude lower, except at Site B with 4.2 mg/m^3. It therefore seems likely that the benzene is associated with petroleum and solvent-based wastes found at higher concentrations in industrial sites. Site F1 was releasing 23 mg/m^3 benzene in the gas from waste deposited five years earlier, which suggests that initial releases may require considerable dilution to reach the TLV adopted by NIOSH of 3.25 mg/m^3. The highest release observed was 114 mg/m^3 from Site D3.

Unlike benzene, which is a carcinogen, the other aromatic hydrocarbons' toxic significance is based on irritant and narcotic properties. The TLVs of toluene, xylene, and the propyl benzenes are consequently higher by about 100-fold, and, although more widespread, these compounds only exceeded their TLVs at some locations at Site D. The origins of toluene and the xylenes appear to be similar to that of benzene, with higher concentrations occurring in industrial sites. However, increasing quantities of higher-molecular-weight alkyl benzenes occur in municipal waste sites, which suggests that increasing proportions of these compounds are derived from degradation of larger molecules.

Vinyl chloride exceeds the TLV of 10 mg/m^3 at some industrial sites but is also widely present at much lower levels. Possible sources of vinyl chloride include direct deposition as a solvent impurity or by-product and release through degradation of PVC, by either release of encapsulated monomer or production of the monomer from polymer breakdown. The presence of tetrachloroethylene at a concentration of 350 mg/m^3 at Site E almost certainly arose from the liquid wastes deposited at that site. In general, the halocarbons are released in greater quantities from younger wastes, reflecting their high volatility and indicating that volatilization is a very important "sink" for these compounds.

The presence of methanethiol and butanethiol in Table 3 reflects TLVs based on protection of workers from intolerable odors rather than any health hazard. Alcohols were released from the younger wastes in considerable quantities, and the concentration of methanol at Site C actually exceeded the TLV of 260 mg/m^3.

The 30 most abundant components from Site D, Probe 3, are listed in Table 4. The total loading of organic vapors was over 5 g/m^3. No TLV is appropriate for comparison with this figure, since the different classes of compounds have widely different physiological effects. Any response, such as headaches, will oc-

TABLE 3—*Components detected at concentrations above recommended long-term toxicity thresholds (TLVs).*

Compound	TLV, mg/m^3	Dilution Factor Required to Reach TLV[a]									
		Domestic Refuse Sites			Sites Containing Industrial Waste						
		A	B	C	D1	D2	D3	E	F1	F2	G
Benzene	3.25	D	1	D	7	4	35	2	7	D	3
Toluene	375	D	D	D	D	1	>2	D	D	D	D
Xylenes	435	D	D	D	D	1	D	D	D	D	D
Propyl benzenes	245	D	D	D	D	1	D	D	D	D	D
Vinyl chloride	10	D	ND	D	D	D	3	2	D	D	D
Tetrachloroethylene	335	D	D	D	ND	D	D	1	ND	D	D
Methanethiol	1	ND	D	87	ND	ND	ND	60	ND	ND	1
Butanethiols	1.5	ND	ND	2	ND	ND	ND	ND	ND	ND	ND
Methanol	260	ND	D	1	ND	D	ND	ND	ND	D	D

[a]A factor of one indicates that the concentration is about equal to the toxicity threshold. D indicates that the compound was below the TLV but was detected, and ND indicates that it was not detected.

TABLE 4—*Site D: principal components detected in Probe 3 by GC/MS.*

Order by Abundance	Compound	Concentration in Probe, mg/m^3
1	octanes	675
	n-octane	95
2	heptanes	481
3	toluene	>460
4	hexanes	358
	n-hexane	96
5	1,2-dichloroethylenes	302
6	nonanes	226
	n-nonane	116
7	xylenes	221
8	di-isopropyl ether	220
9	1,1,1-trichloroethane	177
10	trichloroethylene	170
11	ethyl benzene	170
12	octenes	144
13	1,1-dichloroethane	130
14	decanes	128
	n-decane	51
15	benzene	114
16	methyl cyclohexane	96
17	dichlorofluoromethane	93
18	butanes	90
19	butenes	90
20	cyclohexane	88
21	tetrachloroethylene	84
22	limonene	84
23	propyl benzenes	75
24	1,1,1-trichlorotrifluoroethane	70
25	nonenes	70
26	dimethyl cyclopentanes	68
27	dimethyl cyclohexanes	48
28	chloroethane	46
29	decenes	42
30	other terpenes	40

cur at widely different concentrations for different individuals. However, it is unlikely that any of the cumulative effects will be of greater concern than that known for the presence of individual components, particularly the carcinogens benzene and vinyl chloride. A dilution factor of 100-fold was recommended for the landfill gas from Site D to alleviate any long-term health hazards. Most mixed waste sites in the United Kingdom are likely to require much less dilution, and this was supported by the results from the other industrial sites investigated.

The results from the metal analysis revealed the presence of 24 elements, although none exceeded its TLV. The metals present appeared to be associated with aerosol particles rather than volatile compounds, and the variation between sites was considerable. Although the results suggest that general releases of metals to atmosphere from landfills are of no significance, the deposition of wastes containing arsenic, lead, mercury, or tin may lead to significant releases. Sites containing excessive quantities of these elements were excluded from this initial study.

The bulk components of landfill gas require dilution of about 20-fold to render the air safe and free from explosive hazard. The dilution recommended on the basis of the toxicity of the minor components is therefore only five times greater. Preliminary site investigations suggest that landfill gas is only rarely present at concentrations in excess of 1% of the air above the landfill and that sufficient dilution will always be available over adjoining property.

Particularly Odorous Components

Odor thresholds are not only extremely difficult to estimate, but they are also widely dependent on the individual and on the presence of other compounds. The ASTM odor and taste threshold values book [*12*] was supplemented by other sources [*13-15*], to obtain as extensive a list of odor thresholds as possible. These odor thresholds were then used as the primary criteria for assessing the influence of a given component on the landfill gas smell of the site. Although no odor threshold could be found for some of the components detected, most of these are not thought to be particularly odorous.

Hydrogen sulfide is popularly blamed for landfill odors, but this was not found to be the case at the sites investigated. The bulk gas analysis results are shown in Table 5. The range of detection limits for hydrogen sulfide is due to the various analytical methods employed. The results from Site A were unique and probably unrepresentative. The concentration of hydrogen sulfide definitely varied during sampling, and was not discernible in the odor in the gas above Site A. It seems likely that the probe itself caused a short-term local release which was not representative of the conditions of the site, although the

TABLE 5—*Summary of bulk gas analysis.*

Site Code	Methane, %	Carbon Dioxide, %	Hydrogen, %	Hydrogen Sulfide, mg/m^3
A	66.0	28.3	...	6 to 86
B	37.5	58.5	0.22	<7
C	0.05	79.3	21.1	<1
D1	68.6	26.2	<0.1	<21
D2	69.4	26.0	<0.1	<2
D3	71.0	25.5	<0.1	<21
E	67.0	20.6	11.9	?<1
F1	71.7	23.2	<0.2	<7
F2	57.3	26.4	<0.1	0.36
G	77.0	18.2	<0.1	0.19
H	8.5	87.6	0.34	<0.3

mechanism of this release 3 m below ground level remains uncertain. At the other seven sites, hydrogen sulfide was only detected at levels below 1 mg/m^3, requiring less than a 1000-fold dilution to remove the odor. On the basis of these results, hydrogen sulfide does not appear to play a dominant role in the odor. Table 6 lists all the components that would individually require more than a 1000-fold dilution to render them odorless. The major contributions to the odor appear to come from two groups of compounds. Compounds in the first group are widespread and appear to be produced consistently by the waste. They are the alkyl benzenes and limonene, which along with many other hydrocarbons are probably responsible for the "typical" landfill gas background smell. The dilution required to render the gas odorless appears to be in the range of about 10^3 to 10^4, which suggests that the smell is likely to be detectable off site under some weather conditions.

The second group is dominated by the esters and organosulfurs but also includes specific solvents and butan-2-ol, which appears to have a particularly low odor threshold for an alcohol. These compounds are not widespread and were relatively insignificant at Sites A, F, and D (Probe 1). The esters and organosulfurs appear to be associated with the early decomposition phases. Concentrations were particularly high at Sites C and E, which were both producing significant quantities of hydrogen (Table 5). Methanethiol is the most important, and it is probably this species that local residents mistake for hydrogen sulfide. Not only is the greatest dilution required of methanethiol to remove the odor but it is also one of the more objectionable smells, reminiscent of the typical "bad egg" organosulfur odors. The esters also had important, objectionable odors, with ethyl butanoate making the largest contribution. These compounds are partially responsible for the "sweeter" odors typical of fresh refuse.

Of great importance to the odor from landfill gas is the ability of the components to be removed prior to dispersion in the atmosphere, particularly through top cover material. No work was directed specifically toward this objective, but the dilution factors above ground at Site A provided some limited indications of different attenuation factors. A second analysis was carried out about 15 cm above ground level in the vicinity of a large natural vent where gas was emerging. The bulk gases were present at about 30% of the concentrations in the probe. The 20 most abundant components in the gas from the probe are compared with the aboveground concentrations in Table 7.

Table 7 gives some indication that hydrocarbons may be more readily absorbed as the gas permeates through to the surface. This may be particularly significant both for the longer term odor releases, when alkyl benzenes may be the most odorous components, and for attenuating benzene before it is released to the atmosphere. The effect of different cover materials may be considerable in attenuating the highly odorous esters and organosulfurs, but data presented here can only indicate the scale of attenuation which should be desired to, for example, reduce the smell to levels similar to that of five-year-old waste. It can

TABLE 6—*Highly odorous components.*

	Dilution Factor Required to Reach Odor Threshold, All × 10^{-3}									
	Municipal Waste Sites			Sites Containing Industrial Waste						
Compound	A	B	C	D1	D2	D3	E	F1	F2	G
Limonene	0.4	4.0	1.9	3.3	1.9	1.5	4.2	<0.1	<0.1	0.5
Xylenes	<0.1	0.2	0.1	0.3	1.2	0.5	<0.1	<0.1	<0.1	0.2
Ethyl benzene	<0.1	0.1	<0.1	0.2	1.7	0.9	<0.1	0.4	<0.1	0.4
Propyl benzenes	1.7	2.4	3.0	2.1	7.3	1.9	0.9	2.0	<0.1	<0.5
Butyl benzenes	1.4	0.6	0.3	0.9	0.3	0.2	<0.1	0.2	<0.1	<0.1
Methanethiol	<5	1	2000	<50	<5	<50	1500	<2.5	<12	25
Dimethyl sulfide	<0.1	0.4	0.4	<0.1	0.3	1.2	6	<0.1	<0.1	<0.2
Butan-2-ol	<0.1	>0.2	2.1	<0.1	<0.1	<0.1	1.1	<0.1	0.2	0.3
Methyl butanoate	<0.1	<0.1	0.8	<0.1	<0.2	<0.4	3.0	<0.1	<0.1	2.6
Ethyl propionate	<0.1	<0.1	<0.1	<0.1	<0.1	<0.1	1.4	<0.1	<0.1	0.1
Ethyl butanoate	<0.1	1.3	5.0	<0.2	<0.3	8.7	120	<0.1	0.6	3.3
Propyl propionate	<0.1	<0.1	<0.1	<0.1	<0.1	0.1	2.1	<0.1	<0.1	<0.1
Butyl acetate	<0.1	<0.1	<0.1	<0.1	<0.1	<0.1	2.1	<0.1	<0.1	<0.1
Propyl butanoate	<0.1	<0.1	<0.1	<0.1	<0.1	<0.1	1.0	<0.1	<0.1	<0.1
Dipropyl ethers	<0.1	<0.1	<0.1	<0.1	<0.1	3.0	<0.1	<0.1	<0.1	<0.1

TABLE 7—*Site A: principal components detected by GC/MS.*

		Concentration		Dilution Factor	
Order by Abundance	Compound	Probe, mg/m^3	Above Vent, mg/m^3	At Vent Sample	To Reach the Odor Threshold
1	butyl benzenes	138	15.8	9	1400
2	decanes	137	37.6	4	...
	n-decane	45	8.5	5	45
3	propyl benzenes	69	8	9	1700
4	other terpenes	63	13.5	5	60
5	decenes	62	29.9	2	1.5
6	nonanes	57	10.6	5	...
	n-nonane	37	2	18	19
7	undecanes	48	11	4	...
	n-undecane	21	2.1	10	26
8	xylenes	36	6.6	5	90
9	undecenes	34	13.3	3	...
10	limonene	21	7.6	3	370
11	ethyl benzene	17	2.5	7	85
12	nonenes	12	1.6	7	odorless
13	octanes	10	1.9	5	...
	n-octane	6	0.9	7	6
14	toluene	8	5	2	11
15	heptanes	4	1.4	3	odorless
16	octenes	4	0.6	7	odorless
17	propyl cyclohexanes	4	1.3	3	...
18	tetrachloroethylene	3	0.3	10	odorless
19	?camphor/fenchone	3	1.4	2	3
20	hexenes	2.6	0.79	3	32

be seen that the attenuation required at Sites C and E to meet this objective is considerable, perhaps a factor of 1000.

The absence of carboxylic acids in the samples from Sites A, C, D, and E where condensate and Porapak Q traps were used was also an important result. The alkyl carboxylic acids have a strong characteristic odor, butyric acid has one of the lowest recorded odor thresholds, which has been quoted to be as low as 2×10^{-6} mg/m^3. These acids are plentiful in leachate and undoubtedly are the major contributors to the typical leachate smell. However, their complete absence from the landfill gas analysis suggests that they do not represent an important component of the gas odor. This implies that they are relatively easily absorbed inside the landfill and have a strong affinity for the aqueous phase.

The compounds observed in the gas from mature municipal waste (Site A) were mainly a mixture of hydrocarbons, which agreed well with other gas analysis data available in the literature [*16,17*]. The industrial and fresher municipal wastes widened the range considerably [*18*]. Work in Germany [*17*] has suggested that methane can be used as a tracer gas for the minor components when dilution and mixing with the atmosphere occur. Current work is aimed at validating this approach and examining the diurnal range of gas concentrations above and around a landfill site. Initial results are in broad agreement with the dilution factors of 10^4 observed in Germany. Further work is also planned to include an assessment of cover materials as attenuating media for gaseous releases, since the odor problem of a site may be reduced considerably by the use of appropriate cover.

Conclusions

The industrial waste gas was significantly richer than the municipal waste gas in petroleum-based hydrocarbons and industrial solvents, including a wide range of halocarbons. The quantity of minor components reached over 5 g/m^3 at Site D, and a hazard assessment of the gas from one probe suggests that a dilution factor of 100-fold would be desirable to ensure that the gas does not represent any health hazard for long periods of exposure. Although widely present, vinyl chloride and benzene levels were particularly elevated at industrial sites which had received liquid wastes. The elevated levels of trace components in the industrial wastes does not, however, exacerbate the odor significantly.

The results from the limited number of sites examined suggest that the most important odors are common to many sites and are dependent more on the age and decomposition rate than on the nature of the waste. Methanethiol and ethyl butanoate are particularly significant, whereas hydrogen sulfide and the carboxylic acids are relatively unimportant to the total odor. The presence of particular groups of compounds such as the thiols, alcohols, and esters may well be associated with known phases in the decomposition process, for example, the carbon dioxide or hydrogen-rich stages. Some compounds are apparently produced with great consistency, for example, limonene and some of the alkyl benzenes. These may play an increasing role in the odor of a site as other more potent species subside and the odor strength decreases. The long-term odor appears to be dominated by hydrocarbons and is consequently less strong and less objectionable.

Although there are likely to be considerable interactions between the different compounds, a first estimate of the odor strength can be made by comparing each component with its individual odor threshold. This method suggests that some landfill gases require dispersion in air to over a millionfold dilution to remove the odor, and consequently extensive odor plumes are possible from some sites. The effect of cover materials and decomposition rates will be critical in controlling the magnitude of the odor problem associated with each site. The use of cover materials may also be desirable at sites that receive industrial wastes which are likely to give rise to toxic vapors, to minimize the required dilution of the gas to reach the TLV.

Acknowledgments

The authors would like to thank the Department of the Environment for instigating and funding this work. They are also grateful to those people associated with the sites sampled. Particular mention should be made of Dr. B. Brookes who carried out the GC/MS analysis. Gas analysis was also carried out by D. J. Darke, W. Swanson, and S. Wright. The condensate analysis for metals was carried out by Dr. C. Pickford.

References

[*1*] Rees, J. F., "Major Factors Affecting Methane Production in Landfills," Paper No. 2, *Proceeding*, the Landfill Gas Symposium, 6 May 1981, Harwell Laboratory, Didcot, Oxfordshire, United Kingdom.

[*2*] Rees, J. F. and Viney, I., "Leachate Quality and Gas Production from a Domestic Refuse Landfill. The Implication of Water Saturated Refuse at Aveley Landfill," Report AERE-R-10328, Her Majesty's Stationery Office, London, February 1982.

[*3*] "Waste Tip under Merseyside Scrutiny after H_2S Incident," *Municipal Engineer* (London), Vol. 153, No. 43, 1976, pp. 1581–1582.

[*4*] "Hazardous Waste Disposal Damage Reports," Report No. EPA-530-SW-151.3, U.S. Environmental Protection Agency, Office of Solid Waste Management Programs, Cincinnati, Ohio, 1976.

[*5*] Markle, R. A., Iden, R. B., and Sliemers, F. A., "A Preliminary Examination of Vinyl Chloride Emissions from Polymerization Sludges, during Handling and Land Disposal," *Proceedings of Residual Management by Landfill Disposal*, Hazardous Waste Research Symposium, EPA Report No. EPA-600-9-76-015, U.S. Environmental Protection Agency, Municipal Environmental Research Laboratory, Cincinnati, Ohio, 1976, pp. 186–194.

[*6*] Jones, C. J. and McGugan, P. J., "An Investigation of the Evaporation of Some Volatile Solvents from Domestic Waste," *Journal of Hazardous Materials*, Vol. 2, No. 3, 1978, pp. 235–251.

[*7*] Shen, T. T. and Tofflemire, T. J., "Air Pollution Aspects of Land Disposal of Toxic Wastes," *Journal of the Environmental Engineering Division*, American Society of Civil Engineering, Vol. 106, No. EE1, 1980, pp. 211–226.

[*8*] Dietzmann, H. E., Smith, L. R., Parness, M. A., and Fanick, E. R., "Analytical Procedures for Characterizing Unregulated Pollutant Emissions from Motor Vehicles, Report No. EPA-600/2-79-017, interim report, South West Research Institute, San Antonio, Texas. Sponsored by Environmental Sciences Research Laboratory, Research Triangle Park, N.C., 1979.

[*9*] Brookes, B. and Young, P. J., "The Development of Sampling Methods and GC/MS Analytical Procedures to Identify and Determine the Minor Organic Components of Landfill Gas," *Talanta*, Vol. 30, No. 9, September 1983, pp. 665–676.

[*10*] *Pocket Guide to Chemical Hazards*, F. W. Mackison, R. S. Stricloff, and L. J. Partridge, Eds., DHEW (NIOSH) Publication No. 78-210, U.S. National Institute for Occupational Safety and Health/Occupational Safety and Health Administration (NIOSH/OSHA), Washington, D.C., 1980.

[*11*] "Threshold Limit Values for Chemical Substances in Work Air Adopted by ACGIH for 1982," American Conference of Governmental Industrial Hygienists, Cincinnati, Ohio, 1982.

[*12*] *Compilation of Odor and Taste Threshold Values Data, ASTM Data Series DS48A*, American Society for Testing and Materials, Philadelphia, Pa., 1978.

[*13*] Verscheuren, K., *Handbook on Environmental Data on Organic Chemicals*, Van Nostrand Reinhold Co., New York, 1977.

[*14*] Bedborough, D. R., "Control of Odours and Noise in the Process Industries," Report, Warren Spring Laboratory, Stevenage, Herts, United Kingdom, 1978.

[*15*] "The Identification and Measurement of Refinery Odours," Report No. 8/75, CONCAWE Special Task Force on Odours, Stichting CONCAWE, The Hague, Netherlands, December 1975.

[*16*] Giuliani, A. J., "Application of Conventional Oil and Gas Drilling Techniques to the Production of Gas from Garbage," *Proceedings*, Transmission Conference, Salt Lake City, Utah, 5-7 May 1980, American Gas Association, Policy Evaluation and Analysis Group, Arlington, Virginia.
[*17*] Bruckman, P. and Muelder, W. "Der Gehalt an organischen Spurenstoffen in Deponiegasen, *Schriftenreihe der Landesantalt Immissionsshutz Landes Nordrhein, Westfalen*, Essen, 1982, Vol. 55, pp. 21-28.
[*18*] Young, P. J. and Parker, A., "The Identification and Possible Environmental Impact of Trace Gases and Vapours in Landfill Gas," *Waste Management and Research*, Vol. 1, 1983, pp. 213-226.

Frederic Belkin[1] *and Richard W. Bishop*[1]

Analysis of Chemicals for Waste Disposal

REFERENCE: Belkin, F. and Bishop, R. W., **"Analysis of Chemicals for Waste Disposal,"** *Hazardous and Industrial Waste Management and Testing: Third Symposium, ASTM STP 851*, Larry P. Jackson, Alan R. Rohlik, and Richard A. Conway, Eds., American Society for Testing and Materials, Philadelphia, 1984, pp. 42–47.

ABSTRACT: The U.S. Army Environmental Hygiene Agency (USAEHA) is responsible for investigation of problems associated with the management of hazardous wastes for the U.S. Army. One type of disposal problem arises when an unlabeled container of material is located and disposal is desired. The sample is evaluated in the field by obtaining a history of the material to be disposed, the types of operations performed on the installation, and the length of time and conditions of storage. A proper sample is required, sufficient for analysis, but not so large as to create shipping and disposal problems for the laboratory. Qualitative test results, sample history, physical properties, required detection limits, regulations concerning disposal, and sample preparation are considered when choosing an appropriate method of analysis. An analytical scheme using a variety of instrumental techniques has been developed for the analysis of waste disposal samples. Quality control and good laboratory practices are employed in ensuring the accuracy of the complex analysis.

KEY WORDS: toxic materials, bulk sampling, hazardous wastes, industrial wastes

The U.S. Army Environmental Hygiene Agency (USAEHA) is responsible for investigation of problems associated with the management of hazardous wastes for the U.S. Army.[2] The Organic Environmental Chemistry Division (OECD) and other laboratories at USAEHA perform analyses in support of this mission.

There are two basic types of analytical problems involving chemicals for waste disposal which USAEHA encounters. The first type involves confirmation of the presence and quantity of a suspected chemical. For example, USAEHA performs numerous analyses on transformer fluids to determine the

[1]Chemists, U.S. Army Environmental Hygiene Agency, Aberdeen Proving Ground, Md. 21010.

[2]The opinions or assertions contained herein are the private views of the authors and are not to be construed as official or as reflecting the views of the Department of the Army or Department of Defense.

type and concentration of polychlorinated biphenyls (PCBs), which may be present. The other problem concerns those situations in which an unlabeled drum or container of unknown material is located and disposal is desired.

It is important that an "unknown" be evaluated in the field before the laboratory is sent a sample for analysis [*1*]. This evaluation consists primarily of obtaining a history of the material to be disposed of. The history should include a list of the chemicals suspected to be present, the types of operations performed on the installation, and the length of time and conditions of storage. A proper sample, sufficient for analysis but not so large as to create a shipping and disposal problem for the laboratory, is taken. USAEHA recommends the use of 40-mL glass vials with Teflon-lined screw caps for submission of samples for qualitative and quantitative analysis. Flash point analysis requires an additional 100 mL of liquid sample submitted in suitable glass containers. If there is an indication that a sample may be especially toxic, either chemically or biologically, it should be so noted.

Once a waste sample arrives at USAEHA, it is logged into the sample management computer and assigned a unique laboratory number. The sample description is then recorded on the bulk analysis form (Fig. 1). This description includes the color of the sample, the number of liquid phases, the percentage of each phase, the percentage of sediment, and any characteristic odor. Samples submitted for analysis for specific chemicals such as PCBs, or metals, are then routed to the appropriate analytical group for confirmation and quantitation. The other samples for unknowns' analysis require special treatment.

Each batch of samples submitted from a particular installation is analyzed as a set. This aids the analyst in identifying unknowns and increases productivity of analyses. There are often duplicate types of samples within a set; thus the identification of one sample may help to reveal the components of the look-alikes. For example, if a sample is shown to be a combination of motor oil and mineral spirits, it could indicate that some of the waste samples are from a motor pool area. This enables the analysts to key in on the types of products usually found in this kind of area. This example also demonstrates why a sample history is important to obtain; it can simplify and shorten the analysis considerably. It is usually easier to confirm the presence of a chemical than to identify a complete unknown.

The first step in analyzing unknowns is the performance of some simple analytical screens. These include a check for water solubility, Beilstein's test [*2*], the ignition test, and determination of the percentage of nonvolatiles. The water solubility test aids in determining if the material is aqueous or a water-soluble type of organic compound. The Beilstein's test is needed to ascertain the presence of halogenated compounds, since these compounds often require a different disposal procedure than nonhalogenated organic compounds. The percentage of nonvolatiles indicates the amount of compounds present which are not sufficiently volatile to be chromatographed by gas chromatography (GC), such

Sample Description, Physical and Chemical Tests

Lab No.	Field No.	Odor	# Liquid phases	% Each phase	%Sediment	Color	Soluble in water	Beilstein	Burn in flame	Water present	pH	%Non-volatile	Qualitative organic tests	Flash point (°F)		GC scan

FIG. 1—*Waste disposal sample analysis form.*

as resins and heavy oils. Flash points are determined on organic samples to characterize the ignitability of the sample material [*3*].

The pH is measured on all aqueous samples. Where the pH is outside the range of 5 to 9, the responsible acid or base is identified. Qualitative flame tests for sodium and potassium are determined first on basic samples, since sodium and potassium hydroxides are the most common bases used. If ammonia is present, it can usually be detected by its odor and the pH of the evolving gas. For acidic samples, qualitative tests for chloride, sulfate, and acetate are performed, since hydrochloric, sulfuric, and acetic acids are three of the more common acids and these tests are relatively simple. When these tests prove negative, it is usually necessary to perform the anion tests for phosphoric, nitric, chromic, and hydrofluoric acids. Organic acids are confirmed by either ion chromatography (IC) or GC. Aqueous samples are also analyzed for cyanide content if their odor or history indicate the possible presence of cyanide compounds.

Preliminary qualitative tests for functional groups such as alcohols and ketones are performed on organic samples. The liquid sample is then injected into a gas chromatograph utilizing a thermal conductivity detector (TCD). The presence of water as well as most volatile organic compounds such as solvents can thereby be verified. Additional GC analysis and quantitation is usually done using a flame ionization detector (FID), although other common GC detectors such as electron capture (EC), nitrogen-phosphorus (NP), and Hall electrolytic conductivity detector (HECD) are occasionally used. These detectors, plus a variety of chromatographic columns, can aid in the identification of many common organic compounds. One can get a false negative, however, by failing to run a percent nonvolatile test, because materials such as oils cannot be easily chromatographed. Thus, what appears on a chromatogram represents only the volatile portion of an organic sample. When compounds not easily identified are detected by GC, gas chromatography-mass spectrometry (GC-MS) or gas chromatography-infrared spectrometry (GC-IR) can be utilized for qualitative analysis. Infrared spectroscopy (IR) is usually the analytical instrument of choice for unknown nonvolatile organic compounds. High-performance liquid chromatography (HPLC) can also be used for specific analyses on some samples.

Finally, the metal content of waste disposal samples may be determined in several ways, when necessary. Atomic absorption spectroscopy (AAS) and in-

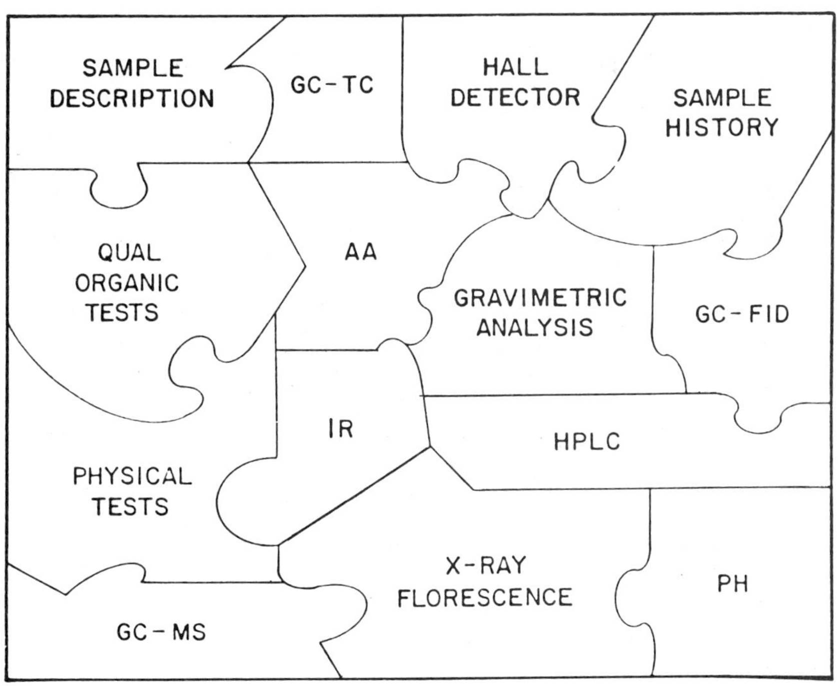

FIG. 2—*Bulk sample analysis, "putting together the pieces of the puzzle."*

ductively coupled plasma (ICP) are routinely used to analyze for the EPA designated extraction procedure toxicity (EPT) metals and can be used for many others. An X-ray fluorescence (XRF) scan is useful for detecting the presence of metals with atomic weights greater than calcium and can be used to estimate quantities. Microscopic analysis is helpful for fiber identification.

Quality control and good laboratory practices are always important aspects in the analysis of unknowns [*4,5*]. Failure to ensure good data can result in incorrect decisions being made in regard to sample disposal. Quality control procedures, such as having two analysts analyze the same sample independently, are employed. All the data are reviewed by the quality control coordinator to ensure that the individual test results are consistent with each analytical test performed and that appropriate analytical tests and quality control techniques are employed.

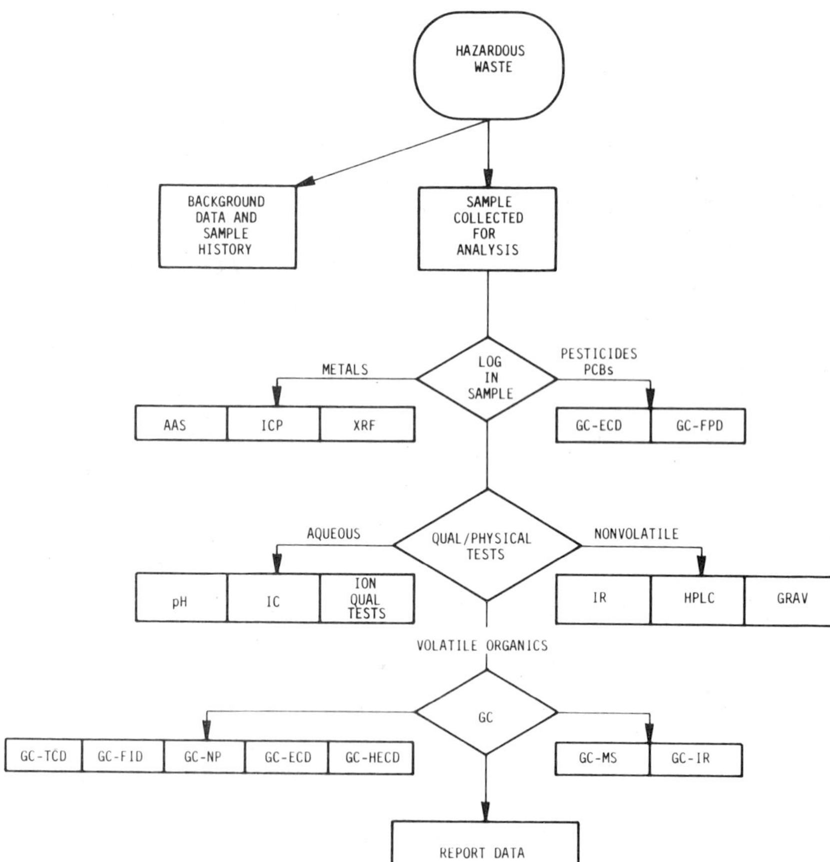

FIG. 3—*Diagram of analysis scheme.*

Matrix effects are important in many waste samples. The matrix could mask a compound of interest or increase the detection limit. A cleanup procedure may be required, or an alternative analytical method may be employed to reduce or eliminate the interference. Spiking the sample with the suspected compounds of interest aids in determining the matrix effects.

After the series of analytical tests is performed on each sample, the analytical data are evaluated, like putting together pieces of a puzzle (Fig. 2). The data from all of the tests should be consistent with the conclusions. Regulations concerning waste disposal must be considered in choosing the analytical procedures and reporting of sample results [*6*]. No sample is analyzed for every possible parameter; the time and expense make this impractical. For example, samples are not routinely screened for radiological hazards unless there is some reason to suspect their presence. Further, there is always a possibility that a particular chemical may be "missed" in the analysis. It is hoped, however, that the analytical scheme that the USAEHA uses (Fig. 3) will minimize the possibility of overlooking the presence of a hazardous waste.

References

[*1*] Meiggs, Theodore O., Wills, Carroll G., and Gallagher, Thomas P., *Conducting Hazardous Waste Site Investigations," ASTM Standardization News*, August 1981, pp. 8–11, 54.

[*2*] Shriner, Ralph L., *The Systematic Identification of Organic Compounds*, 6th ed., Wiley, New York, 1980, pp. 81–83.

[*3*] Sax, N. Irving, *Dangerous Properties of Industrial Materials*, 5th ed., Van Nostrand Reinhold, New York, 1979, pp. 234–235.

[*4*] Inhorn, Stanley L., *Quality Assurance Practices for Health Laboratories*, American Public Health Association, Washington, D.C., 1978, pp. 3–126.

[*5*] MacDougall, Daniel M., *Analytical Chemistry*, Vol. 52, 1980, pp. 2242–2249.

[*6*] Code of Federal Regulations, "Identification and Listing of Hazardous Waste," Title 40, Part 261, Revision of July 1, 1983, U.S. Government Printing Office, Washington, D.C., pp. 360–391.

Pierre L. Côté[1] *and Denis Isabel*[2]

Application of a Dynamic Leaching Test to Solidified Hazardous Wastes

REFERENCE: Côté, P. L. and Isabel, D., "**Application of a Dynamic Leaching Test to Solidified Hazardous Wastes,**" *Hazardous and Industrial Waste Management and Testing: Third Symposium, ASTM STP 851*, Larry P. Jackson, Alan R. Rohlik, and Richard A. Conway, Eds., American Society for Testing and Materials, Philadelphia, 1984, pp. 48-60.

ABSTRACT: A dynamic leaching test using a fixed leachant renewal schedule developed for testing nuclear waste forms was not found to be widely applicable to solidified hazardous wastes because of the large range of leachability encountered and because of limitations in the analytical detection of the chemical species in the leachate. The data generated could be interpreted using a simple diffusion model. The range of measured diffusivities, from $10^{-15.5}$ to $10^{-8.4}$ cm^2/s, was covered by using a variable leachant renewal schedule. The precision associated with any single determination was approximately one order of magnitude.

KEY WORDS: hazardous wastes, toxic wastes, solidification, stabilization, fixation, waste form, standard test, leaching, dynamic test, diffusion, release, industrial wastes, solid wastes

Addition of inorganic materials such as lime, cement, and fly ash to aqueous wastes to produce solid matrices decreases the waste leachability through reduction of the contact surface between the leaching medium and the waste. Advantage can be taken of the fact that solidified waste can be molded into regular geometrical shapes to measure leachability using mass transport principles. In a dynamic leaching test, a solidified waste specimen of known surface area is immersed in a leaching medium at a specified ratio of leachant volume to sample surface area. The leaching medium is replaced at frequent intervals, and the masses of the species of interest leached during each inter-

[1] Process development engineer, Environment Canada, Environmental Protection Service, Wastewater Technology Centre, Burlington, Ontario, Canada L7R 4A6.

[2] Graduate student, Institut national de la recherche scientifique, INRS-eau, Ste-Foy, Quebec, Canada GIV 4C7.

val are determined. The test is nonagitated because it is assumed that the migration rate of the contaminants in the leachate is several orders of magnitude larger than in the solid matrix, thus eliminating concentration gradients in the leaching medium.

Diffusion has been shown to be the factor governing leaching rates in short-term leachability studies of solidified low-level radioactive wastes [*1–3*] and of solidified hazardous wastes [*4,5*]. A standard method was proposed by the American Nuclear Society (ANS) [*6*] to evaluate leachability, using a short-term test and interpret the results following a diffusion model. This method, developed for measurement of radionuclide leachability, is not readily applicable to solidified hazardous wastes. Hazardous wastes may contain a variety of contaminants, such as heavy metals, anions, and toxic organics, with significantly different leachabilities. Furthermore, the analytical detection limits and measurement precision and accuracy of these contaminants are not generally as good as for radionuclides. The work reported in this paper was undertaken to investigate the use of a dynamic leaching test to evaluate the leachability of solidified hazardous wastes.

Background

Diffusion Model

The solution of a semiinfinite medium diffusion model with uniform initial concentration and zero surface concentration is presented by Godbee and Joy [*7*] in this form:

$$\left(\frac{\Sigma a_n}{A_0}\right)\left(\frac{V}{S}\right) = 2\left(\frac{D_e}{\pi}\right)^{1/2} t_n^{1/2} \tag{1}$$

where

a_n = contaminant loss during leaching period, n, mg,
A_0 = initial amount of contaminant present in the specimen, mg,
V = volume of specimen, cm^3,
S = surface of specimen, cm^2,
t_n = time (end of the leaching period), and,
D_e = effective diffusion coefficient, cm^2/s.

Assuming that the leaching rate is linear over a leachant renewal period, the following solution can also be derived [*7*]:

$$\left(\frac{a_n}{A_0}\right)\left(\frac{V}{S}\right)\left(\frac{1}{\Delta t_n}\right) = \left(\frac{D_e}{\pi}\right)^{1/2} \frac{1}{T_n^{1/2}} \tag{2}$$

where

Δt_n = duration of the leaching period, s, and
T_n = time (middle of the leachant period), s.

The relationship between the three time variables of Eqs 1 and 2 is illustrated in Fig. 1. Equations 1 and 2 provide two ways of calculating the effective diffusion coefficient, D_e. With Eq 1, D_e can be calculated from the slope of a plot of the cumulative fraction leached, $\Sigma a_n/A_0$, versus the square root of time, $t_n^{1/2}$. Using Eq 2, an averaged D_e value can be derived from the incremental fraction, a_n/A_0, leached at each period, n. The diffusion coefficient calculated from these equations is termed "effective" because diffusion takes place in the liquid filling the interstices of a porous body. The liquid path length can therefore be much longer than that assumed by the model [*8*]. Furthermore, A_0 includes both the mobile and nonmobile species of the contaminant of interest [*7*].

Considering the variables in Eqs 1 and 2, it can be seen that calculation of D_e values could be performed under different test conditions. A_0 varies with the size of the specimen; a_n is the product of leachate volume and concentration measured; (V/S) is a shape factor function of the geometry of the specimen; and, finally, the time variables are dependent on the leachate renewal schedule. For a given solidified waste specimen and contaminant, any combination of test conditions that approach the initial and boundary conditions of the sample diffusion model should allow calculation of a unique D_e value.

The condition of uniform initial concentration can be respected by carefully preparing the specimens in the laboratory. The assumption of a semi-infinite medium is approximated by selecting a specimen size or test duration, so that a negligible fraction of the initial amount A_0 will be leached during the test. These two conditions are discussed in detail in the draft ANS [*6*] standard.

The third condition, of zero surface concentration, can be approached by selecting a combination of the sample size and shape, the ratio of the leachant volume to sample geometrical surface area, and the leachant renewal frequency that would minimize the concentration reached during each leaching interval yet allow analytical detection. Practical considerations (availability of waste, molds, leaching vessels, and so on) set limits to the sample size and shape and to the volume of the leachant. The ANS standard [*6*] specifies a

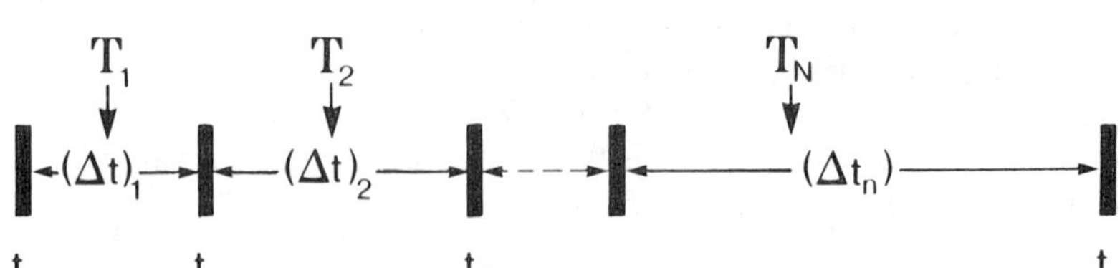

FIG. 1—*Relationship between the three time variables of a dynamic leaching test.*

leachant volume to geometrical surface area ratio of 10 cm in order to minimize the leachant composition changes during short leaching intervals and still provide sufficient amounts of extracted species for analysis.

Under these conditions, an optimal leachant renewal frequency, for the leaching of a given species from a particular matrix, could be defined as that frequency that would compromise between the model assumption of zero concentration in the leachate and the analytical measurement requirement. If the constraint is imposed that a_n is to be the same for all leaching periods, it can be readily demonstrated from Eq 1 that

$$t_n = n^2 t_1 \tag{3}$$

If t_1 is known, this equation can be used to schedule an optimal leachant renewal series for diffusion-controlled leaching. Figure 2, constructed from Eq 1, can be used to calculate the duration of the first leaching period, t_1, provided that the approximate value of D_e is known. For example, considering the desired contaminant amount to be measured in the leachate, a_1, and the initial amount in the specimen, A_0, it is determined that $(a_1/A_0) = 10^{-4}$. Furthermore, if $D_e \simeq 10^{-12}$, use of Fig. 2 shows that $t_1 \simeq 10^{3.75}$ s or 1.5 h. The corresponding optimal renewal schedule, as given by Eq 3, is 1.5, 6.0, 13.5, 24.0, and so forth.

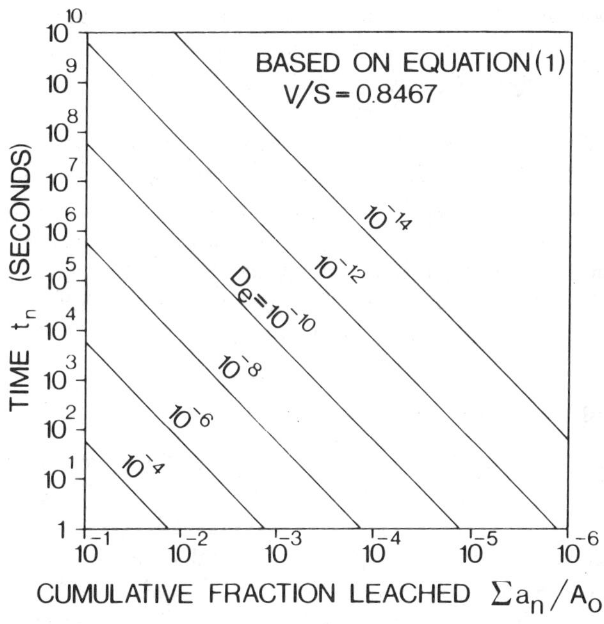

FIG. 2—*Relationship between the leaching time and the cumulative fraction leached.*

Test Protocols

Measurements of the leachability of solidified low-level radioactive wastes have been performed for the last 25 years using various protocols [*2*]. A first step toward standardization was made by the International Atomic Energy Agency [*9*]. The proposed test procedure was not widely accepted for several reasons analyzed by the American Nuclear Society [*6*].

The standard protocol proposed by the ANS [*6*] lasts five days, during which the leachant is renewed seven times. The data are analyzed using the incremental leaching rate approach (Eq 2), and a leachability index is calculated as follows:

$$LX = \frac{1}{m} \sum_{n=1}^{m} \log \left[\frac{\beta}{D_e} \right]_n \tag{4}$$

where

LX = leachability index,
β = constant, 1 cm^2/s, and
m = number of leaching periods (7).

Following this approach, the leachability index is a dimensionless "figure on merit" that qualitatively describes leachability. The LX value can be interpreted as the negative logarithm of the effective diffusivity only provided long-term leaching data support the hypothesis that diffusivity controls leachability. It is important to note that this approach would not be applicable if Eq 1 was used to analyze the data, since calculation of the D_e value requires that diffusion be the controlling leaching mechanism [a straight line must be obtained when plotting $(\Sigma a_n/A_0)$ versus $t_n^{1/2}$].

A standard method was recently proposed for testing of solidified hazardous waste by the U.S. Environmental Protection Agency (EPA) [*10*]. This method is an adaptation of the ANS protocol [*6*], with the main difference being the duration of the test, which was increased to 14 days to include ten leachant renewals. To the authors' knowledge, this method has not been validated to address the concerns about its applicability to solidified hazardous wastes raised in the introduction of this paper.

Experimental Design and Methods

Experimental work was conducted to measure the effect of leachant renewal frequency and waste strength, A_0, on calculated diffusivities of a variety of chemical species from a solidified waste matrix.

A synthetic waste solution was prepared from distilled water by addition of 0.04 *M* each of arsenic [as sodium arsenite ($NaAsO_2$)], cadmium (as cadmium nitrate [$Cd(NO_3)_2$]), chromium [as chromic chloride ($CrCl_3$)], lead (as lead nitrate [$Pb(NO_3)_2$]), and phenol. This resulted in a 3.94% total solids solution with a pH of 2.2. Part of this solution was diluted by a factor of 5 to provide a second waste strength (0.008 *M*) for experimentation.

The synthetic waste solutions were solidified by adding two parts of fly ash (from the Lakeview Power Generating Station, Mississauga, Ontario) and one part of Type I portland cement to the synthetic waste solution (weight-to-weight ratios). The additives and the synthetic wastes were mixed to a homogeneous slurry in a Hobart mixer. Two-inch cubes (5.08 cm) were prepared as per the ASTM Test for Compressive Strength of Hydraulic Cement Mortars (C 109-80). An alternate sample preparation method (for samples that would not retain their physical integrity when immersed in water) would be to place the slurry in a Soxhlet extraction thimble. The cubes, unmolded after 24 h, were cured for 28 days at 20°C and 100% humidity before the leaching experiments were initiated. Addition of fly ash and cement increased the volume of the original waste by a factor of 2.3. This volume increase factor is higher than those obtained when the minimum amount of additives are used [*11*]. It was used here to avoid sample disintegration during the test. The resulting initial amounts, A_0, of the chemical species monitored are indicated in the second column of Table 1 for the 0.04-*M* waste strength.

The experimental design for the leaching tests is summarized in Table 2. All the test conditions were duplicated. Six specimens of each of the full-strength synthetic waste formulation (0.04 *M*) and the fivefold diluted formulation (0.008 *M*) were leached. In addition, two specimens were leached as blanks, with solidified blocks prepared with distilled water in lieu of the synthetic waste. All the specimens were leached for 28 days following three leachant renewal schedules. The leachants were changed at 2, 7, 24, 48, 72, 96, 120, 192, 336, 504, and 672 h for the standard leachant renewal schedule. The half and double-frequency schedules were derived from the standard schedule by, respectively, dropping every second leachant renewal and inserting a leachant renewal between every two existing ones. Note that the first

TABLE 1—*Analytical conditions.*

Parameter	Initial Amount in a 5-cm Cube (Waste Strength 0.04 *M*), mg	Method of Analysis in the Leachate	Detection Level, mg/L	Pseudo-Equilibrium Concentration, mg/L[a]
As	170.8	AA,[b] graphite furnace	0.002	0.5
Cd	256.3	AA, graphite furnace	0.001	<0.001
Cr	118.6	AA, graphite furnace	0.001	0.05
Na	52.4	AA, flame	0.1	60.0
Pb	472.4	AA, graphite furnace	0.001	0.01
Phenol	214.3	steam distillation, colorimetry	0.03	6.0
NO_3	565.4	technicon, colorimetry	0.01	3.0

[a] After 28 days of leaching without leachant renewal.
[b] AA = atomic absorption.

TABLE 2—*Experimental design.*[a]

Leachant Renewal Schedule	Waste Strength		
	Blank	0.008 *M*	0.04 *M*
Double	...	7, 8	9, 10
Standard	1, 2	3, 4	5, 6
Half	...	11, 12	13, 14

[a] The numbers in the table identify the specimens. The leachant renewal schedules are described in the text.

seven leachant renewals of the standard schedule corresponds to the ANS-proposed standard test [*6*] and also follows closely the optimal schedule (for diffusion-controlled leaching) given by Eq 3. The first ten leachant renewals of the standard schedule correspond to the EPA-proposed standard test [*10*].

The ANS-proposed test procedure [*6*] was followed for leaching. The specimens were suspended at the centroid of the leachant using a harness made from fishing line. The suspenders of the harness were secured between the lid and the 2-L plastic bottle. At the end of each leaching period, pH and conductivity of the leachates were measured and aliquots were analyzed for arsenic, cadmium, chromium, sodium, lead, phenol, chloride, and nitrate (NO_3) using the analytical methods listed in Table 1.

Results

Data reduction and interpretation will be illustrated considering arsenic release from Specimens 4, 6, 8, 10, 12, and 14. The results of the other specimens and species will then be presented in the form of calculated leachability indexes. A complete set of results in tabular and graphical forms for all the specimens and species is available from the authors upon request.

The concentrations of arsenic measured in the leachates for each leaching period for Specimens 4, 6, 8, 10, 12, and 14 are shown in Table 3. As expected, the concentrations increase with an increase in the waste strength and a decrease of the leachant renewal frequency. These data also illustrate that spacing the leachant renewal periods following Eq 3 ensures a more or less constant concentration in the leachates.

Cumulative fractions leached were calculated and are plotted as a function of the square root of time in Fig. 3. The graph shows that, after an initial washing period of about 24 h ($t^{1/2} = 300\ s^{1/2}$), leaching is controlled by diffusion. The diffusion coefficients, calculated from the diffusion-controlled part of the curves, using Eq 1 and expressed as $pD_e = -\log D_e$, are reported in the first column of Table 4. It should be noted that pD_e values are, on average, lower (higher leachability) for the lower waste strength. This is because, for a given renewal frequency, the lower waste strength data sets approximate

TABLE 3—*Arsenic leaching as a function of waste strength and leachant renewal frequency.*

End of Leaching Period, h	Arsenic Concentration, mg/L					
	0.04-*M* Waste Strength			0.008-*M* Waste Strength		
	Double (10)[a]	Standard (6)	Half (14)	Double (8)	Standard (4)	Half (12)
1[b]	0.025			0.014		
2	0.022	0.045		0.010	0.022	
4.5	0.027			0.025		
7	0.011	0.025	0.053	0.015	0.042	0.049
15.5	0.012			0.028		
24	0.009	0.017		0.025	0.060	
36	0.010			0.005		
48	0.007	0.012	0.030	0.004	0.005	0.010
60	0.007			0.002		
72	0.006	0.009		0.002	0.002	
84	0.011			0.002		
96	0.010	0.014	0.040	0.002	0.005	0.007
108	0.008			0.002		
120	0.014	0.012		0.003	0.003	
156	0.014			0.005		
192	0.014	0.018	0.030	0.003	0.005	0.007
228	0.014			0.005		
264	0.014	0.016		0.006	0.010	
300	0.010			0.008		
336	0.012	0.016	0.022	0.005	0.010	0.015
420	0.015			0.009		
504	0.026	0.040		0.009	0.012	
588	0.014			0.008		
672	0.025	0.044	0.058	0.014	0.018	0.030

[a] Specimen identification number (see Table 2).
[b] The end of the leaching period indicated was the target schedule. The actual times were used in the diffusivities calculations.

more closely the model boundary condition of zero concentration in the leachate (as can be noticed from the concentration data of Table 3). Furthermore, the pD_e values also decrease with increasing renewal frequency in accordance with a better approximation of the same boundary condition.

The same data were also used to calculate leachability following the approach proposed by the ANS [6]. A D_e value was calculated from the incremental leach rate of each period using Eq 2. The arithmetic average from all the leaching periods (including those in the initial washing phase) was then calculated using Eq 4 to generate the *LX* values presented in the second column of Table 4. The *LX* values are lower than the pD_e values for every specimen (higher diffusivity). This effect, due to initial washing, is more pronounced with the lower waste strength, as can be seen in Fig. 3.

Examination of the cumulative fraction leached versus the square root of time plots for all the species and specimens showed that, after the initial

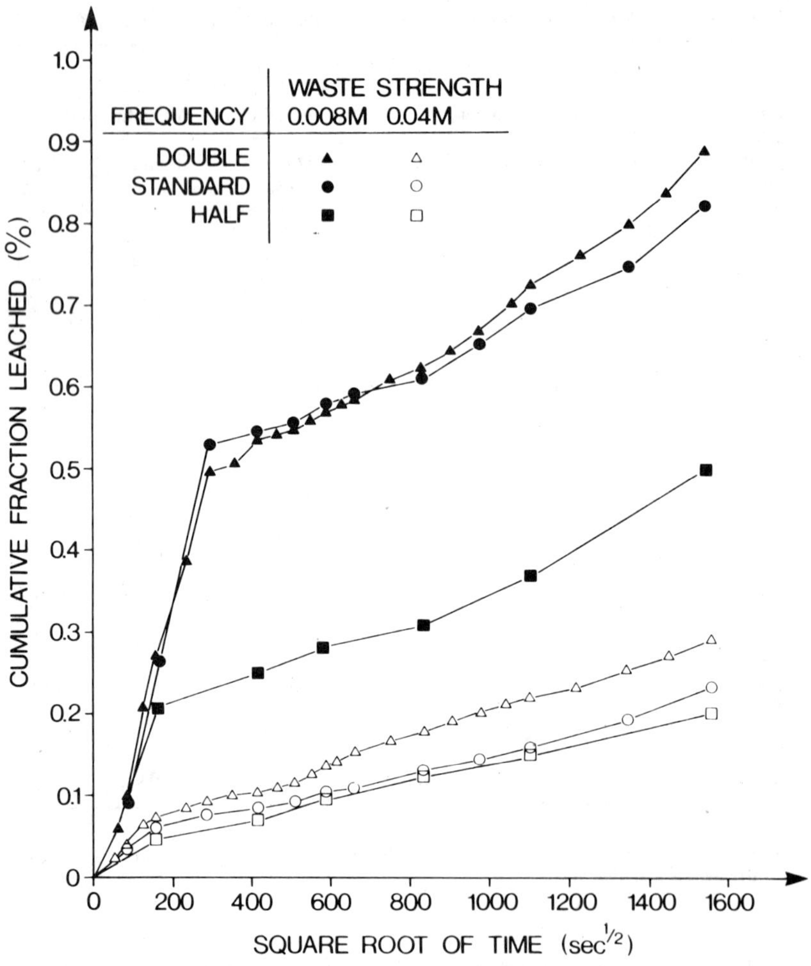

FIG. 3—*Arsenic data interpreted following a simple diffusion model for 28 days of leaching.*

TABLE 4—*Diffusivity of arsenic.*

	pD_e	*LX*
Specimen	Calculated with Eq 1	Calculated with Eqs 2 and 4
8 (0.008 *M*, double)	11.28	10.93
4 (0.008 *M*, standard)	11.52	11.12
12 (0.008 *M*, half)	11.63	11.35
10 (0.04 *M*, double)	11.82	11.69
6 (0.04 *M*, standard)	12.10	11.93
14 (0.04 *M*, half)	12.17	11.96

washing period, diffusion controlled leaching during the 28 days of the experiment. Leachability indexes, *LX*, were calculated using Eqs 2 and 4 rather than Eq 1 because, as explained in the Background section, this approach is more generally applicable to short-term leaching tests. The results are presented in Table 5. For some species and test conditions, *LX* could not be calculated because of analytical detection limitations. It was arbitrarily decided that *LX* values would be calculated only with those data sets in which at least 90% of the concentrations measured were above the detection level specified in Table 1.

Another factor indirectly influences calculation of *LX* values. Pseudo-equilibrium concentrations, obtained after one month of continuous leaching (without leachant renewal), are presented in the last column of Table 1. The difference between these concentrations and the detection level concentrations constitutes a range that determines the test conditions that can be used to measure the leachability. For most of the data reported in Table 5, the measured concentrations did not approach the saturation limit.

Discussion

The dynamic leaching tests based on a simple diffusion model proposed by the ANS [*6*] and the EPA [*10*] cannot be widely applied to solidified hazard-

TABLE 5—*Leachability index,* LX, *after 28 days of testing.*

Leachant Renewal Schedule	Species	Waste Strength	
		0.008 *M*	0.04 *M*
Double	As	10.96, 10.93	11.67, 11.69
	Cd	..., ...	..., ...
	Cr	..., ...	..., ...
	Na	8.66, 8.67	8.41, 8.43
	Pb	..., ...	..., ...
	phenol	..., ...	9.46, 9.49
	NO_3	..., ...	10.78, 11.05
Standard	As	11.15, 11.12	11.96, 11.93
	Cd	..., ...	..., ...
	Cr	..., ...	..., ...
	Na	8.60, 8.64	8.38, 8.40
	Pb	..., ...	..., ...
	phenol	10.10, 10.14	9.42, 9.74
	NO_3	..., ...	11.16, 10.78
Half	As	11.28, 11.35	11.91, 11.96
	Cd	..., ...	..., ...
	Cr	..., ...	13.22, 13.15
	Na	8.63, 8.61	8.34, 8.34
	Pb	..., ...	15.36, 15.48
	phenol	10.03, 10.10	9.38, 9.64
	NO_3	..., ...	10.78, 10.68

ous wastes. These two protocols provide for leaching following a fixed leachant renewal schedule, which will result in measurement of an effective diffusion coefficient in a limited number of cases only. This can be demonstrated by considering Eq 3 and Fig. 4. Since the leachant renewal schedules of the previously proposed procedures approximate that described by Eq 3 (with $t_1 = 2$ h), the same fraction, a_n/A_0, will be leached during each period. Figure 4 was constructed from Eq 1 for the test conditions used in the authors' leaching experiments ($V/S = 0.8467$; $V = 1.5$ L) and with a value of $t_1 =$ 2 h. Figure 4 gives the expected concentration in the leachate (provided diffusion controls) as a function of two intrinsic properties of the fixed waste: the initial amount of a contaminant, A_0, and its effective diffusion coefficient, D_e. This figure can be used to illustrate how analytical considerations limit the applicability of the method. The range of concentrations, between analytical detection and saturation levels, determines a range of measurable diffusivities for a given A_0. However, as the measured concentration increases, the simple diffusion model assumption of zero surface concentrations become less valid, and the diffusivities calculated using Eq 1 or 2 are underestimated.

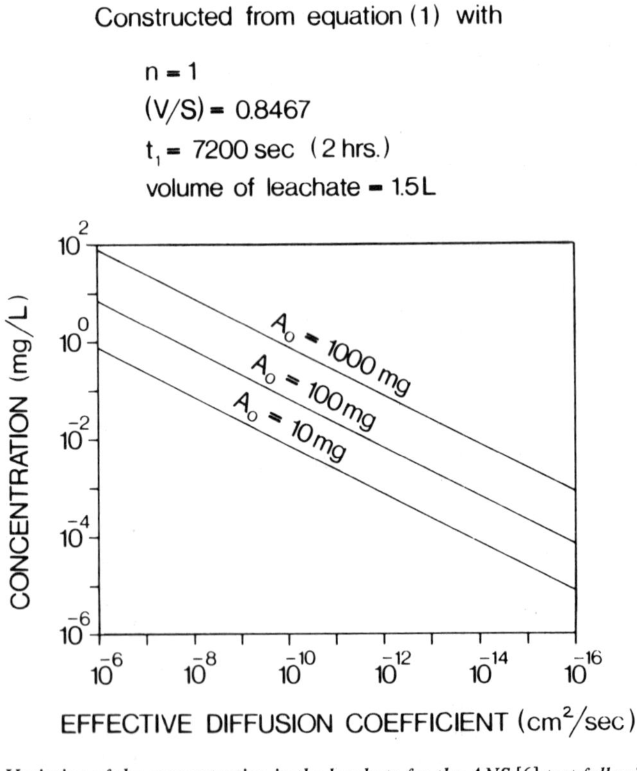

FIG. 4—*Variation of the concentration in the leachate for the ANS* [6] *test following a simple diffusion model.*

As an illustration of these facts, consider the arsenic and lead leaching results. For arsenic, the concentration ranged from 0.002 to 0.5 mg/L, and $A_0 = 170.8$ mg (Table 1). Under these conditions, Fig. 4 shows that the range of measurable diffusivities (as pD_e) is from 8.8 to 13.6. Table 5 results (standard frequency, 0.04-m waste strength) indicate that arsenic leachability falls within that range and was indeed successfully measured. For lead, on the other hand, with a concentration range from 0.001 to 0.01 mg/L and $A_0 = 472.4$ mg (Table 2), Fig. 4 shows a range of measurable diffusivities from 13.1 to 15.1. The results of Table 5 (standard frequency, 0.04-m waste strength) indicate that lead leachability falls outside those limits. It could be measured, however, using the "half" leachant renewal schedule and was found equal to 15.42.

The leachate concentration effect on calculated diffusivities can be evaluated with the arsenic data for which pD_e could be calculated under six sets of test conditions (Table 4, first column). The effective diffusion coefficient increases with decreasing waste strength and increasing leachant renewal frequency (that is, calculated diffusivities increase with decreasing concentrations in the leachant). This suggests that the lowest calculated $pD_e = 11.28$ would be closest to the real value, since these test conditions more closely approximate the zero surface concentration condition. The diffusivity determined under the six tests conditions ranged from 11.28 to 12.17, a factor of almost 10, which indicates that the precision of any single determination would be of one order of magnitude.

Conclusions

A simple diffusion model was successfully used to interpret one-month leaching data of several chemical species from a solidified hazardous waste.

A standard test using a fixed leachant renewal schedule developed for testing of nuclear waste forms [*6*] was not widely applicable to a solidified hazardous waste because of the large range of leachability encountered and because of limitations in the analytical detection of the chemical species in the leachate.

It was shown that different leachant renewal schedules can be used to cover the range of leachabilities and more closely approach the assumptions of the diffusion model used for data interpretation.

Pseudo-equilibrium data should always be generated together with the leaching test to ensure that saturation does not limit the leaching process.

The results of short-term tests should be reported using the leachability index (Eq 4) proposed by the ANS [*6*] until more fundamental studies ascertain that diffusion is the controlling leaching mechanism.

The precision of the results, when expressed as a diffusion coefficient, is approximately one order of magnitude.

A test of this type has the potential of being used as a regulation tool, since it measures the mobility of contaminants in a solidified waste matrix. It is currently considered for inclusion in a solidified waste assessment protocol being developed at Environment Canada's Wastewater Technology Centre.

References

[*1*] Godbee, H. W., Compere E. L., Joy, D. S., Kibbey, A. H., Moore, J. G., Nestor, Jr., C. W., Anders, O. U., and Neilson, R. M., Jr., "Application of Mass Transport Theory to the Leaching of Radionuclides from Waste Solids," *Nuclear and Chemical Waste Management*, Vol. 1, 1980, pp. 29–35.

[*2*] Mendel, J. F., "A Review of Leaching Test Methods and the Leachability of Various Solid Media Containing Radioactive Wastes," BNWL-1765, Battelle Pacific Northwest Laboratories, Richland, Wash., 1973.

[*3*] Moore, J. G., Godbee, H. W., Kibbey, A. H., and Joy, D. S., "Development of Cementitious Grouts for the Incorporation of Radioactive Wastes, Part 1: Leach Studies," ORNL-4962, Oak Ridge National Laboratory, Oak Ridge, Tenn., 1975.

[*4*] Johnson, J. C., Lancione, R. L., and Sanning, D. E., "Stabilization Testing and Disposal of Arsenic Containing Wastes," in *Toxic and Hazardous Waste Disposal*, Vol. 2, R. B. Pojasek, Ed., Ann Arbor Science Publishers, Ann Arbor, Mich., 1979, Chap. 11.

[*5*] U.S. Army Corps of Engineers, "Solidification/Stabilization of Sludge and Ash from Wastewater Treatment Plants," draft report, Environmental Laboratory, U.S. Army Engineer Waterways Experiment Station, Vicksburg, Mich., 1982.

[*6*] "Measurement of the Leachability of Solidified Low-Level Radioactive Wastes," second draft of a standard, Working Group ANS-16.1, American Nuclear Society, 1981.

[*7*] Godbee, H. W. and Joy, D. S., "Assessment of the Loss of Radioactive Isotopes from Waste Solids to the Environment, Part 1: Background and Theory," TM-4333, Oak Ridge National Laboratory, Oak Ridge, Tenn., 1974.

[*8*] Treybal, Robert E., *Mass Transfer Operations*, 2nd ed., McGraw-Hill, New York, 1968.

[*9*] Hespe, E. D., "Leach Testing of Immobilized Radioactive Waste Solids," *Atomic Energy Review*, Vol. 9, No. 1, 1971, pp. 195–207.

[*10*] U.S. Army Corps of Engineer, "Guide to the Disposal of Chemically Stabilized and Solidified Wastes," U.S. Environmental Protection Report SW-872, Cincinnati, Ohio, September 1980.

[*11*] Côté, P. L. and Hamilton, D. P., "Leachability Comparison of Four Hazardous Waste Solidification Processes," Paper presented at the 38th Annual Purdue Industrial Waste Conference, West Lafayette, Ind., 10–12 May 1983.

Howard W. Kilau[1] *and Ishwarlal D. Shah*[1]

Chromium-Bearing Waste Slag: Evaluation of Leachability When Exposed to Simulated Acid Precipitation

REFERENCE: Kilau, H. W. and Shah, I. D., **"Chromium-Bearing Waste Slag: Evaluation of Leachability When Exposed to Simulated Acid Precipitation,"** *Hazardous and Industrial Waste Management and Testing: Third Symposium, ASTM STP 851*, Larry P. Jackson, Alan R. Rohlik, and Richard A. Conway, Eds., American Society for Testing and Materials, Philadelphia, 1984, pp. 61-80.

ABSTRACT: Waste slag dumps constitute a large potential source of chromium contamination of groundwaters, particularly under the enhanced leaching action of acid precipitation. Laboratory leaching experiments were carried out to determine the composition factors controlling the chromium leachability of waste slags when exposed to acidic solutions designed to simulate acid precipitation.

Two critical slag composition factors were found to control the chromium leachability: (1) the calcium oxide (CaO)/silicon dioxide (SiO_2) ratio and (2) the magnesium content. It was determined that above a 2.0 CaO/SiO_2 ratio, chromium exists in slag as calcium chromite, which can be vulnerable to leaching by acid precipitation, especially if it is oxidized to calcium chromate when exposed to the environment over an extended period.

Experimental results showed that maximum chromium leachability from industrial slags occurred when the composition of the slag was at about a 2.0 CaO/SiO_2 ratio. Between a 1.0 and 2.0 CaO/SiO_2 ratio, in the presence of sufficient magnesium, picrochromite ($MgO \cdot Cr_2O_3$) was formed, which is very stable towards oxidation and was resistant to dissolution by simulated acid precipitation.

KEY WORDS: slags, slag disposal, wastes, chromium, chromium steels, chromium oxides, chromates, leaching, water pollution, precipitation (meteorology), acidity, hazardous wastes, industrial wastes

Metal producers in the United States discard an estimated 88.9 Gg (98 000 tons) of chromium each year in waste slags [*1*]. This constitutes a large potential source of chromium contamination of groundwaters, particularly under the enhanced leaching action of acid precipitation on waste slag dumps. Num-

[1]Research chemist and metallurgist, respectively, U.S. Department of the Interior, Bureau of Mines, Twin Cities Research Center, Minneapolis, Minn. 55417.

erous investigations into the effect of acid precipitation on vegetation, surface waters, animal life, and stone architecture are to be found in the literature [*2–7*]. The potential effects of acid precipitation on waste dump sites have not been addressed. Many sites, such as those from waste ferrous slag disposal, are not considered hazardous under normal conditions. However, the prevalence of abnormal acid precipitation conditions could increase unabated into the future, leading to the possible leaching of hazardous elements from these normally innocuous wastes.

Of the numerous toxic elements that could be leached from waste slag, chromium appears to be of major concern to human health. Chromium in the hexavalent (VI) oxidation state is very toxic, and several of the chromates have been determined to be carcinogenic [*8*]. The U.S. Environmental Protection Agency (EPA) National Interim Primary Drinking Water Standards (proposed in 1976) restrict chromium to a maximum allowable concentration of 50 mg/m^3 [*9*]. Therefore, even moderate leaching of chromium from waste slag landfills into groundwater supplies would potentially exceed this low permissible level. Under certain conditions, there is also the possibility of runoff resulting in the bioaccumulation of chromium in algae or fish, which could cause the movement of the element through the food chain where it could subsequently become available for human consumption [*7*].

Very little has been reported in the literature regarding the leaching of chromium from waste slag [*10–12*]. This report describes research on the leachability of chromium from a variety of industrial slags from stainless steel and chrome alloy operations. The work attempts to assess the potential for chromium leaching from waste slag exposed to acid precipitation and to determine the slag compositions that will predispose a slag to harm the environment through the presence of leachable soluble chromium compounds.

Experimental Procedure

Two methods were used to determine the chromium leachability of the specimen slags. The first was a long-term test using a lysimeter-type apparatus. The second method was a simple, short-term batch test using stirred slurries.

The lysimeter apparatus (diagrammed in Fig. 1) continuously circulated leachate through a bed of -100-mesh (<147-μm particle size) slag. The pH was measured in the upper reservoir and controlled to a selected constant value of 4.5 by the automated addition of doser acid. The doser acid was usually a mixture of 0.25 $kmol/m^3$ (0.5 *N*) in sulfuric acid (H_2SO_4) and 0.34 $kmol/m^3$ (0.34 *N*) in nitric acid (HNO_3), although the acid strength was changed on occasion depending on the slag basicity.

Each lysimeter experiment was begun by adding 0.1 kg of slag sample to a cone of filter paper in the separatory funnel of the apparatus in Fig. 1. Beginning with 250 cm^3 of distilled water in the lower reservoir, the peristaltic pump was started to give a solution circulation rate of 0.217 cm^3/s (13.0 mL/min).

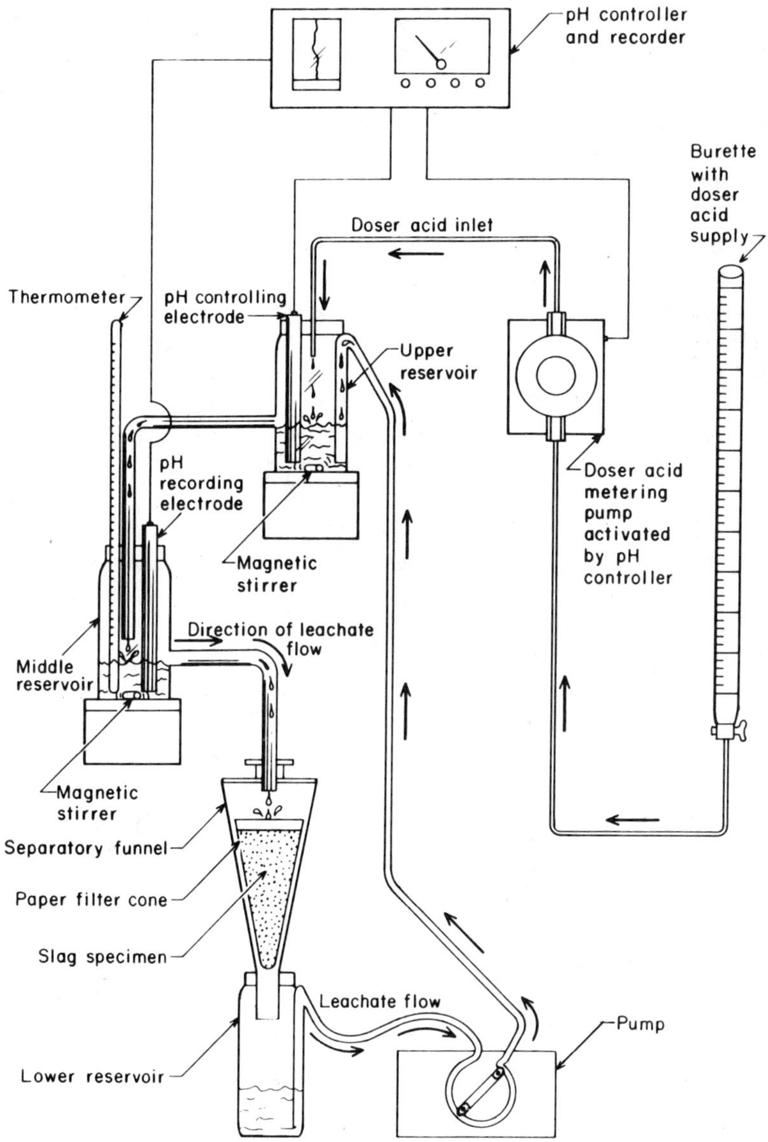

FIG. 1—*Schematic diagram of lysimeter-type apparatus.*

Doser acid immediately dripped into the system to control automatically the pH at 4.5 in response to the controlling electrode of the upper reservoir. The leachate circulation rate of 0.217 cm^3/s enabled a shallow pool of liquid leachate to be maintained above the slag sample, thereby providing a more uniform percolation through the slag bed. The middle reservoir recording electrode monitored the pH of the circulating leachate just before it descended to the slag sample. A Chemtrix pH controller (Model 44R)[2] was used in combination with an FMI acid-metering pump (Model RPG6) to control and measure the pH. The slag specimen was leached 605 ks (7 days) at about a 2.5:1 liquid/solid ratio. In the slurry test, 0.01 kg of −100-mesh slag was stirred for 18 ks (5 h) or 86.4 ks (24 h) in 1000 cm^3 of dilute sulfuric acid leachant (100:1 liquid/solid ratio).

The slags were obtained from industrial stainless steel producers and from a chromium waste recycling operation producing pig metal for stainless steel production. The chemical analyses of the slags as supplied (base Slags A, B, C, and D) are given in Table 1. The compositions of base Slags A, B, and C were altered for many tests by adding reagent-grade calcium oxide (CaO), silicon dioxide (SiO_2), or magnesium oxide (MgO), followed by melting to form a homogeneous mixture. After holding the slag mixture 57.6 ks (16 h) above the fusion point at 1723 K (1450°C) in an air atmosphere, the slag temperature was raised to 1773 K (1500°C), and the slag was poured onto a clean, cold steel plate. In many cases, calcium fluoride (CaF_2) and ferric oxide (Fe_2O_3) were added to aid the pouring of the viscous slag melts from the platinum-rhodium alloy melting crucibles. All the slags were ground to −100 mesh for the leaching tests. The analyses for these altered slags are given in Table 1 (Slags A1 through A6, B1 through B3, and C1 through C10).

The slags and leachates were analyzed using standard procedures. Analyses for silicon, calcium, magnesium, aluminum, and fluorine were accomplished by fusing the slag with sodium peroxide in an iron crucible followed by perchloric acid dehydration and filtration. The filtrate was analyzed by atomic absorption spectrometry (AAS) to determine calcium, magnesium, and aluminum levels. The fluorine level was determined with an ion selective electrode, while the silica residue was determined gravimetrically. The chromium, iron, manganese, and nickel in the slags were analyzed using acid dissolution in hydrochloric acid (HCl), hydrofluoric acid (HF), and HNO_3 followed by AAS. The leachates from the slag leaching experiments were analyzed using AAS except for fluorine, for which the ion selective electrode technique was used.

Experimental Results

Base Slags A, B, and C and several altered slags (A1 and C1 through C5) of Table 1 were leached for 605 ks (7 days) using the lysimeter apparatus to simulate environmental leaching by acid rain. The simulated acid rain solution was controlled at a constant 4.5 pH using a doser acid containing H_2SO_4 and HNO_3 in a 1.16:1 weight ratio. Also, for comparison, a physical mixture containing 4% chromium [as chromic oxide (Cr_2O_3)] in inert polystyrene powder

[2] Reference to specific equipment does not imply endorsement by the Bureau of Mines.

TABLE 1—*Chemical analyses of slags obtained from industrial chrome alloy operations and altered slags prepared by adding reagents and melting in a laboratory furnace.*

Slag	Analyses, weight %								
	Chromium	CaO	SiO_2	MgO	Al_2O_3	Fluorine	Manganese	Iron	Nickel
Base Slag A	0.9	60.4	22.7	8.1	3.5	1.8	0.5	0.4	0.023
A1	0.6	38.8	51.7	2.5	2.0	1.0	...	...	...
A2	0.6	41.1	31.0	16.7	2.3	1.0	...	5.0	...
A3	0.6	41.8	30.4	18.1	2.3	1.0	...	3.9	...
A4	0.7	42.8	30.0	16.7	2.3	0.8	...	4.7	...
A6	1.0	56.8	22.5	7.3	3.0	3.0	...	3.1	...
Base Slag B	0.7	54.0	32.8	4.7	2.4	2.4	1.0	0.2	0.022
B1[a]	0.9	49.7	33.6	6.9	3.2	2.3	1.3	0.7	...
B2[b]	1.0	53.3	30.6	6.0	2.5	3.0	1.0	0.6	...
B3	0.7	57.1	33.5	6.2	2.5	2.1	0.8	...	...
Base Slag C	4.8	29.7	24.2	19.7	6.6	2.3	4.4	1.9	...
C1	0.7	33.5	34.9	18.7	4.2	1.3	1.7	1.5	...
C2	1.2	38.2	28.6	...	...	...	...	...	...
C3	2.3	40.5	24.0	14.0	6.4	1.9	4.8	2.1	...
C4	4.6	39.4	18.9	15.9	5.4	5.1	3.2	1.3	...
C5	2.1	39.2	11.1	9.4	3.3	3.9	2.1	16.5	...
C6	2.4	39.6	24.5	16.1	5.9	2.9	3.6	1.8	...
C7	2.5	47.5	25.1	11.5	5.4	1.9	3.8	1.3	...
C8	2.5	46.4	23.4	9.6	5.6	2.2	3.2	1.6	...
C9	2.1	49.6	18.7	13.2	4.8	2.2	2.9	1.0	...
C10	2.2	52.2	19.3	14.3	4.2	2.1	3.4	1.2	...
Base Slag D	8.1	18.0	31.6	11.9	9.4	3.3	8.4	1.9	...

[a]Slags B1, B2, and B3 were not prepared from the base Slag B given in the table but from a different chunk of base Slag B of higher chromium content.

[b]Slag was not held at melting temperature for 57.6 ks (16 h), but was poured immediately upon melting.

(200 to 400 mesh) was leached under the same conditions. The leachates obtained at the end of the 605-ks (7-day) period were analyzed, and the results are reported in Table 2.

The base Slags A, B, C, and D and several altered slags (A2 through A6, B1 through B3, and C6 through C10) of Table 1 were leached for 18 ks (5 h) or 86.4 ks (24 h) using the slurry method. The chromium leachability results for several leachant sulfuric acid concentrations are given in Table 3.

Discussion

Leaching Methods

The EPA method [*9*] for determining the toxicity of solid waste has several elements in common with the lysimeter and slurry tests used in this work. However, no effort was made to duplicate the EPA extraction procedure. The EPA test has little relevancy when applied to waste ferrous slags. For example, the use of acetic acid to control the pH in the EPA test is not applicable practically or logically to waste slags. Many waste slags contain large amounts of CaO, which consume excessive volumes of a weak acid such as acetic acid to maintain the required pH level. The EPA test is modeled on the codisposal of toxic wastes in an actively decomposing municipal landfill which overlies a groundwater aquifer. Slag dumps are inorganic, relatively sterile wastes compared with the biologically active municipal landfills the EPA had in mind when devising its test.

The waste slag tests developed in this report are based on a leachant composed of the major components of acid precipitation, that is, sulfuric and nitric acids. In the lysimeter tests, these acids were mixed in a weight ratio of H_2SO_4 to HNO_3 of 1.16:1, a value reported in the literature for a sampling of acid rain in the eastern United States in 1975 [*13*]. This ratio varies throughout the United States, but the 1.16 value was considered typical for simulating acid rain percolating through a waste slag heap in the northeastern industrial area of the United States.

The lysimeter tests of base Slags A, B, and C showed these slags to have good resistance to chromium leachability under the test conditions (Table 2). The lysimeter results demonstrate that the solubility of chromium from these slags is less than that of pure Cr_2O_3 when the slags are freshly discharged from the steel-making furnace and before they are subjected to weathering or other environmental conditions. During the leaching, gypsum ($CaSO_4 \cdot 2H_2O$) precipitated from the leachate. Gypsum was deposited as a white powder in various regions of the leaching apparatus, particularly on top of the slag bed, and was identified by X-ray diffraction.

Influence of CaO/SiO_2 Ratio

Although the chromium leachability differences between as-supplied slags were not detectable in the lysimeter testing, the slurry testing revealed marked

TABLE 2—*Chemical analyses of leachate liquids obtained after leaching industrial chrome slags and altered slags for 605 ks (7 days) in the lysimeter.*

	Concentration of Elements in Leachate, mg/l[a]										
Slag	Chromium	Magne- Calcium	Alumi- sium	Silicon	num	Fluorine	Manga- nese	Iron	Sulfur	Nickel	Chromium Leached, %[b]
Cr_2O_3[c]	21	...	...	...	...	...	...	...	626	...	0.1
Base Slag A	<2	1800	3	9	<7	4	<1	<1	370	...	<0.1
A1	4	792	8	41	<7	...	<1	<1	328	...	0.2
Base Slag B	<2	1692	4	40	<7	6	<1	<1	416	0.4	<0.09
Base Slag C	<2	1233	21	7	<7	...	...	...	...	...	<0.01
C1	31	1658	8	18	<7	10	<1	<1	442	...	1.5
C2	16	2263	12	14	<7	9	<1	<1	417	...	0.5
C3	505	4026	77	39	...	...	...	...	314	...	9.8
C4	1198	4275	11	36	...	1	...	<1	355	...	12.8
C5	298	3095	13	27	<7	15	...	<1	333	...	5.7

[a]0.001 × milligrams per/litre = kilograms per cubic metre (SI units).
[b]Percentage of initial chromium in slag leached into liquid as calculated from leachate analysis.
[c]A physical mixture (0.1kg) of reagent Cr_2O_3 + polystyrene powder was leached.

TABLE 3—*Chemical analyses of chromium in leachate liquids obtained after leaching industrial chrome slags and altered slags using the slurry method.*

Slag	Hours[a] Leached	H_2SO_4 in leachant, g/L[b]	Chromium in Leachant, mg/L[c]	Chromium Leached, %
Base Slag A	5	0[d]	<0.5	<0.5
	5	10.56	35	37.8
	5	21.12	85	90.6
	5	42.24	62	67.0
A2	24	3.52	4	6.8
	24	8.80	12	20.2
A3	24	3.52	3	4.9
	24	8.80	6	10.9
A4	24	3.52	6	8.7
	24	8.80	17	24.5
A6	24	3.52	34	33.6
	24	8.80	58	57.6
Base Slag B	5	8.80	24	36.3
	5	17.60	29	43.0
	5	35.20	39	59.1

B1	5	8.80	50	56.8
B2	5	8.80	50	49.6
B3	5	8.80	46	69.9
Base Slag C	5	7.04	2	0.8
	5	14.08	12	5.5
	5	28.16	11	5.1
C6	5	7.04	34	14.2
	5	14.08	95	39.6
C7	5	8.80	96	38.4
C8	5	8.80	180	81.8
C9	5	8.80	150	70.5
C10	5	8.80	133	60.4
Base Slag D	5	8.80	96	11.6
	5	17.60	102	12.3
	5	35.20	106	12.9

[a] 3.6 × hours = kiloseconds (SI units).
[b] Grams per/litre = kilograms per cubic metre.
[c] 0.001 × milligrams per/litre = kilograms per cubic metre.
[d] Leached with distilled water.

differences. Under the conditions of slurry testing with H_2SO_4 leachant, base Slags A and B showed considerable vulnerability to chromium leaching, base Slag D showed moderate leachability, and base Slag C demonstrated high resistance to the sulfuric acid solutions (Table 3). The leaching susceptibility of base Slags A and B appeared to be influenced by the slag CaO/SiO_2 ratio (the components of the ratio, CaO and SiO_2, being in concentration units of weight percent).

Figure 2 plots the ratio of CaO/SiO_2 versus the percentage of chromium extracted during a 18-ks (5-h) period of slurry leaching at various concentrations of sulfuric acid leachant. The percentages of chromium leached from Slags A and C in the figure are approximate values calculated by interpolating between experimental results in Table 3 for adjacent leachants of different H_2SO_4 concentrations. The figure reveals a minimum chromium leachability occurring somewhere between a 0.57 and 1.65 CaO/SiO_2 ratio for all concentrations of leachant acid. For the weakest concentration, 8.8 kg/m^3 H_2SO_4, a maximum chromium leachability appeared to be between 1.65 and 2.66 CaO/SiO_2.

This pattern of chromium leachability dependence on the CaO/SiO_2 ratio was duplicated in both the lysimeter and slurry tests when base Slags A, B, and C were altered and melted in air. The chromium leachability data from the lysimeter and slurry testing for alterations of Slag C (slags C1 through C5, Table

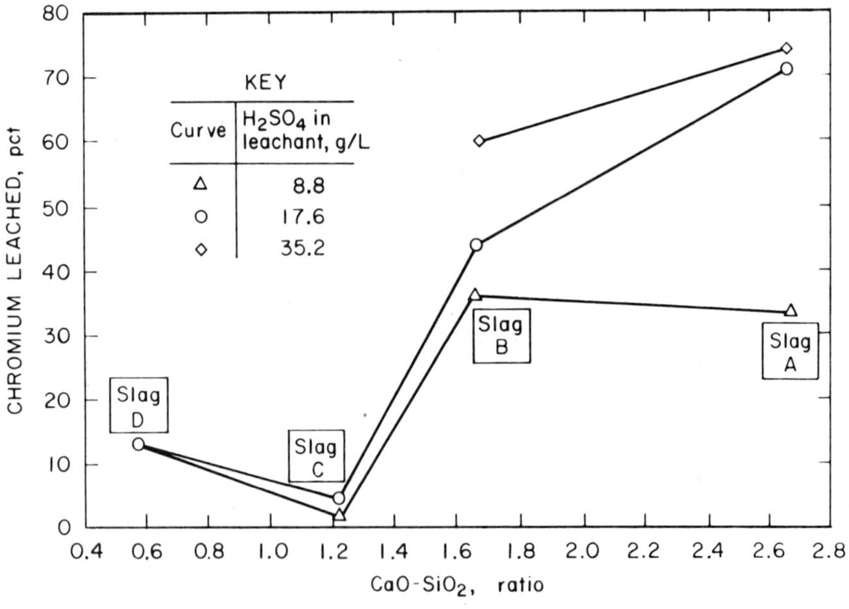

FIG. 2—*Chromium leachability dependence on CaO/SiO_2 ratio for base slags when leached with dilute sulfuric acid solutions using the slurry method (g/L = kg/m^3).*

2; and slags C6 through C10, Table 3) were plotted against CaO/SiO_2 to give Fig. 3. In this case, the maximum leachability is more pronounced and occurs near the 2.0 CaO/SiO_2 ratio. Although the percentages of chromium leached at any particular CaO/SiO_2 ratio are different under the two testing conditions (slurry and lysimeter) the pattern of minima and maxima are the same, that is, a minimum occurs at about a ratio of 1.3 CaO/SiO_2 and a maximum at about a 2.0 ratio. The results indicate that waste chromium-bearing slags are particularly susceptible to leaching of chromium by simulated acid precipitation when they contain calcium in amounts sufficient to give a CaO/SiO_2 ratio of about 2.

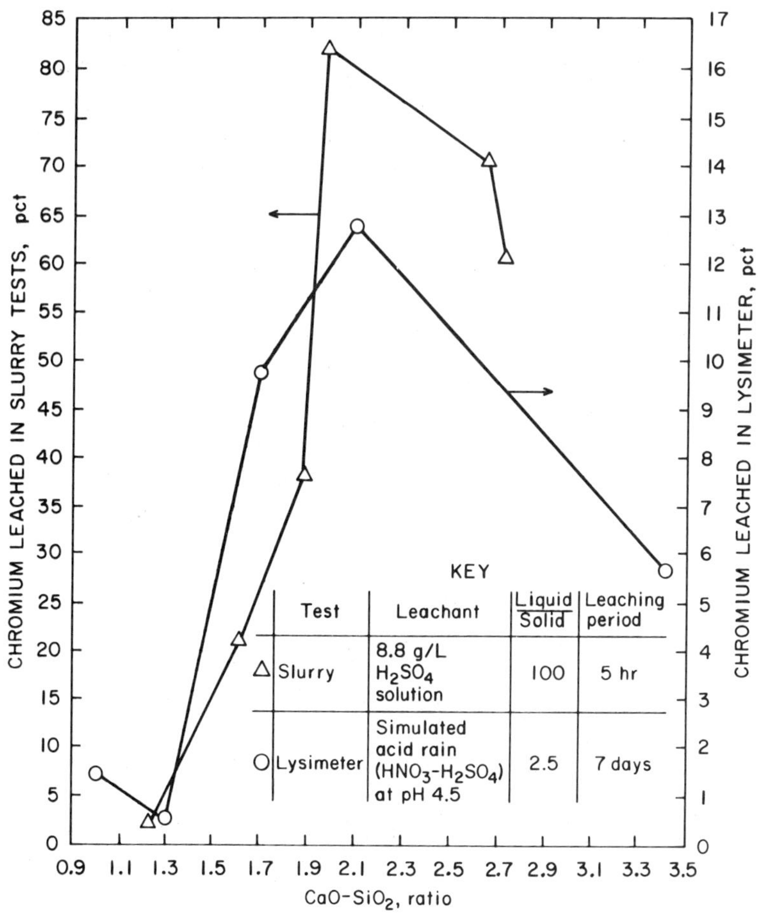

FIG. 3—*Comparison of chromium leachability from altered Slag C in slurry and lysimeter tests ($g/L = kg/m^3$, $3.6 \times h = ks$, and $86.4 \times days = ks$).*

Phase Diagram

An examination of the CaO-SiO_2-Cr_2O_3 system delineated by Glasser and Osborn [*14*] indicates the importance of the 2.0 ratio to waste slag chromium leachability. Figure 4 presents the subsolidus phase diagram for the ternary system at 1623 K (1350°C). At the low chromium concentrations (less than 8.1%) encountered in the tested waste slags, the diagram shows a phase transformation near the 2.0 CaO/SiO_2 ratio depending on the slag's chromium content. If lime is added to the system in the 1.5 to 2.0 CaO/SiO_2 region, two phase boundaries are crossed near the 2.0 ratio and Cr_2O_3 is converted to calcium chromite ($CaO \cdot Cr_2O_3$).

Oxidation to Calcium Chromate

Calcium chromite is not greatly soluble in acid [*15*], but in the presence of air and CaO at elevated temperatures it is readily oxidized to acid-soluble calcium chromate ($CaCrO_4$) [*15-17*]. The relative insolubility of calcium chromite is evident in Table 2, where the calcium chromite in base Slag A (calcium chromite presumed present, as derived from the phase diagram of Fig. 4 for a 2.66 CaO/SiO_2 ratio) did not leach under the lysimeter test conditions. How-

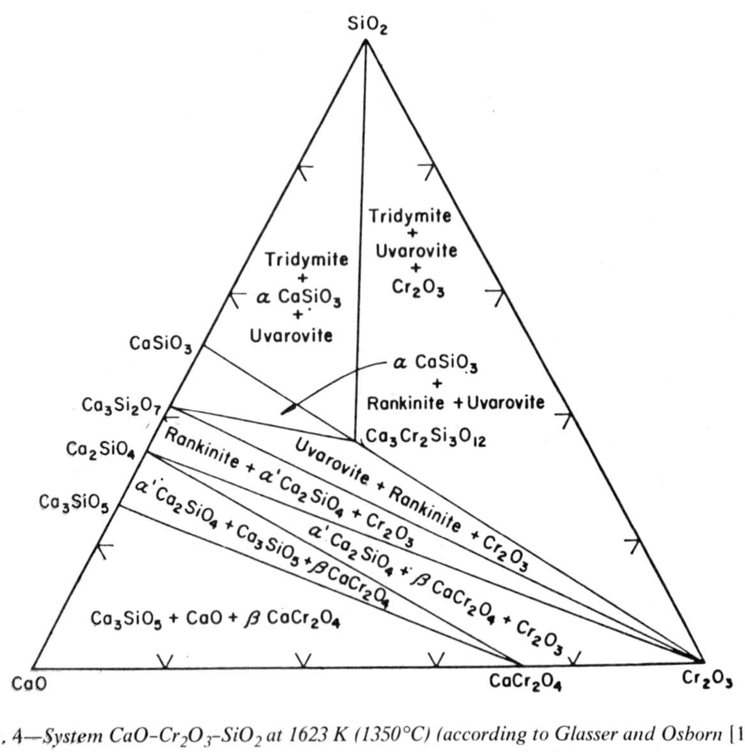

FIG. 4—*System CaO-Cr_2O_3-SiO_2 at 1623 K (1350°C) (according to Glasser and Osborn* [14]*).*

ever, the greater solubility of calcium chromite than of other chromium phases below the 2.0 CaO/SiO_2 ratio was revealed when the base slags were subjected to the stronger acid and higher liquid/solid ratio of the slurry testing (Table 3 and Fig. 2).

The oxidation of calcium chromite to $CaCrO_4$ renders the slag's chromium content very susceptible to leaching, even under the milder conditions of lysimeter testing (Table 2, Slags C3 through C5). The preparation of altered slags by melting for an extensive period under air atmosphere ensures the oxidation of calcium chromite to calcium chromate and explains the greater chromium leachability of altered slags in the composition range of calcium chromite formation.

It should be pointed out that $CaCrO_4$ cannot exist in the slag in the steelmaking furnace. Before it is tapped, the slag is in contact with the molten iron bath, which would reduce $CaCrO_4$ by such slag/metal reactions as

$$3Fe^0_{(metal)} + 2CaCrO_{4(slag)} \rightarrow CaO \cdot Cr_2O_{3(slag)} + 3FeO_{(slag)} + CaO_{(slag)} \quad (1)$$

$$3Fe^0_{(metal)} + 2CaCrO_{4(slag)} \rightarrow FeO \cdot Cr_2O_{3(slag)} + 2FeO_{(slag)} + 2CaO_{(slag)} \quad (2)$$

The poor leachability of base Slag A in the lysimeter test compared with those of altered slags of similar CaO/SiO_2 ratio probably indicates that Slag A has not oxidized appreciably. Apparently, when the slag was tapped from the industrial furnace and cooled, time for oxidation of calcium chromite was insufficient, despite the severance of slag/metal bath contact. Entrained metallic particles in the slag may also inhibit oxidation during this cooling period.

The question arises as to whether $CaO \cdot Cr_2O_3$ in waste slag would oxidize to yield soluble $CaCrO_4$ under conditions of disposal in the environment (long exposure to air, summer heat, and acid precipitation). The present work does not answer this question with certainty because of short leaching periods used. The free energy change for the oxidation of $CaO \cdot Cr_2O_3$ was not calculated since a free energy value for $CaO \cdot Cr_2O_3$ could not be located in the thermodynamic literature. However, thermodynamic calculations show that the oxidation reaction

$$2CaO + Cr_2O_3 + 1.50_2 \rightarrow 2CaCrO_4 \quad (3)$$

is feasible at ambient temperatures. The standard free energy of this reaction was calculated to be 299.6 kJ (−71.6 kcal). Dufau [*16*] found experimentally that the oxidation of $CaO \cdot Cr_2O_3$ begins below 100°C in air. Therefore, waste slag containing chromium in the form of calcium chromite is believed particularly vulnerable to oxidation to soluble $CaCrO_4$ when exposed for a long period in the environment. Hence, a slag composition having a CaO/SiO_2 ratio near 2.0 should be generally avoided, since it predisposes the slag to the formation of oxidizable calcium chromite.

This recommendation might be expanded with caution to include the range well above 2.0 ratio, except that the experimental data do not correspond well with the phase diagram information. The large drop off in chromium leachability for altered slags having ratios greater than about 2.6 (Fig. 3) is not completely understood. The moderately reduced chromium leachability between the ratios 2.0 and 2.6 is reasonable, since lime addition beyond the maximum point of 2.0 CaO/SiO_2 does not create more calcium chromite but simply dilutes the slag and consumes acid. The severe drop-off between 2.6 to 3.4, however, requires a different explanation. A possible explanation may be in the limitations of the leaching system. High lime addition could cause excessive precipitation and adsorption of gypsum on the slag residue. This adsorbed layer could act as a barrier to prevent reaction between the H_2SO_4 leachant and calcium chromate in the slag. Figure 2 indicates that the phenomenon is influenced by acid strength. The unaltered slags of Fig. 2 display no drop off of chromium leachability at high CaO/SiO_2 ratios for the stronger acid concentrations. The result suggests that the gypsum barrier, if it exists, dissolves in the stronger acid to allow normal chromium leaching.

Influence of MgO

When the chromium leachability of various alterations of Slags A and B were compared with those of altered Slag C, numerous deviations indicated that the CaO/SiO_2 ratio was not the sole composition factor controlling chromium leachability. The variation seemed random for slags with CaO/SiO_2 ratios higher than 1.9, but in the region between ratios 1.2 and 1.9 the leachability appeared to follow a sequence related to slag magnesium concentration. In this region, chromium leached less readily as the magnesium content in the slag was increased. This result is illustrated in Fig. 5, which plots the chromium leachability of altered slags in slurry testing (Slags A2 through A4, B1 through B3, and C6 through C7) against the slag MgO content in the region of interest. The chromium leachability as a function of magnesium concentration appears to have approximately linear characteristics in the 6 to 20% MgO range examined. The dependence of chromium leachability on magnesium content was further supported by a similar plot (Fig. 6) of the slurry test data from base Slags A, B, C, and D. As in Fig. 2, the percentages of leached chromium for Slags A and C are interpolated values. Figure 6 shows that increasing the magnesium content in slag again depresses the slag chromium leachability, with the exception of Slag A at the stronger acid concentrations. In this case, the high CaO/SiO_2 ratio (2.66) rather than slag MgO content is the dominating factor influencing chromium leachability.

Phase Diagrams

The inhibiting effect of increased slag magnesium content on chromium leachability, in the range below 2.0 CaO/SiO_2, appears to originate from the

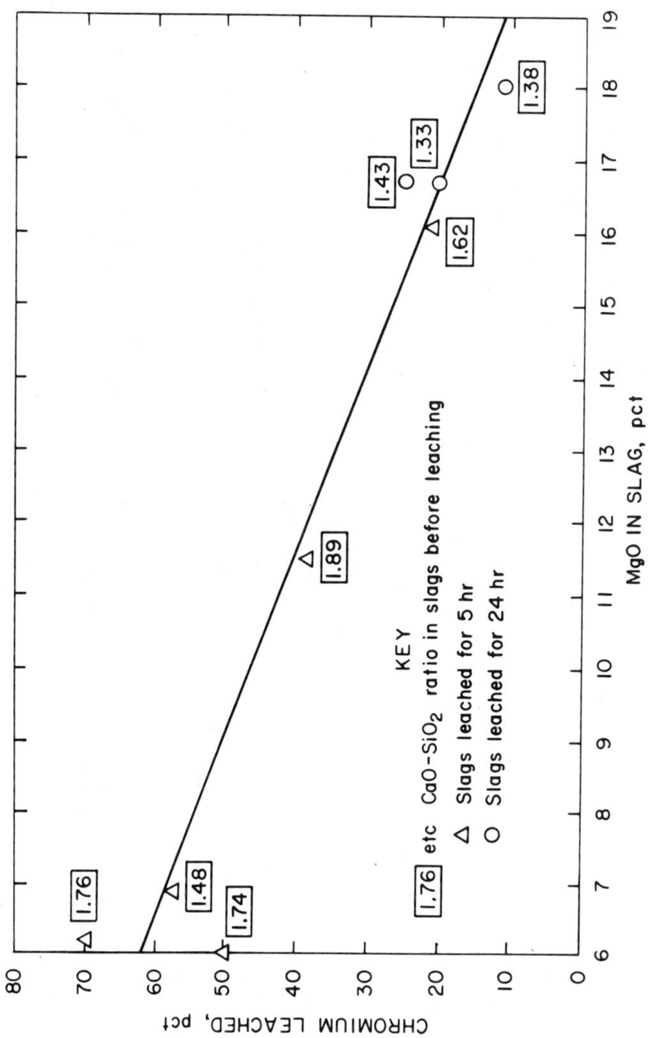

FIG. 5—*Dependence of chromium leachability of altered slags on slag magnesia content in the range 1.33 to 1.89 CaO/SiO_2. All the slags were slurry tested in 8.8 kg/m^3 H_2SO_4 solution (3.6 × h = ks).*

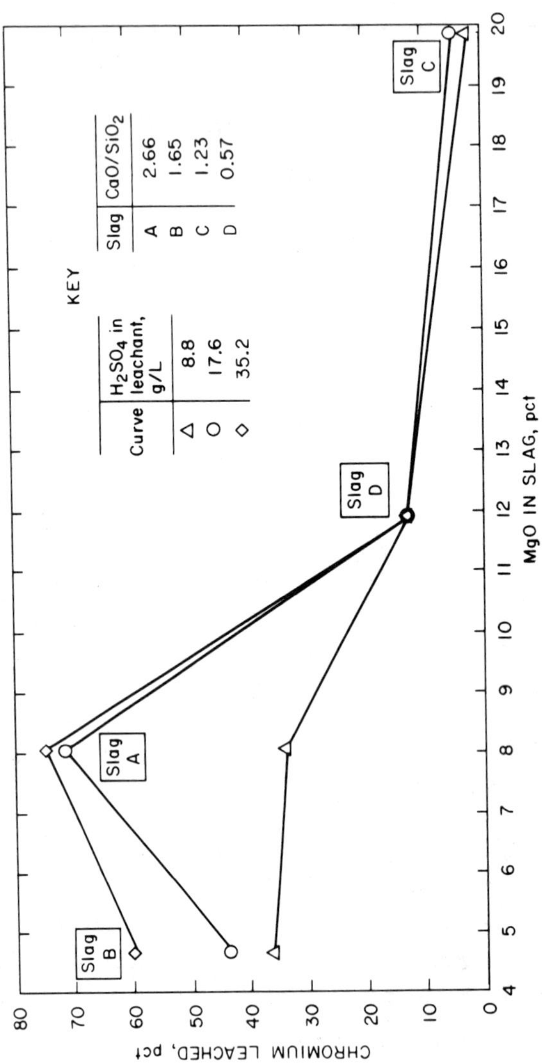

FIG. 6—*Effect of magnesium concentration in slag on the slurry testing of base Slags A, B, C, and D in leachants containing various concentrations of* H_2SO_4 *(g/L = kg/m*^{3}*).*

formation of insoluble spinel compounds. White's research into refractory-slag systems reports a fact remarkably pertinent to waste slag chromium leachability [*18–19*]. He showed that the CaO/SiO_2 ratio of 2.0 was a critical factor in determining the phase combinations present in chrome-magnesite refractory. White determined that the phase combinations up to a ratio of 2.0 CaO/SiO_2 contained all the chromium in a spinel, solid-solution phase with MgO, FeO, Fe_2O_3, and alumina (Al_2O_3). At ratios greater than 2.0, the spinel structure gave way to formation of calcium chromites, ferrites, and aluminates.

The quaternary phase diagram for the system CaO-SiO_2-Cr_2O_3-MgO, shown in Fig. 7, was developed by Berezhnoi and Gul'ko [*20*]. Examination of this phase diagram indicates that formation of the $MgO \cdot Cr_2O_3$ spinel (picrochromite) is likely responsible for the diminished chromium leachability for slags of high magnesium content. As the tetrahedron of the system is traversed through its internal, elementary tetrahedra enclosing various phase domains, the compositional requirements for existence of picrochromite become evident. Thus, if a traverse is made in the direction of increasing lime content at relatively low and constant Cr_2O_3 and MgO levels, it is evident that the formation of $MgO \cdot Cr_2O_3$ spinel ends on crossing the phase boundary at the 2.0 CaO/SiO_2 ratio. Above the 2.0 ratio, chromium exists wholly in the $CaO \cdot Cr_2O_3$ state, which, as determined earlier, is readily oxidized to acid-soluble $CaCrO_4$ under appropriate conditions.

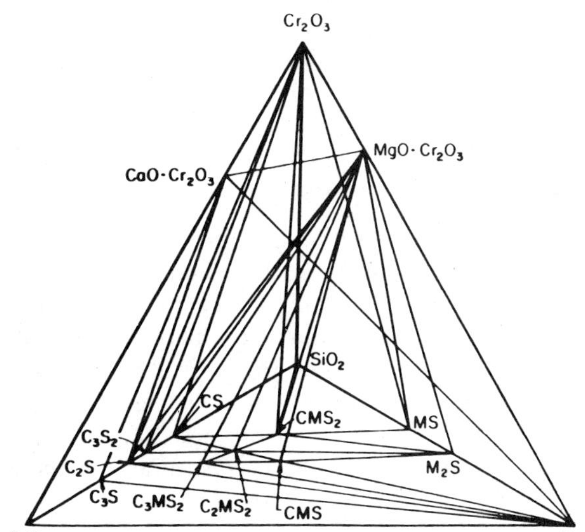

FIG. 7—*System CaO-MgO-Cr_2O_3-SiO_2, showing subsolidus compatability relations. C = CaO, M = MgO, S = SiO_2 (according to Berezhnoi and Gul'ko* [20]*).*

Stability of $MgO \cdot Cr_2O_3$

Picrochromite is an exceedingly stable compound that is very resistant to dissolution by mineral acids. Dufau [*16*] observed that the compound was attacked with difficulty by HF and HCl, while Ebelman [*21*] said it was not attacked at all by concentrated HCl. Doerner [*15*] found picrochromite to be insoluble even in nitric acid. Figure 6 likely reveals the superior resistance of picrochromite to acid leaching over other chromium species in the slag, such as Cr_2O_3, $FeO \cdot Cr_2O_3$, and mixed spinel $(Mg,Fe,Al)O \cdot Cr_2O_3$. The greater leachability of base Slag A (Fig. 6) probably demonstrates the greater solubility of calcium chromite in strong sulfuric acid solution in comparison with the spinel and other chromium compounds present in Slags, B, C, and D.

The picrochromite, unlike calcium chromite, is very resistant to oxidation. Bayer and Wiedemann [*22*] studied the reaction of MgO with Cr_2O_3 in air using differential thermal analysis. They found the stable spinel $MgO \cdot Cr_2O_3$ was formed preferentially. Oxidation to soluble magnesium chromate ($MgCrO_4$) required high oxygen pressures. This resistance of picrochromite to oxidation under normal conditions was confirmed by Ford and Rees [*23*]. Therefore, the picrochromite is a most desirable form for chromium to be disposed to the environment since it likely possesses excellent resistance both to oxidation and solubility under acid rain conditions.

Ford and Rees [*24*] also reported that the system CaO-SiO_2-Cr_2O_3-MgO is analogous to the systems CaO-SiO_2-Cr_2O_3-Fe_2O_3 and CaO-SiO_2-Cr_2O_3-Al_2O_3. The oxidation product $CaCrO_4$ would again be expected to form in these systems when the CaO/SiO_2 ratio achieved a 2.0 value. Therefore, in more complex actual slag systems containing both iron oxide and alumina, one may still expect the chromium leachability to increase remarkably near the 2.0 CaO/SiO_2 ratio.

Conclusions

The authors conclude that certain compositions of chromium-bearing waste slag are susceptible to leaching of their chromium content into the environment under conditions of acid precipitation. There are two composition factors which predispose slag to chromium leachability: (1) the CaO/SiO_2 ratio and (2) the magnesium content. When the ratio is above 2.0, chromium is present in normal ferrous slags (where CaO and SiO_2 are the majority constituents) as calcium chromite regardless of the magnesium content. Calcium chromite is slightly soluble in acid and might leach significantly under severe acid rain conditions. Of more concern is the possibility that calcium chromite may oxidize to calcium chromate under conditions of disposal in the environment. Calcium chromate would be readily soluble when exposed to acid precipitation in the industrialized areas of the United States where slags are normally discarded. The chromium leached would be in the hexavalent state, the condition most dangerous to man and vegetation. This potential problem can be minimized if the slag composition is well below a 2.0 CaO/SiO_2 ratio, with enough magnesium present to form the insoluble picrochromite ($MgO \cdot Cr_2O_3$).

References

[1] "Industrial Wastes: An Unexplored Source of Valuable Minerals," GAO Report EMD-80-45, Comptroller General's Report to the Congress, U.S. General Accounting Office, Washington, D.C., 15 May, 1980.

[2] Kish, T., "Acid Precipitation: Crucial Questions Still Remain Unanswered," *Journal of the Water Pollution Control Federation*, Vol. 53, No. 5, May 1981, pp. 518-521.

[3] Glass, N. R., Glass, G. E., and Rennic, P. J., "Effects of Acid Precipitation," *Environmental Science and Technology*, November 1979, Vol. 13, No. 11, pp. 1350-1355.

[4] Gauri, K. L. and Holdren, G. C., Jr., "Pollutant Effects on Stone Monuments," *Environmental Science and Technology*, Vol. 15, No. 4, April 1981, pp. 386-390.

[5] Hilleman, B., "Acid Precipitation," *Environmental Science and Technology*, Vol. 15, No. 10, October 1981, pp. 1119-1124.

[6] *Acid Rain*, EPA Report 600/9-79-036, U.S. Environmental Protection Agency, Washington, D.C., July 1980.

[7] Kincannon, D. F., Bates, M. H., and DeVries, R. N., "Control of Pollution in Waters Containing Heavy Metals," Technical Completion Report, Project No. A-087-OKLA, Oklahoma Water Resources Research Institute, Oklahoma State University, Stillwater, Oklahoma, September 1980.

[8] Hartford, W. H., in *Encyclopedia of Chemical Technology*, 3rd Ed., Vol. 6, Wiley, New York, 1979, pp. 55-102.

[9] U.S. Environmental Protection Agency, "Hazardous Waste Management System," *Federal Register*, Vol. 45, No. 98, 19 May 1980, pp. 33127-33128, 33257.

[10] Nagaya, K. "Innoculation of Chrome Ore Slag," Japanese Patent 77, 27, 097, 1 March 1977; *Chemical Abstracts*, Vol. 87, 1977, p. 26589v (English summary).

[11] Eguchi, M. and Uchida, R., "Study of the Fixing of Chromium with Iron-Silicate Slag, II: Effects of Oxygen Potential and Lime on the Fixing with Melting Treatment," *Tohoku Daigaku Senko Seiren Kenkyusho Iho*, Vol. 34, Nol. 1, 1978, pp. 35-44 (Japanese); *Chemical Abstracts*, Vol. 90, 1979, p. 26827h (English summary).

[12] Gemmell, R. P., "Revegetation of Derelict Land Polluted by a Chromate Smelter, Part 1: Chemical Factors Causing Substrate Toxicity in Chromate Smelter Waste," *Environmental Pollution*, Vol. 5, No. 3, 1973, pp. 181-197.

[13] Likens, G. E., "Acid Precipitation," *Chemical and Engineering News*, Vol. 54, 22 November 1976, pp. 29-44.

[14] Glasser, F. P. and Osborn, E. F., "System CaO-Cr_2O_3-SiO_2" in *Phase Diagrams for Ceramists*, E. M. Levin, C. R. Robbins, and H. F. McMurdie, Eds., American Ceramic Society, Columbus, Ohio, 1964, pp. 226-227.

[15] Doerner, H. A., "Roasting of Chromite Ores to Produce Chromates," BuMines RI 2999, U.S. Bureau of Mines, Department of the Interior, Washington, D.C., 1930.

[16] Dufau, E., "Properties of Calcium Chromite and Magnesium Chromite" in *A Comprehensive Treatise on Inorganic and Theoretical Chemistry*, J. V. Mellor, Wiley, New York, 1931, pp. 198-199.

[17] Ford, W. W. and White, J., "CaO-Cr_2O_3 System" in *Phase Diagrams for Ceramists*, E. M. Levin, C. R. Robbins, and H. F. McMurdie, Eds., American Ceramic Society, Columbus, Ohio, 1964, pp. 226-227.

[18] White, J., "Phase Distribution in Ceramics" in *Ceramic Microstructures*, R. M. Fulbrath and J. A. Pask, Eds., Wiley, New York, 1968, Chapter 35, pp. 728-762.

[19] White, J., "Magnesia-Based Refractories" in *High Temperature Oxides, Part I: Magnesia, Lime, and Chrome Refractories*, A. M. Alper, Ed., Academic Press, New York, 1970, Chapter 2, pp. 77-141.

[20] Berezhnoi, A. S. and Gul'ko, N. V., "The System CaO-MgO — Cr_2O_3-SiO_2" in *Phase Dia-*

grams for Ceramists, E. M. Levin, C. R. Robbins, and H. F. McMurdie, Eds., American Ceramic Society, Columbus, Ohio, 1975 Supplement, p. 279.

[*21*] Ebelmen, J. J., "Properties of Magnesium Chromite" in *A Comprehensive Treatise on Inorganic and Theoretical Chemistry*, J. V. Mellor, Wiley, New York, 1931, p. 199.

[*22*] Bayer, G. and Wiedemann, H. G., "Formation and Decomposition of Calcium and Cadmium Chromate During Solid State Reactions Between Cr_2O_3 and Carbonates and Hydroxides" in *Proceedings*, Fifth International Conference on Thermal Analysis, H. H. Chihara, Ed., Heyden, Bellmawr, N.J., 1977, pp. 437-439.

[*23*] Ford, W. F. and Rees, W. J., "System CaO-MgO-Cr_2O_3" in *Phase Diagrams for Ceramists*, E. M. Levin, C. R. Robbins, and H. F. McMurdie, Eds., American Ceramic Society, Columbus, Ohio, 1964, p. 209.

[*24*] Ford, W. F. and Rees, W. J., "The CaO-MgO-SiO_2-Al_2O_3-Cr_2O_3-Fe_2O_3 System," *Transactions of the British Ceramic Society*, Vol. 57, No. 5, 1958, pp. 233-241.

Kau-Fui V. Wong,[1] *Subrata Sengupta,*[1] *and Dipanker Dasgupta*[1]

Prediction of Cation Mobility in Soils from the Linear Distribution Coefficient

REFERENCE: Wong, K.-F. V., Sengupta, S., and Dasgupta, D., **"Prediction of Cation Mobility in Soils from the Linear Distribution Coefficient,"** *Hazardous and Industrial Waste Management and Testing: Third Symposium, ASTM STP 851*, Larry P. Jackson, Alan R. Rohlik, and Richard A. Conway, Eds., American Society for Testing and Materials, Philadelphia, 1984, pp. 81-91.

ABSTRACT: The ability to predict pollutant migration in the soil should be one of the objectives in disposal operations because the soil is the ultimate receptacle of the waste. The objective of the present work is to determine the relative mobility of different cations in a mixture traveling through the soil. The method used here is to derive a relationship from the exact solution to the one-dimensional, convective-dispersive-adsorptive equation which describes the motion of a cation within a complex mixture moving through the soil. It has been found that the length of time a cation in solution mixture takes to travel a certain depth of soil is directly proportional to $(1 + K')$ where $K' = [K\rho(1 - \epsilon)]/\epsilon$. K is the linear distribution coefficient.

KEY WORDS: pollutant, migration, distribution coefficient, adsorption isotherm, hazardous wastes, industrial wastes

Since the soil is the ultimate receptacle of waste, disposal operations are an environmental concern. Pollutant migration from a disposal site is an important consideration in disposal operations. Thus, the ability to predict pollutant migration in the soil is of environmental interest.

The objective of the present work is to determine the relative mobility of different cations in a complex mixture traveling through the soil. The method used is to derive a relationship from the solution to the convective-dispersive-adsorptive equation which describes the motion of a cation within the complex mixture.

[1]Associate professor, professor, and research assistant, respectively, Department of Mechanical Engineering, University of Miami, Coral Gables, Fla. 33124.

For the one-dimensional flow of a cation in a mixture through a soil column, the concentration of Ion A is given by the following transport equation.

$$\frac{\partial C}{\partial t} + V_0 \frac{\partial C}{\partial Z} = D_0 \frac{\partial^2 C}{\partial Z^2} - \frac{(1-\epsilon)\rho}{\epsilon} \frac{\partial S}{\partial t} \tag{1}$$

where

C = concentration of Ion A in solution,
V_0 = average pore velocity,
Z = longitudinal coordinate along flow direction,
t = time,
D_0 = dispersion coefficient,
ϵ = porosity, and
S = amount of ion A reacted per unit weight of soil.

Equation 1 may be normalized to

$$\frac{\partial X}{\partial t} + V_0 \frac{\partial X}{\partial Z} = D_0 \frac{\partial^2 X}{\partial Z^2} - \frac{K\rho(1-\epsilon)}{\epsilon} \frac{\partial X}{\partial t} \tag{2}$$

where

$X = \dfrac{C}{C_0}$,
$S = KC$,
C_0 = initial concentration of Ion A in solution,
K = linear distribution coefficient of Ion A between the solution and the soil, and
ρ = soil density.

This equation reduces to

$$\frac{\partial X}{\partial t} + V \frac{\partial X}{\partial Z} = D \frac{\partial^2 X}{\partial Z^2} \tag{3}$$

where

$$D = \frac{D_0}{1 + \dfrac{K_\rho(1-\epsilon)}{\epsilon}} \tag{4a}$$

$$V = \frac{V_0}{1 + \dfrac{K_\rho(1-\epsilon)}{\epsilon}} \tag{4b}$$

Appropriate initial and boundary conditions are

$$X = 1.0 \qquad Z = 0 \tag{5a}$$

$$\frac{\partial X}{\partial A} = 0 \qquad Z = L \tag{5b}$$

$$X = 0 \qquad t = 0 \text{ in } 0 \leq Z \leq L \tag{5c}$$

where L is the total length of the soil column. Actually, a practical condition at $Z = 0$ would be the third type (Robin) of boundary condition [1]. However, this complicates the problem and, for most cases, gives results that are essentially the same as the first type (concentration specified) of boundary condition results. Lai and Jurinak [2,3] used a first-type boundary condition.

The exact analytical solution to Eq 3, subject to Eq 5, is given by Cleary and Adrian [1]. This solution is reproduced in Eq 6.

$$X = 1 - 2 \sum_{m=1}^{\infty} \exp\left[\left(\frac{VZ}{2D}\right) - \left(\frac{V^2 t}{4D}\right) - D\beta_m^2 t\right]$$
$$\times \frac{\beta_m^2 + \left(\frac{V}{2D}\right)^2}{L\left\{\beta_m^2 + \left(\frac{V}{2D}\right)^2\right\} + \left(\frac{V}{2D}\right)} \sin \beta_m Z$$
$$\times \frac{\exp\left(-\frac{VL}{2D}\right)\left[-\frac{V}{2D}\sin \beta_m L - \beta_m \cos \beta_m L\right] + \beta_m}{\frac{V^2}{4D^2} + \beta_m^2} \tag{6}$$

where the eigenvalues, β_m, are the positive roots of

$$\beta_m \cot \beta_m L = -\left(\frac{V}{2D}\right) \tag{7}$$

Derivation

The exact solution in Eq 6 may be written as

$$X = 1 - 2 \sum_{m=1}^{\infty} \exp\left[\left(\frac{VZ}{2D}\right) - \left(\frac{V^2 t}{4D}\right) - D\beta_m^2 t\right]$$
$$\times f\left(\frac{V}{D}, \beta_m, Z\right) \tag{8}$$

It is evident that the function $f(V/D, \beta_m, Z)$ is independent of the parameter K. The variation of X with time t is dependent on K if the other parameters are kept constant. Consider two ions, A and B, with distribution coefficients, K_1 and K_2. Let X_1 = concentration profile of Ion A after time t_1 and X_2 = concentration profile of Ion B after time t_2.

$$X_1 = 1 - 2 \sum_{m=1}^{\infty} \exp\left[\left(\frac{V_1 Z}{2D_1}\right) - \left(\frac{V_1^2 t_1}{4D_1}\right) - D_1 \beta_m^2 t_1\right] \times f_1\left(\frac{V_0}{D_0}, \beta_m, Z\right) \quad (9)$$

$$X_2 = 1 - 2 \sum_{m=1}^{\infty} \exp\left[\left(\frac{V_2 Z}{2D_2}\right) - \left(\frac{V_2^2 t_2}{4D_2}\right) - D_2 \beta_m^2 t_2\right] \times f_2\left(\frac{V_0}{D_0}, \beta_m, Z\right) \quad (10)$$

If the concentration profile of Ion A after time t_1 is identical to that of Ion B after time t_2, then

$$X_1 = X_2 \quad (11)$$

One solution of Eq 11 is

$$\exp\left[\frac{V_1 Z}{2D_1} - \left(\frac{V_1^2 t_1}{4D_1}\right) - D_1 \beta_m^2 t_1\right] = \exp\left[\frac{V_2 Z}{2D_2} - \left(\frac{V_2^2 t_2}{4D_2}\right) - D_2 \beta_m^2 t_2\right]$$

Hence

$$\frac{V_1^2 t_1}{4D_1} + D_1 \beta_m^2 t_1 = \frac{V_2^2 t_2}{4D_2} + D_2 \beta_m^2 t_2 \quad (12)$$

$$\frac{t_2}{t_1} = \frac{V_1 \cdot \dfrac{V_0}{4D_0} + D_1 \beta_m^2}{V_2 \dfrac{V_0}{4D_0} + D_2 \beta_m^2}$$

$$= \frac{\dfrac{1}{1 + \dfrac{K_1 \rho(1-\epsilon)}{\epsilon}\left[V_0 \dfrac{V_0}{4D_0} + D_0 \beta_m^2\right]}}{1 + \dfrac{K_2 \rho(1-\epsilon)}{\epsilon}\left[V_0 \dfrac{V_0}{4D_0} + D_0 \beta_m^2\right]}$$

$$= \frac{1 + \frac{K_2\rho(1-\epsilon)}{\epsilon}}{1 + \frac{K_1\rho(1-\epsilon)}{\epsilon}} = \frac{1 + K_2'}{1 + K_1'} \tag{13}$$

where

$$K' = \frac{K\rho(1-\epsilon)}{\epsilon}$$

Equation 13 gives the relationship between the relative migration rates for the ions and their distribution coefficients. It states that a large distribution coefficient gives rise to a slower migration rate, that is, a longer length of time. The mobility or the migration rate of an ion may be interpreted as inversely proportional to the length of time the ion takes to migrate a certain distance under a fixed set of physical conditions.

Experimental Method

The Florida soil samples used were the Hallandale and the Plantation soil series. Table 1 lists the particle-size distribution analysis of the two soil series [*1*]. The chemical analyses and physical properties of the soils are listed in Table 2.

From Tables 1 and 2, it is evident that the Hallandale soil series does not show much variation from 0 to 40.6 cm. This is the soil used for the present work and referred to as Hallandale fine sand. Also, at least three chemically distinct layers are evident in the Plantation soil series—Plantation muck top layer (Horizon I A1), Plantation muck middle layer (Horizon II A2), and Plantation muck bottom layer (Horizon II C1).

The liquid mixture used was a composite from the Pompano Beach, Florida, anaerobic digestion plant. The composite liquid was obtained by mixing portions of the filtrate collected from the plant over a period of several months.

The equilibration time was first determined experimentally. This is the length of time needed to equilibrate the soil and the liquid mixture, so that the ionic concentration in the liquid remains constant with respect to time.

The chemical distribution coefficients were determined by the following procedure:

1. In five separate 250-mL bottles, 0.2 g, 1 g, 2 g, 5 g, and 10 g of soil were mixed with 100 mL of the characteristic filtrate. The pH of the filtrate was kept at its natural level of 6.8.
2. The bottles were shaken for the length of the equilibration time, as determined previously.

TABLE 1—*Particle-size distribution analysis of selected soils.*[a]

Soil Series and Sample Number	Horizon	Depth, cm	Particle-Size Distribution, %						
			Very Coarse Sand (2 to 1 mm)	Coarse Sand (1 to 0.5 mm)	Medium Sand (0.5 to 0.25 mm)	Fine Sand (0.25 to 0.10 mm)	Very Fine Sand (0.10 to 0.05 mm)	Silt (0.05 to 0.002 mm)	Clay (Less Than 0.002 mm)
Hallandale									
S6-3-1	A1	0 to 10.2	0.1	1.0	16.2	71.1	2.4	5.0	4.2
S6-3-2	A2	10.2 to 25.4	0	1.0	15.4	79.0	2.6	0.8	1.2
S6-3-3	B1	25.4 to 35.6	0	1.3	16.3	77.3	1.8	0.8	2.5
S6-3-4	B2	35.6 to 40.6	0	1.4	17.3	72.5	2.5	1.0	5.3
Plantation									
S6-8-3	IIA1	0 to 15.2	0.1	1.7	17.7	76.7	2.2	0.4	1.2
S6-8-4	IIA2	15.2 to 45.7	0.1	1.7	17.5	77.4	2.3	0	1.0
S6-8-5	IIC1	45.7 to 58.4	0.1	2.0	18.4	74.8	2.0	0.7	2.0

[a] Analyzed by the Soil Characterization Laboratory, Department of Soil Science, University of Florida Agricultural Experiment Station, Gainesville, Florida.

TABLE 2—*Chemical and physical properties of selected soils.*

	Hallandale				Plantation		
	S6-3-1	S6-3-2	S6-3-3	S6-3-4	S6-8-3	S6-8-4	S6-8-5
Horizon	A1	A2	B1	B2	IIA1	IIA2	IIC1
Depth, cm	0 to 10.2	10.2 to 25.4	25.4 to 35.6	35.6 to 40.6	0 to 15.2	15.2 to 45.7	45.7 to 58.4
SOIL REACTION, pH							
H_2O 1:1	5.8	6.2	6.3	7.4	6.3	6.3	7.1
0.01 *M* $CaCl_2$ 1:1	5.2	5.6	6.4	7.4	5.8	6.2	7.2
1 *N* KCl 1:1	5.1	5.7	6.0	7.1	5.9	6.2	7.7
CATION EXCHANGE CAPACITY, meq/100 g							
	15.7	2.3	1.8	13.3	3.5	1.5	6.6
WATER CONTENT AT VARIOUS PRESSURES (BARS), %							
0.10 bar	27.9	5.1	3.2	11.1	5.6	2.8	3.6
0.33 bar	24.0	3.4	2.1	9.6	3.1	1.5	2.0
15.00 bars	6.6	1.8	1.5	1.4	1.5	1.4	1.0

3. The contents of the bottle were filtered through 2-μm millipore filter paper in an ultrafiltration unit.
4. The filtrate was evaporated on a steam table, dried in an oven at 150°C, and finally ashed in a muffle furnace at 550°C.
5. The residue was dissolved in 25 mL of 15 *N* nitric acid and filtered through Whatman 3 paper. The volume of the filtrate was brought up to 100 mL with distilled water.
6. The solution was analyzed on an atomic absorption spectrometer for the cations.

The experimental results were used to determine, by least-squares fit, the linear distribution coefficients of the ions between the soils and the characteristic liquid.

Results

In Tables 3 through 6 are listed the values of the distribution coefficients experimentally determined for nine different cations in a complex mixture with soil samples from Florida, [*4,5*]. A linear distribution coefficient has been used [*6*].

The results for Hallandale fine sand are given in Table 1. The distribution coefficients for the cations between the complex mixture and the Plantation muck series are given in Tables 4 through 6. In each table, the ions are listed in order of increasing K' value. K_s' is the dimensionless distribution coefficient parameter for the most mobile ion. In Table 1, the most mobile ion is manganese. The most mobile ions are iron, manganese, and manganese in Tables 2, 3, and 4, respectively. The length of time for the most mobile ion to

TABLE 3—*Relative migration rates of ions in Hallandale fine sand.*[a]

Metal	K'	$\frac{t}{t_s} = \frac{1 + K'}{1 + K_s'}$, $K_s' = 302$
Mn	302	1.000
Cd	307	1.017
Co	755	2.495
Cu	870	2.875
Ni	2570	8.485
Fe	2620	8.650
Zn	3710	12.248
Ca	4730	15.614
Cr	7370	24.327

[a] Subscript "*s*" denotes the most mobile ion in the group.

TABLE 4—*Relative migration rates of ions in plantation muck top layer.*[a]

Metal	K'	$\frac{t}{t_s} = \frac{1 + K'}{1 + K'_s}$ $K'_s = 220$
Fe	220	1.000
Mn	260	1.181
Cu	370	1.679
Cd	452	2.050
Cr	484	2.195
Zn	590	2.674
Ca	3020	13.670
Ni	3250	14.710

[a]Subscript "*s*" denotes the most mobile ion in the group.

TABLE 5—*Relative migration rates of ions in plantation muck middle layer.*[a]

Metal	K'	$\frac{t}{t_s} = \frac{1 + K'}{1 + K'_s}$ $K'_s = 65$
Mn	65	1.000
Cu	137	2.091
Cd	348	5.288
Zn	670	10.167
Cr	1560	23.652
Ni	1960	29.712
Fe	2030	30.773
Co	3600	54.561
Ca	5020	76.076

[a]Subscript "*s*" denotes the most mobile ion in the group.

travel a certain depth is designated t_s. The corresponding lengths of time, t, for the other ions to travel that same depth is expressed in terms of t_s. The predicted values of t are given in the third column of each table. It is evident that the ions are also in order of decreasing mobility. This is a direct consequence of Eq 13.

Numerical solutions to Eq 3 with the boundary conditions of Eq 5 were computed for the most mobile ion up to a predetermined time, t_s. (The hydrodynamic dispersion coefficient for a similar type of soil was found from the literature [*8, 9*] and used in the computation. Representative values were chosen for the pore water velocity [*9*]. Numerical solutions were then computed for the other ions up to predetermined times, t, where $t = (1 + K')t_s/$

TABLE 6—*Relative migration rates of ions in plantation muck bottom layer.*[a]

Metal	K'	$\frac{t}{t_s} = \frac{1 + K'}{1 + K'_s}$ $K'_s = 730$
Mn	730	1.000
Cu	800	1.096
Co	834	1.142
Zn	1450	1.985
Cd	1640	2.245
Fe	1690	2.313
Ca	2400	3.285
Ni	2890	3.955
Cr	9410	12.874

[a] Subscript "*s*" denotes the most mobile ion in the group.

$(1 + K'_s)$. The concentration profiles for every ion were compared and found to be the same, within computational accuracy. This demonstrates that Eq 13 is, indeed, a solution to Eq 11. Numerical solutions were obtained for all the ions listed in Tables 1 through 4.

Summary

The relative lengths of time for different cations in a mixture to travel through a certain depth of soil have been expressed in terms of their distribution coefficients. The integral transform method of solution, as obtained by Cleary and Adrian [*1*], was used for the one-dimensional, convective-dispersive-adsorptive partial differential equation, which describes the motion of a cation within a complex mixture moving through the soil.

It has been found that the length of time a cation in solution takes to travel a certain depth of soil is directly proportional to $(1 + K')$, where $K' = K\rho(1 - \epsilon)/\epsilon$ and K is the distribution coefficient. This has been demonstrated to be true using the numerical solutions.

References

[*1*] Cleary, R. W. and Adrian, D. D., "Analytical Solution of the Convective-Dispersive Equation for Cation Adsorption in Soils," *Soil Science Society of America, Proceedings*, Vol. 37, 1973, pp. 197–199.

[*2*] Lai, S., and Jurinak, J. J., "Numerical Approximation of Cation Exchange in Miscible Displacement Through Soil Columns," *Soil Science Society of America, Proceedings*, Vol. 35, 1971, pp. 894–899.

[*3*] Lai, S. and Jurinak, J. J., "Cation Adsorption in One-Dimensional Flow Through Soils: A Numerical Solution," *Water Resources Research*, Vol. 8, No. 1, 1972, pp. 99–107.

[4] Sengupta, S., Farooq, S., Gerrish, H., Wong, K., and Daly, E., "Characterization and Environmental Studies of Pompano Beach Anaerobic Digestion Facility," Annual Report, D.O.E. Contract No. EV 78-S-05-6072, University of Miami, Florida, August 1979.
[5] Sengupta, S., Dasgupta, D., Wong, K. V., Daly, E., Farooq, S., and Gerrish, H. P., "Numerical Simulation of Chemically Reacting Flow Through Soils: A Parametric Study," *International Journal of Simulation*, Vol. 1, No. 1, 1981, pp. 23–31.
[6] Lapidus, L. and Amundson, N. R., "Mathematics of Adsorption in Beds, VI. The Effect of Longitudinal Diffusion in Ion Exchange and Chromatographic Columns," *Journal of Physical Chemistry*, Vol. 56, 1952, pp. 984–988.
[7] Calhoun, F., Caldwell, R. E., and Carlisle, V. W., "Soil Survey of Broward County Area, Florida," U.S. Department of Agriculture, in cooperation with the University of Florida, Gainesville, Fla., 1972.
[8] Cherry, J., *Groundwater*, Prentice-Hall, Englewood, N.J., 1979.
[9] Tarver, G. R., "Hydrology of the Biscayne Aquifer in the Pompano Beach Area, Broward County, Florida," Florida Geological Survey Report on Investigations No. 36, 1964.

James K. Petros,[1] G. Michael Alsop,[1] and Richard A. Conway[1]

Rapid Extraction Methods for Organics in Soil

REFERENCE: Petros, J. K., Alsop, G. M., and Conway, R. A., **"Rapid Extraction Methods for Organics in Soil,"** *Hazardous and Industrial Waste Management and Testing: Third Symposium, ASTM STP 851*, Larry P. Jackson, Alan R. Rohlik, and Richard A. Conway, Eds., American Society for Testing and Materials, Philadelphia, 1984, pp. 92–99.

ABSTRACT: This article presents rapid methods for the extraction and subsequent analysis of organics in soils potentially contaminated from chemical wastes. A methylene chloride extraction, on both a micro scale in a syringe and a macro scale in a shaken flask, with subsequent gravimetric analysis are described. On wet soils, anhydrous sodium sulfate is added to bind the water before adding methylene chloride and mixing with a vortex touch mixer. A baseline procedure is the direct gravimetric determination of volatile organics corrected with a Karl Fischer determination of water. Estimates of time saved by using the rapid methods, as well as precision and accuracy data, are presented.

KEY WORDS: soil, extraction, chemical solvent extraction, microextraction, organic chemicals, chemical wastes, gravimetric analysis, hazardous wastes, industrial wastes

Passage of the Resource Conservation and Recovery Act (RCRA) in 1976 and the Comprehensive Environmental Response, Compensation and Liability Act (CERCLA, Superfund) in 1980 led to increased public awareness and regulatory effort in the arena of solid and hazardous waste management. In all sectors increased emphasis was placed on characterizing solid wastes of all types and evaluating existing land disposal facilities, especially if leakage were suspected. The need for analyzing a large number of samples (such as, waste samples from a sludge lagoon or soil samples from a field investigation) in order to evaluate adequately the variability of their organic content caused the authors to assess the adequacy of the methods available (circa 1979) to measure various parameters. One of the apparent major problems was the lack of

[1] Project scientist, development scientist, and corporate development fellow, respectively, Union Carbide Corp., Engineering and Technology Services Division, Research and Development Dept., South Charleston, W.Va. 25303.

readily applicable methods available to evaluate a wide variety of organic constituents in solid wastes, sludges, soils, and sediments.

Published methods that were available at that time were limited in their applicability [*1*]. No approved, standardized methods [U.S. Environmental Protection Agency (EPA), American Society for Testing and Materials (ASTM)] were available, although several groups around the country were actively developing such methods. Although a variety of methods have now been described in various documents [*2,3*], the internal methods developed at Union Carbide are still appropriate for use in internal investigations and still offer advantages over some of the now standardized methods.

The basic test available at the start of this work was the classic Soxhlet extraction [*1*]. Although at that time the Soxhlet was largely an empirical test, it (or a variation of it) was typically used to determine the "oil and grease" or organic content of aqueous sludges. As the authors' in-house work evolved, three objectives were established in developing an improved method or methods:

1. The method should be *simple* and *rapid.* Because of the large numbers of samples, it was desirable to develop a less equipment-intensive method that was easy for technicians to follow and troubleshoot without the aid of an analytical chemist; also because of the quantity of samples, a rapid turnaround of equipment and data was desirable.
2. The method should provide a reasonable degree of *precision and accuracy.* Since the data generated were mainly for internal use, a rigorous quality assurance (QA) program was not always needed; however, it was necessary that any test developed provide an indication of contamination/no contamination or, hopefully, an accurate assessment of the degree of contamination. Some examples of data generated, and guidelines used to evaluate the data, are discussed later in the text.
3. The method should be *adaptable* to various solvents, different types of wastes, soils, or both, and a wide variety of other background matrices. Examples of materials to be tested include organic residues and sludges, biological sludges, clay soils, sandy soils, and dredged sediment. Different solvents may be appropriate (for example, dichloromethane or methanol), depending on what analyses might be performed on the extract.

The methods that were subsequently developed were not intended to determine trace quantities (< 100 mg/kg) of organics in soil or wastes. In most cases, the levels of organics were one to three orders of magnitude greater than this.

Methods

Three different extraction methods have been developed and used by Union Carbide for various internal studies. These methods are presented in detail in Table 1 and briefly summarized here.

TABLE 1—*Description of analytical methods.*

SHAKE FLASK EXTRACTION METHOD

1. Place a known weight of dry sample (103°C), usually 1 to 3 g, into a 125-mL Erlenmeyer flask with a glass stopper.
2. Add 50 mL methylene chloride, stopper, and shake vigorously for 1 min.
3. Filter the entire sample through a preweighed GF/C filter pad.
4. Transfer the pad and solids to a tared aluminum weighing dish. Wash any solids adhering to the filter equipment into the dish with a small amount of solvent.
5. Dry the pan and filter pad with the remaining solids at 103°C to drive off the residual solvent.
6. Weigh the dried, extracted solids, and calculate the extractable organic content (EOC) as follows:

$$\text{Percent EOC} = \left(1 - \frac{\text{weight of extracted solids}}{\text{weight of initial solids}}\right) \times 100$$

SYRINGE MICROEXTRACTION METHOD

Sample Bottle Preparation

1. Seal a 5-mL serum bottle with a Teflon septum and aluminum cap.
2. Evacuate the sample bottle with a syringe needle inserted through the septum, using 60 mm Hg vacuum. Heat the bottle for 15 min with an infrared heat lamp from a distance of about 5 in. (12.7 cm).
3. Allow the sample bottle to cool in the dessicator and get a tare weight. Be sure that any fingerprints, or other marks, are removed from the bottle prior to weighing.

Sample Extraction

1. Place a GF/C filter disk and a small piece of glass wool in a 5-cm^3 glass syringe.
2. Add a predetermined amount of the sample to syringe barrel.

Note: Sample size is determined by organic load in sample. Use 10-mg sample for high organic concentrations; 5-g sample for very low organic concentrations.

3. Attach a No. 27½ syringe needle capped with a small septum, and introduce 3 mL methylene chloride ($MeCl_2$) into syringe barrel.
4. Let the syringe stand for 1 min to extract the sample, remove the septum, and insert the needle into the evacuated serum bottle (through septum cap) until all $MeCl_2$ is drawn into the serum bottle.
5. Inject an additional 3 mL of $MeCl_2$ into the syringe barrel to complete the extraction.
6. Reevacuate the sample bottle and heat with an infrared light (Step 2, sample bottle preparation) until the $MeCl_2$ has evaporated.
7. Continue vacuum and heat for an additional 10 min.
8. Allow the sample bottle to cool in the dessicator and get a tare weight. Be sure that any fingerprints, and other marks, are removed from the bottle prior to weighing.
9. The difference between the tare weight and the gross weight of the serum bottle is the weight of extractable organic material. Calculate the weight percent of extractable organic content as follows:

$$\text{Percent EOC} = \frac{A - B}{C} \times 100$$

where
A = gross weight of serum bottle,
B = tare weight of serum bottle, and
C = initial weight of soil sample.

TABLE 1—*Continued.*

VORTEX MIXER EXTRACTION METHOD

1. Weigh 10 g of representative sample into a 100-mL centrifuge tube. Record the sample weight to the nearest 0.01 g.
2. Add 15.0 g of sodium sulfate (Na_2SO_4) anhydrous, or 1.5 times the sample weight in Step 1, to bind up the water in the sample.
3. Mix the sample and Na_2SO_4 thoroughly in the centrifuge bottle. If a dry sample is not obtained, add more Na_2SO_4 as needed. It may be desirable to grind the sample and Na_2SO_4 in a mortar and pestle in order to obtain adequate mixing. Add 50.0 mL of dichloromethane to the centrifuge tube. Cap.
4. Mix on a vortex touch mixer for 2 min.
5. Allow the solids to settle; centrifuge (or filter) if necessary.
6. Remove 25 mL of supernatant methylene chloride from the centrifuge bottle and add to a tared, 200-mL round-bottom flask. Remove the solvent by stripping in a 25°C water bath with 50 mm of vacuum. Strip to dryness (~ 40 min).
7. Reweigh the flask with the organic residue.
8. Calculate the extractable organic content as follows:

$$\text{Percent EOC (g/g)} = \frac{\text{residue weight (g)}}{\text{sample weight (g)}} \times \frac{50\ \text{mL}\ MeCl_2}{25\ \text{mL}\ MeCl_2} \times 100$$

WATER CONTENT (KARL FISCHER METHOD)

1. Add about 5 g of representative sample to 50 mL of anhydrous methanol in a 100-mL glass bottle.
2. Shake the sample vigorously for 5 min on a mechanical mixer to allow the water to be extracted by the methanol. Allow the solids to settle.
3. Withdraw 1 mL of the methanol mixture with a syringe and inject it into a standard Karl Fischer water apparatus. Titrate according to the standard procedure. A Beckman Model KF4B Aquameter was used for titration.
4. Follow the manufacturer's instructions for calibrating the instrument. Run a water content on 1 mL of the anhydrous methanol to correct for background water content.
5. Calculate the percentage of water in the sample, using the formula:

$$\text{Water (\%)} = \frac{AFS}{W} \times 100$$

where

A = millilitres of titrant used,
F = KF factor, milligrams of water per millilitre of titrant,
S = extract volume (50 mL), and
W = extract weight of sample, mg.

SOXHLET EXTRACTION METHOD

1. Extract 10 g of sample (previously dried at 103°C for 16 h) in a Soxhlet extractor using 200 mL of methylene chloride and a 26 to 60-mm thimble. Extract for 3 h.
2. When the extraction is completed, remove the solvent by vacuum distillation at 60 mm Hg and 60°C.
3. Weigh the extract.
4. Compute the extractable organic content as follows:

$$\text{Percent EOC} = \frac{\text{weight of extract}}{\text{weight of initial sample}} \times 100$$

Shake Flask Extraction Method

The shake flask extraction test is used primarily for determining the total extractable organic content (EOC) in samples in which the major constituents are nonvolatile. A sample of oven-dried, weighed material is extracted for 1 min with methylene chloride, the slurry is passed through a glass-fiber filter, and the retained solids are dried and weighed. The gravimetric finish actually weighs the solids not extracted; the EOC is obtained by the difference. This method is appropriate for dry samples and uses much less equipment than a standard Soxhlet apparatus.

Syringe Microextraction Method

In the syringe microextraction method, the sample is extracted twice with methylene chloride; each time the solvent is drawn into a preevacuated and preweighed 5-mL serum vial with a septum cap. The solvent is evaporated under vacuum and low temperature. The residue remaining in the bottle is the EOC. This method uses less equipment and space than the shake flask method.

Vortex Mixer Extraction Method

The vortex mixer extraction method was developed to provide rigorous agitation and to allow the direct extraction of wet samples by adding anhydrous sodium sulfate to the samples. Mixing is accomplished using a standard vortex mixer for 2 min. The advantages to this method are the ability to use a larger sample size than with the prior methods, use of wet "as-is" samples, and same-day laboratory results.

The other method provided in Table 1 is the standard Soxhlet extraction method, which was used for comparative purposes. The method cited is similar to that referenced in SW-846, *Test Methods for Evaluating Solid Waste* [2]. A correction for volatile organic content using the Karl Fischer water method is also described.

Results and Discussion

Some early laboratory data comparing the shake flask method to a Soxhlet extraction (3-h reflux) are given in Table 2. On three different samples of an organic sludge, agreement between the two methods was within a ratio of 1.04 to 1.06. At lower organic concentrations contained in marine sediment, the agreement was adequate on some samples and poor on others. Except for one sample, felt to reflect an analytical error, the shake flask method consistently gave higher results, likely because of better mixing of the solvent with the sorbed organic material.

Table 3 compares the shake flask, syringe microextraction, and Soxhlet ex-

TABLE 2—*Comparison of shake flask to Soxhlet extraction results.*

	Organic Content, Weight %	
Sample	Soxhlet	Shake Flask
ORGANIC SLUDGES (SAMPLE DRIED AT 103°C)		
1	67.0	64.2
2	83.8	80.4
3	45.1	42.4
Average	65.3	62.3
MARINE SEDIMENT (SAMPLE DRIED AT 103°C)		
1	0.78	0.84
2	0.50	0.68
3	1.77[a]	0.63
4	0.18	0.41
5	0.36	0.58
6	0.07	0.22
7	nil	0.44

[a] Possible analytical error.

TABLE 3—*Comparison of three different extraction methods on dry soil.*[a]

	Clay Soil Dry Weight Extractable Organic Content, %	
Method	Mean	Standard Deviation
Soxhlet (16-h reflux)	0.79	0.07
Soxhlet (3-h reflux)	0.75	0.05
Syringe microextraction	0.65	0.02
Shake flask	0.91	0.08

[a] All the analyses were performed in triplicate. The soil was dried at 103°C for 16 h.

traction methods (using two different reflux times). It is clear that with this soil the 3-h reflux period provided essentially the same result as the 16-h period for the Soxhlet. Although this may not hold true for all organic compounds or all sample matrices, it is worth investigation on a case-by-case basis as the savings in labor and sample turnaround time are significant. The shake flask again showed a higher result, with the syringe method giving a slightly lower value than the Soxhlet method. For many types of evaluations, the differences in the results would not be significant.

In order to compare all of the methods on the same sample, a soil containing mainly nonvolatile organics was mixed with anhydrous sodium sulfate and analyzed by all four methods. The results, shown in Table 4, indicate fairly

TABLE 4—*Comparison of four extraction methods on wet clay soil.*[a]

	Wet Weight Extractable Organics, %	
Method	Mean	Standard Deviation
Shake flask extraction	1.40	0.26
Syringe microextraction	1.80	0.20
Vortex method	1.47	0.12
Soxhlet (3-h)	0.48	0.15
Soxhlet (16-h)	1.13	0.13

[a] Na_2SO_4 was added to soils for *all* the methods. Clay soil 77.4% solids, 18.7% water. All the analyses were performed in triplicate. Solvent stripping: vortex and Soxhlet at 25°C; microsyringe at 40 to 50°C.

good agreement among all four methods. In this case, the 3-h Soxhlet extraction was not sufficient to measure the total organic content. The standard deviations are similar for all the methods (done in triplicate). Once again, for a determination of the relative level of contamination, any of the three methods developed by the authors should be suitable.

Values for samples in which volatile organics (boiling point $< 103°C$) are a concern can be obtained by calculating the difference between the shake flask EOC and water content (using the Karl Fischer method cited in Table 1).

The authors have learned from numerous investigations that our methods are sensitive and accurate enough to allow initial decisions to be made regarding the magnitude of contamination. In some cases qualitative or quantitative analyses (such as gas chromatography/mass spectrometry, gas chromatography/flame ionization detection, infrared) may be used on extracts obtained with these methods to provide more definitive results. It is recognized that more exhaustic extraction procedures would need to be followed to perform quantitative analyses at levels of "trace" contamination.

Table 5 shows the relative differences in time of each of the extraction methods discussed herein. The time gained in using the rapid methods allows additional samples to be run each day, thereby increasing the data base and quality

TABLE 5—*Time comparisons for methods evaluated.*[a]

Method	Elapsed Time, h[b]
16-h Soxhlet	19
3-h Soxhlet	6
Syringe microextraction	2
Shake flask	3
Vortex extraction	3

[a] Time for set of six analyses on similar samples.
[b] Projected time from the start of the analysis to the time the results are available.

of the study. In addition, the "same-day" analysis allows us to provide real-time feedback, if warranted, to an ongoing program, whether it be excavation of contaminated soil or soil boring for installation of new groundwater monitor wells.

One additional advantage to using the rapid, less labor-intensive methods is the ability to routinely perform duplicate analyses of all samples. This has proven to be an invaluable asset to interpreting results. Given the nonhomogeneity of most soil samples, it is not uncommon to get an outlier, giving either false-positive or false-negative results. Running duplicate samples generally allows one to avoid these erroneous readings and to have more confidence in the results.

Conclusions

1. Three extraction methods have been developed to provide quickly a measure of organic content of residues and soils at a level of about 0.5% and above.
2. The shake flask extraction method, which measures loss of weight on oven-dried solids, is used primarily for nonvolatile constituents.
3. The syringe microextraction method, which measures weight of organics extracted from dried samples, uses minimum equipment, space, and sample size.
4. The vortex mixer extraction procedure, which uses sodium sulfate to bind water, can handle wet "as-is" samples of larger sizes and provides intimate mixing of solvent and solids.
5. The results from triplicate extractions were suitably precise and accurate (as opposed to the Soxhlet technique) to be useful in preliminary and in-house investigations. Where quantitative determination of microcontamination is of concern or fine lines need to be drawn, more exhaustive and detailed extraction procedures should be employed.

References

[1] *Standard Methods for the Examination of Water and Wastewater*, 15th ed., American Public Health Association, Washington, D.C., 1980.

[2] *Test Methods for Evaluating Solid Waste, Physical/Chemical Methods*, SW-846, 2nd ed., U.S. Environmental Protection Agency, Washington, D.C., 1982.

[3] *Solid Waste: Methodology for Evaluation—Volume II: Analysis of Residual Wastes*, SW-196C.2, U.S. Environmental Protection Agency, Washington, D.C., 1981.

Harley W. Young,[1] *James C. Young,*[1] *Mustafa Boybay,*[2] *and Turgut Demirel*[2]

Effect of Chemical Treatment of Fly Ash on the Leaching of Metals

REFERENCE: Young, H. W., Young, J. C., Boybay, M., and Demirel, T., **"Effect of Chemical Treatment of Fly Ash on the Leaching of Metals,"** *Hazardous and Industrial Waste Management and Testing: Third Symposium, ASTM STP 851,* Larry P. Jackson, Alan R. Rohlik, and Richard A. Conway, Eds., American Society for Testing and Materials, Philadelphia, 1984, pp. 100–117.

ABSTRACT: An investigation was conducted to determine the ability of chemical treatment to reduce the leaching of metals from fly ash. A further objective was to determine the effect of cofiring refuse-derived fuels with coal on leachate composition. Samples of fly ash were treated with portland cement, chemically pure lime, and phosphoric acid and subjected to a column leaching process. The leachates were composited weekly and analyzed using X-ray fluorescence and atomic absorption spectrometry.

Portland cement or lime reduced the quantity of leached arsenic 40 to 80% from the control samples, whereas those samples fixed with phosphoric acid yielded 50 times more arsenic. Generally, the samples treated with portland cement showed the most uniform reduction in leachate concentrations.

Leachates from fly ash produced upon cofiring refuse-derived fuel with coal showed consistently higher concentrations of all elements except for calcium, selenium, and alkalinity.

Compaction and curing of the chemically treated ash samples showed little improvement in the ability to control the amount and types of metals appearing in leachates.

KEY WORDS: fly ash disposal, leaching, heavy metals, chemical treatment, ash disposal, groundwater pollution, hazardous wastes, industrial wastes

Increasing energy costs and problems with the nuclear power industry have compelled the growing use of coal as a fuel in public power production. Also, there has been increasing interest in the use of refuse-derived fuel (RDF), the combustible portion of municipal solid waste, to supplement coal. It has been estimated that 50 million tons (45 metric tons) of fly ash are now produced in

[1]Graduate research assistant and professor, respectively, Civil Engineering Department, University of Arkansas, Fayetteville, Ark. 72701.

[2]Postdoctoral research fellow and professor, respectively, Civil Engineering Department, Iowa State University, Ames, Iowa 50011.

the United States each year. Current trends indicate that this amount may increase as much as fourfold in the near future [*1*].

This increased burning has resulted in the production of huge amounts of waste fly ash. Although some of this material has found uses in the replacement of portland cement and for soil stabilization, much of it is sluiced directly to on-site ponds or placed in sanitary landfills.

The major environmental concern with fly ash disposal is the leaching of heavy metals and toxic elements to underlying groundwater sources [*1,2*]. The heavy metals found in greatest concentration include antimony, cadmium, lead, nickel, thallium, and zinc; other toxic elements include arsenic and selenium [*3*].

Because of the potential for contaminating underground drinking water supplies, new and improved methods of disposal and analysis are needed. One potential method for reducing or eliminating contamination from leaching is to chemically treat the ash prior to or during disposal. The intent of this treatment is to chemically fix or stabilize the metallic elements in an insoluble form or by encasement in a nonporous, solid matrix so that significant leaching does not occur. Consequently, a study was undertaken to assess the effect of chemical treatment on the leaching of metallic elements from fly ash and to evaluate the effect of decreasing the available surface area through compaction and solidification in reducing the concentration of toxic elements in the leachate. A secondary objective was to determine the effect of cofiring RDF fuels with coal on the leachate composition.

Experimental Program

Apparatus and Test Plan

Leaching test methods used to date fall into three groups. First, the field cell test probably yields the most directly meaningful data. However, it suffers from being specific to only one location and entails much time and expense. Second are the batch or shake tests. One of these, the EPA extraction procedure, includes a 24-h mixing time and a large ratio of water to material, and the pH is adjusted to 5 with 0.5 *N* acetic acid over the course of the extraction. This test is based on the premise that the rigor of the extraction will show the worst possible case. The third test method—the type used in this study—is the column leaching test. The use of such columns has been described in detail by Young [*4*] and Rademaker and Young [*5*]. For this study, leachate columns were fabricated from 4-in. (100-mm) diameter Pyrex tubes with 60° glass funnels fused to the lower end (Fig. 1). Glass stopcocks with Teflon stoppers were fused to the stem of each funnel, and a short length of Tygon tubing was attached. This tubing led to glass collection bottles placed beneath each column. Layers of glass beads and acid-washed silica sand were used as a support for the ash samples.

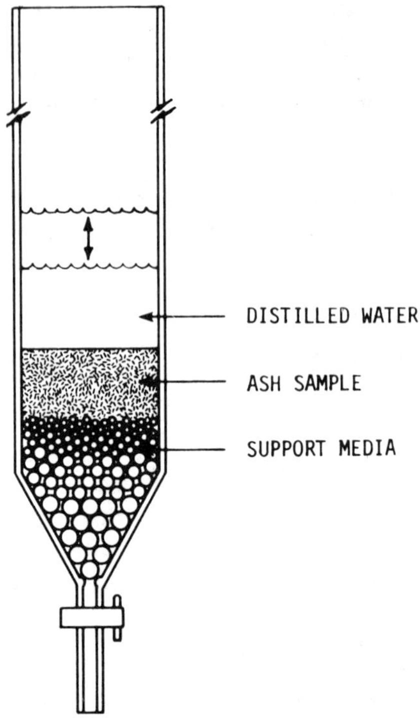

FIG. 1—*Schematic diagram of leaching column showing the placement of the support media and ash sample.*

Prepared samples of fly ash were placed in the columns and saturated initially using 1200 mL of deionized water. Leachate samples were withdrawn daily from the bottom of the column and were composited weekly. Finally, deionized water was added to the top of each column after each withdrawal to maintain a constant liquid volume. The pH was monitored in the collection bottles, and redistilled nitric acid was added as necessary to maintain the pH at less than 2.

Flow rates during removal of leachate from each column were governed by the rate of withdrawal from columns containing unfixed fly ash control samples, which provided a flow rate of about 20 to 30 mL/min. The leachate was removed from all the test columns at one time each day. The amount of leachate removed daily, 321 mL, was based on the least amount that was sufficient for the required analyses. Based on a bulk porosity of 70% in the ash, this was approximately 0.65 pore volumes per day.

Selection of Chemical Agents

Three chemical agents were chosen for testing: portland cement, chemically pure lime, and phosphoric acid. Portland cement and lime were chosen because of their widespread availability, their relatively low cost, and their common use for soil stabilization. Phosphoric acid was selected because of its ability to react rapidly with fly ash and form a solid matrix.

Nomenclature

The labeling convention used in this paper is presented in Table 1. The ash samples are represented by a two or three-character code. Each code corresponds to one ash sample in one column. The control column samples obtained when burning coal only are designated as "FA." The samples collected when burning RDF plus coal fly ash are labeled "GFA." Also, RDF plus coal fly ash will be referred to as "RDF fly ash," whereas coal-only fly ash will be designated simply as "fly ash."

All the codes for ashes treated with portland cement contain the letters "CM," the codes for ashes fixed with lime include the letters "LI," and those for phosphoric acid include the letters "PA." Codes ending in the letter "A" represent RDF fly ash, while those preceded by the letter "F" are for coal-only fly ash. The codes ending in a number represent treated RDF fly ash that had been compacted, crushed, and sorted by particle size. The number represents the average particle size for that ash, as will be described in later sections of this paper.

Characterization and Analysis

Barium, arsenic, lead, and mercury content were determined using atomic absorption spectrometry. Further elemental content was determined using X-ray fluorescence spectroscopy as described by Young, Boybay, Young, and Demirel [*6*]. Alkalinity (ALK), total dissolved solids (TDS), and pH were measured by tests described in *Standard Methods* [*7*].

Ash Preparation

Samples of fly ash were obtained in August 1980 from the electrostatic precipitators at the Ames, Iowa, municipal power plant. This ash was produced

TABLE 1—*Fly ash sample code designations.*

Type	Fly Ash from Coal Only, Uncompacted	Fly Ash from RDF Plus Coal: Uncompacted	Fly Ash from RDF Plus Coal: Compacted, Crushed, and Sized
Portland cement	FCM	CMA, CMO	CM1, CM2, CM3
Lime	FLI	LIA	LI1, LI2, LI3
Phosphoric acid	FPA	PAA	PA1, PA2, PA3
Control	FA	GFA	

when burning a mixture of 55% Iowa coal and 45% Colorado subbituminous coal by weight. Ash samples were also collected when firing a mixture of 65% coal, by weight, and 35% pulverized refuse from the Ames solid waste recovery system. This RDF consisted of combustible items up to 2 in. (50 mm) in diameter remaining after pulverization of municipal solid waste having a typical composition of 35 to 45% paper, 14% food and garbage, 9% metals, 10% glass, and 25% yard refuse, textiles, wood, rubber, and plastics. After removal of noncombustibles, the resulting RDF contributed about 17% to the energy released through burning and amounted to about 40% of the total fuel mix by weight. Table 2 lists the partial composition of these ashes as measured by X-ray fluorescence analysis. In addition to an amorphous silica matrix, carbon residuals were present as well as many trace elements including the heavy metals.

Two groups of ash samples were tested. The first group consisted of ash samples treated with chemical additives and placed in the leaching columns in an uncompacted loose condition. The second group consisted of ash samples that had been treated, compacted, crushed, and sorted by size. These compacted fly ash samples contained a larger amount of chemical additive than did the uncompacted samples. Both groups included RDF fly ash and coal fly ash. The specific gravity of the chemically treated ash samples ranged from 2.5 to 2.8.

Preparation of Treated Ash Samples

Uncompacted Samples—The samples of ash and treatment agent were weighed and mixed. The proportions used in uncompacted samples were made on the basis of a 90:10 weight ratio of ash to chemical. Ten percent deionized water by weight was added. These ash samples were then mixed and allowed to cure in a 100% moisture room for ten days. This curing period was

TABLE 2—*The partial composition of Ames power plant ashes.*

Element[a]	GFA[b]	FA[c]	BA[d]
Mg	0.75	1.22	0.57
Na	0.50	0.97	0.18
Fe	17.75	10.22	18.05
Ti	0.47	0.45	0.22
Si	16.67	11.76	12.55
Ca	4.86	6.57	8.09
Al	10.80	10.81	9.49
K	1.14	0.84	0.72

[a] Concentrations are in percent as the element.
[b] GFA = fly ash from refuse derived fuel and coal.
[c] FA = fly ash from coal only.
[d] BA = bottom ash from coal only.

based on previous work concerning the use of fly ash for soil stabilization [*8*]. After curing, corrections were made for moisture changes and added material, so that 500 g of dry fly ash ultimately were added to each leaching column.

Compacted Samples—Samples to be compacted were mixed in proportions given in Table 3. Higher concentrations of fixing chemicals were used in these samples than in the uncompacted samples to ensure that compacted particles of fly ash would not disintegrate in the leaching columns. The CMO sample (Table 1) contained similar quantities of portland cement and lime but was not compacted. Water was added until the consistency was adequate for compaction. The moist samples, except CMO, were then placed in a 2-in. (50-mm) diameter mold and compacted using ten blows on each side with a 5-lb (23-Hg) hammer over a 12-in. (300-mm) drop. This procedure has been found to yield compaction roughly equivalent to standard Proctor density [*9*]. These samples plus the uncompacted CMO sample were then placed in a moisture room for curing. The samples treated with phosphoric acid were allowed to react for 24 h before water was added. These samples solidified so quickly when water was added that compaction was not possible, and they were placed directly in the moisture room for curing. All the samples were removed from the moisture room after ten days and then were crushed and sieved into three size ranges. Corrections were made for added water and fixing chemical. The samples were weighed and sufficient amounts were placed in the leaching columns to give a total addition of 500 g of original fly ash.

The samples of compacted RDF fly ash treated with portland cement were analyzed with a Quantachrome mercury porosimeter. This analysis indicated a solid particle density of 2.81 kg/m^3. Based on these measurements and bulk volumes in the leaching columns, bulk porosities ranged from 65 to 75% for all the ash samples.

Test Results

The column leachates generated in this study were clear, colorless, odorless, alkaline solutions and—with the exception of phosphoric acid-treated ash

TABLE 3—*Quantities of chemical additives in compacted samples.*

Sample	Chemical Added	Percentage of Chemical[a]	Percentage of Water[a]
Portland cement, CM1 CM2 CM3	Type I PC $Ca(OH)_2$	15 3	14.8
Lime, LI1 LI2 LI3	$Ca(OH)_2$	20	13.2
Phosphoric acid, PA1 PA2 PA3	H_3PO_4	40	25.0

[a] On the basis of the weight of fly ash.

samples—had a pH in the range on 10 to 12. Over the twelve-week leaching period, the pH changed less than 2 pH units. In all cases, the pH change was toward neutral. These leachates, especially those from ashes treated with phosphoric acid or lime, were nearly saturated with the dissolved salts of calcium, potassium, or phosphate complexes of these metals. The leachate from ashes treated with phosphoric acid generally had a pH ranging from 4.0 to 6.0, depending on the quantity of acid used in fixing the ash.

Total Leached Quantities

Concentrations of metals and other chemical parameters were measured in each leachate sample on the first, second, fourth, eighth, and twelfth weeks (C_1, C_2, C_4, C_8, and C_{12}). The data were then plotted as shown for zinc, copper, and calcium in Fig. 2 to provide a visual impression of the leaching pattern for each element. The length of the bar at each data point represents the probable error (approximately 68% of standard deviation).

Total leached quantities (TLQ) were obtained by multiplying the concentrations from each test analysis by the volume of leachate produced. This is equivalent to the area under the time-versus-concentration curve times the total volume of leachate removed over the period of the study. Using the 0.321 L produced each day and assuming these concentrations were constant over test periods of 10.5, 10.5, 21.0, 28.0, and 28.0 days, TLQs were approximated using the following equation:

$$\text{TLQ} = 3.375(C_1 + C_2) + 6.749\,C_4 + 8.999(C_8 + C_{12}) \qquad (1)$$

where

TLQ = total leached quantity, and
C_i = concentration for week i.

The units of the TLQ values depend on the units measured. For most elements it is milligrams.

Table 4 lists both the TLQs and the maximum measured concentrations for the elements tested when using unaltered fly ash. Total dissolved solids (TDS) is a measure of the residue remaining upon evaporation after filtration and consists of all the nonvolatile dissolved material present in the leachate samples. The TLQs of dissolved solids for the control sample leachates ranged from 18 000 to 26 000 mg. For the leachates from phosphoric acid or lime-treated ashes, the TLQs ranged from 25 000 to 27 000 mg. A TLQ of 20 000 mg dissolved solids would correspond to 4% of the original quantity of fly ash placed in the column (500 g). Heavy metals in the control sample leachate typically ranged from 5 to 30 μg TLQ for mercury up to 1 to 2 mg TLQ for lead. The TLQ values for arsenic ranged from 90 to 140 μg, while selenium TLQ values averaged 45 mg.

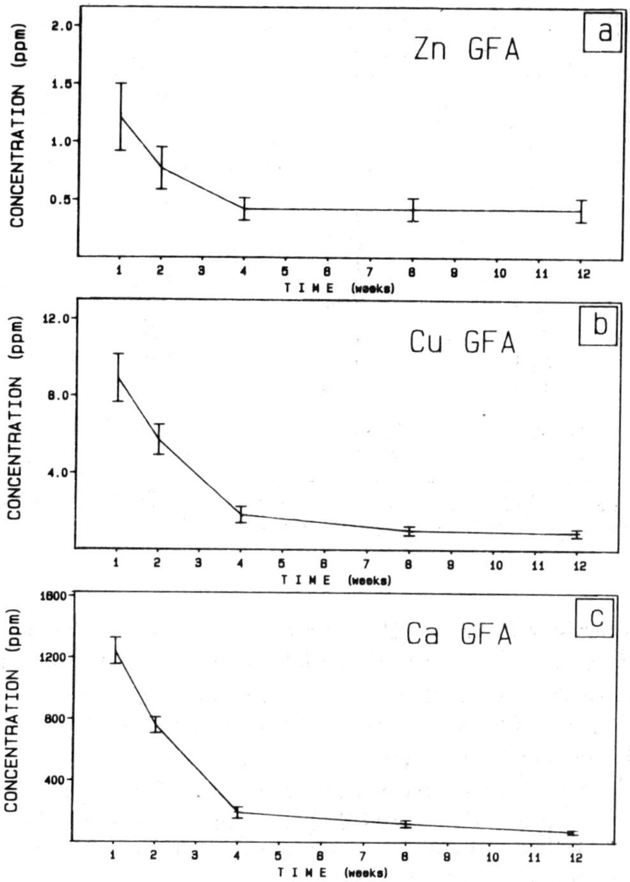

FIG. 2—*Typical concentration-versus-time curves for leachates from fly ash.*

Effect of Chemical Treatment on Uncompacted Ashes

The leachates produced from the uncompacted ash samples (CMA, FCM, and CMO) were compared to assess the relative effect of the chemical stabilizing agent used. The ashes treated with all-chemical agents had a granular texture with an appearance much like an unscreened sand and gravel mixture.

Portland Cement—Portland cement in uncompacted samples (CMA, FCM, and CMO) was found to reduce the TLQs of lead, selenium, copper, calcium, and total dissolved solids from 30 to 50% below the TLQs for unfixed ash samples (GFA and FA) (Fig. 3). The TLQ values for arsenic were inconclusive because the arsenic concentrations in most leachate samples were found to increase throughout the twelve-week test period, as illustrated in the concentration-versus-time plots in Fig. 4. The increase in arsenic concentration is

TABLE 4—*The TLQ and maximum reported concentrations of metals in the leachates from RDF fly ash and coal-only fly ash.*

Element or Test	TLQ[a]		Maximum Measured Concentration, mg/L	
	GFA	FA	GFA	FA
K	619	170	79.4	8.50
Ca	8 220	9 810	917	743
Cu	70.3	68.5	6.64	4.40
Zn	15.1	14.2	1.14	0.60
As	114	...	0.043	...
Se	39.7	41.6	3.64[b]	2.90[b]
Rb	16.6	13.4	1.93	1.10
Ba	54.3	45.2	6.91[b]	3.80
Hg	13.6	5.96	0.001	0.0002
Pb	1.34	1.20	0.19	0.11
TDS	22 800	23 130	3 250	2 150
ALK	16 000	18 850	1 820	1 330

[a] All units are in milligrams except arsenic, lead, and mercury, which are in micrograms.
[b] These values exceed extraction procedure (E.P.) limits.

probably the result of the pH decreasing as the more soluble alkaline salts were dissolved from the fly ash, indicating that arsenic is possibly leached as the arsenate ion, AsO_4^{3-}, rather than As^{3+}. These data suggest that a considerable amount of arsenic may eventually leach from fly ash deposits with or without treatment using portland cement.

The TLQ results for mercury and zinc were mixed (Fig. 3). The uncompacted RDF and coal ash sample (CMA) showed higher values than the control and FCM samples. This situation was reversed for barium, where uncompacted FCM values were above those of the control while the CMA values were much below. In all cases except for mercury, zinc, and barium, the leachates from unfixed control samples (GFA or FA) showed higher TLQ values than for ash samples treated with portland cement. Alkalinity was reduced in the CMA sample only; the values for the FCM and CMO and control samples were similar. The TLQ values for rubidium were essentially the same in leachates from ashes treated with portland cement. Potassium concentrations in the leachates from all three fixed samples were higher than in the control samples.

Although a larger amount (15%) of portland cement was included in the CMO sample (this amount corresponded to that used for compacted samples along with 3% lime), the TLQ values reported for this sample were sometimes higher and sometimes lower than in the other test samples.

Lime—Leachates from ash samples prepared with chemically pure lime showed increases in TLQs for several of the elements tested over the leachates from the control samples (Fig. 5). There was a substantial reduction, however, in the TLQ values for mercury and arsenic. Rubidium concentrations tended to increase in the leachates from lime-treated ashes, and minor increases were

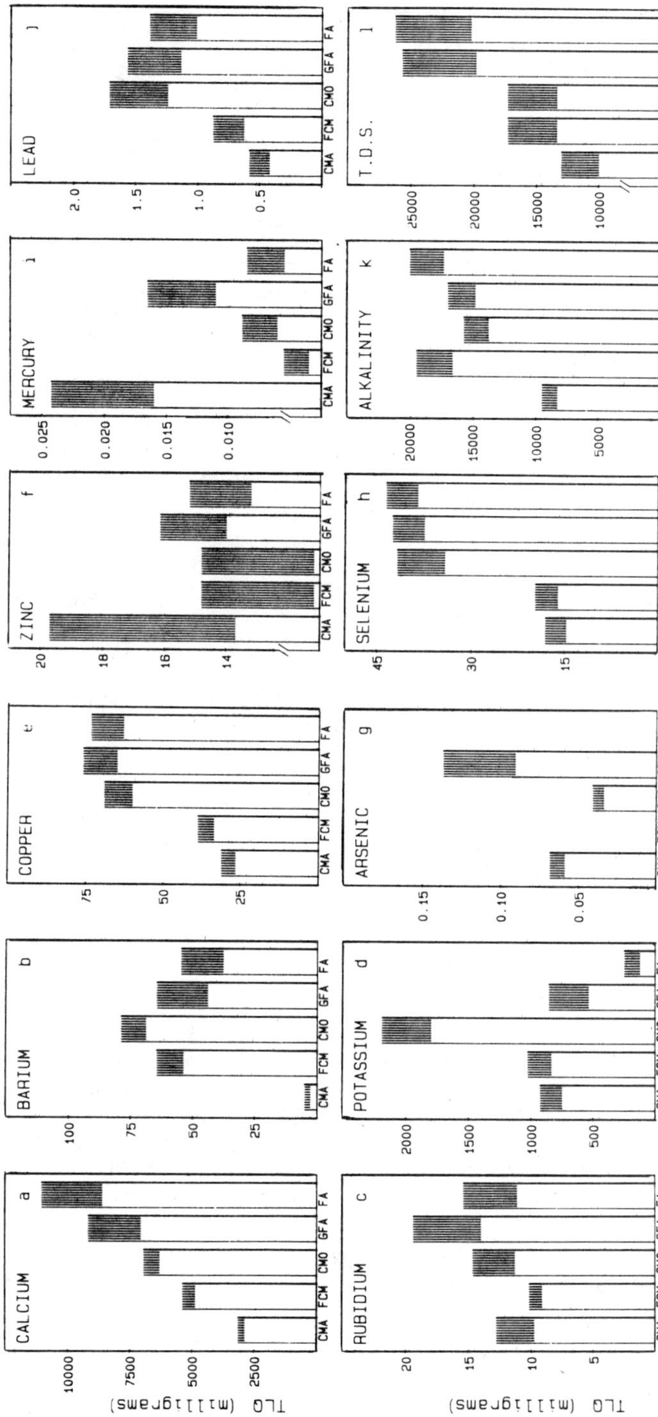

FIG. 3—*Total leached quantities of various elements in leachates from uncompacted ash samples treated with portland cement and from control samples. (The hatched areas represent probable error.)*

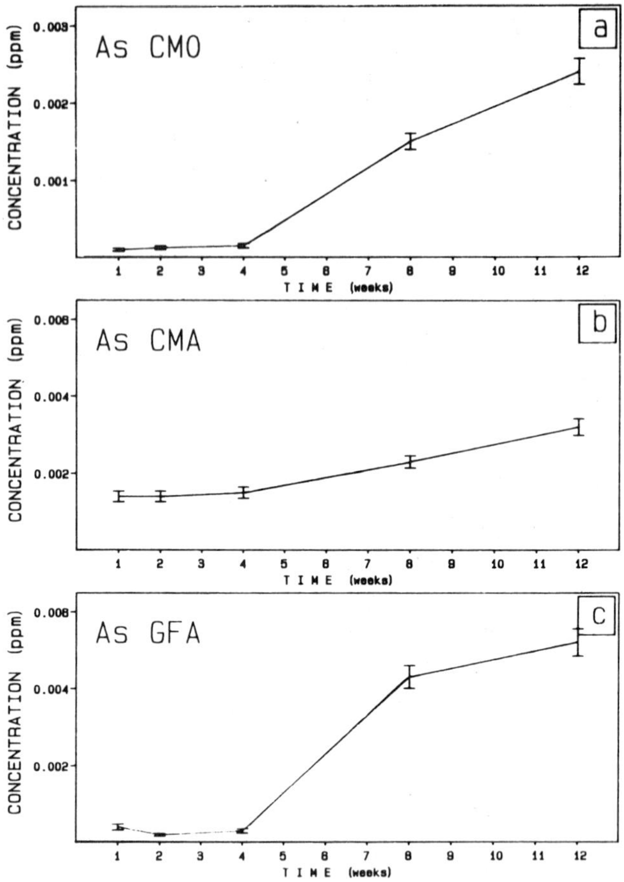

FIG. 4—*Concentration versus time for arsenic in leachates from ash samples receiving various treatments.*

apparent for copper and zinc in some of the leachates. The alkalinity and total dissolved solids were substantially higher for the leachate from the lime-treated ashes than from the two controls. Large increases were observed consistently in barium concentrations in the leachates from all the lime-treated samples. Calcium and potassium were present at increased levels in leachates from fixed ash samples as well as a 20 to 30% greater concentration for selenium.

Generally, leachates from all the ashes fixed with lime showed higher TLQ values than from the controls (GFA and FA) for most of the elements tested, with the notable exception of arsenic and mercury. There also seemed to be a consistent trend toward lower TLQ values in leachates from the fly ash samples not containing RDF material.

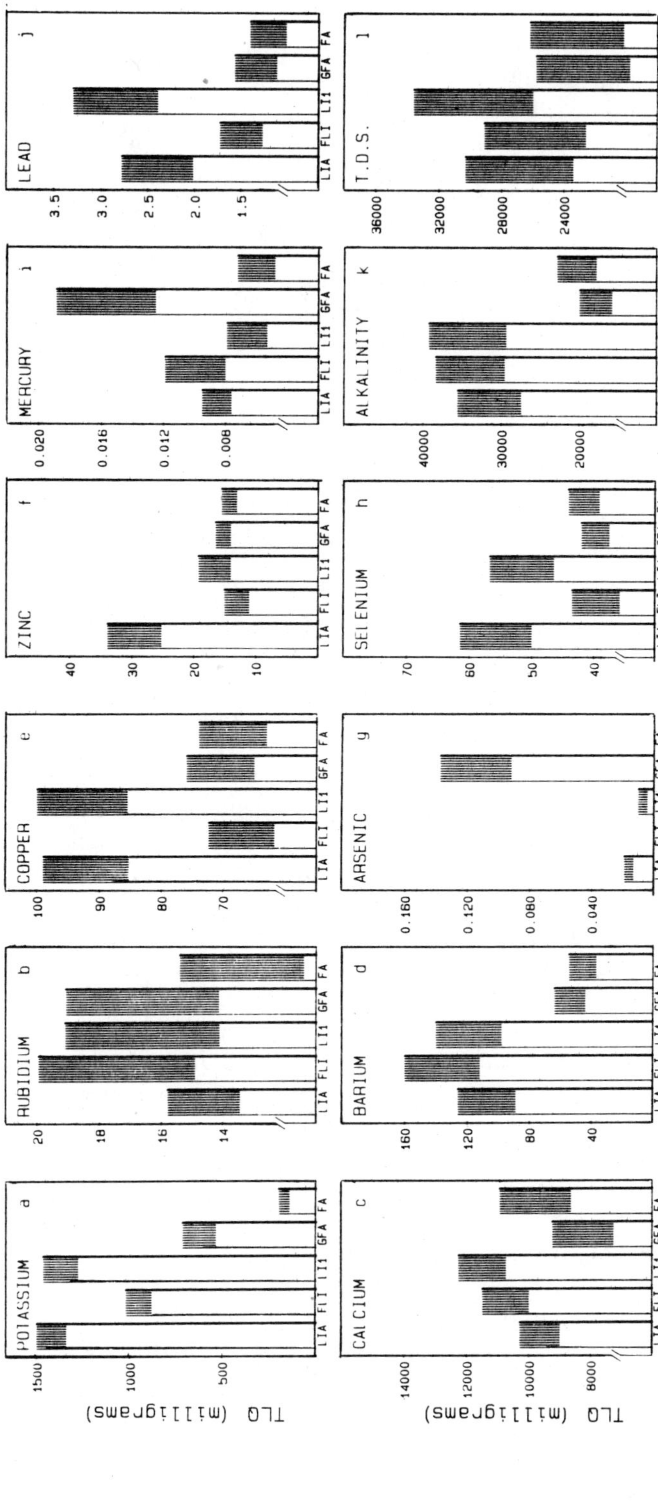

FIG. 5—*Total leached quantities of various elements in leachates from ash samples treated with lime and from control samples. (The hatched areas represent probable error.)*

Phosphoric Acid—Ash samples treated with phosphoric acid produced leachates quite different in character from those to which lime or portland cement were added. The concentration of a number of the test elements increased substantially with time, as was shown in Fig. 4 for portland cement-treated ash samples. Figure 6 shows the TLQ values for all twelve tests for three phosphoric acid-treated ashes and two controls. Among the elements that increased with stabilization were zinc, copper, selenium, lead, potassium, and rubidium. Arsenic concentrations were so high that stocks of phosphoric acid were analyzed to check against possible contamination. Alkalinity was substantially lower than in control ash samples.

The Effect of RDF on Leachate Composition

The effect of RDF on leachate quality was assessed from a comparison of the TLQ for each coal fly ash sample with the TLQ for ash samples containing RDF. This relationship is summarized in Fig. 7. The length of each bar represents the total leached quantity for each test, as noted to the right of Fig. 7.

Untreated control fly ash showed lower TLQs than did untreated RDF-coal ash, with the exception of calcium, selenium, and alkalinity. That is, the RDF caused an increase in the amount of most elements in the leachate. RDF contributed most to increases in potassium (72%), rubidium (20%), barium (20%), mercury (56%), and lead (12%).

Comparison of the data for the uncompacted ash samples treated with portland cement shows lower values for potassium, calcium, copper, barium, lead, and total dissolved solids in the leachates from ash samples containing RDF than from the ash samples collected when burning coal only. Higher TLQ values of mercury, zinc, and rubidium were observed with the RDF ash samples. Of these, only the changes for mercury, rubidium, zinc, barium, and alkalinity were statistically significant. Thus, when using portland cement as a treatment agent, the RDF caused an increase in the amount of some elements and a decrease in others. However, the elements that showed an increase were different from those in the untreated samples containing RDF fly ash.

Phosphoric acid treatment caused leaching of higher concentrations of all metals from RDF fly ash than from coal fly ash.

With lime stabilization, the TLQs for potassium, copper, zinc, selenium, and lead were significantly higher in leachates from the samples containing RDF than from the samples containing coal ash only. There was no change for mercury. In general, the leachates from the non-RDF fly ash samples showed lower concentrations of metals than those from RDF fly ash.

This evidence indicates that there was a general increase in the total level of contaminants in leachates from fly ashes produced when cofiring RDF material with coal. There were no consistently significant lower values for any one parameter in all four treatments.

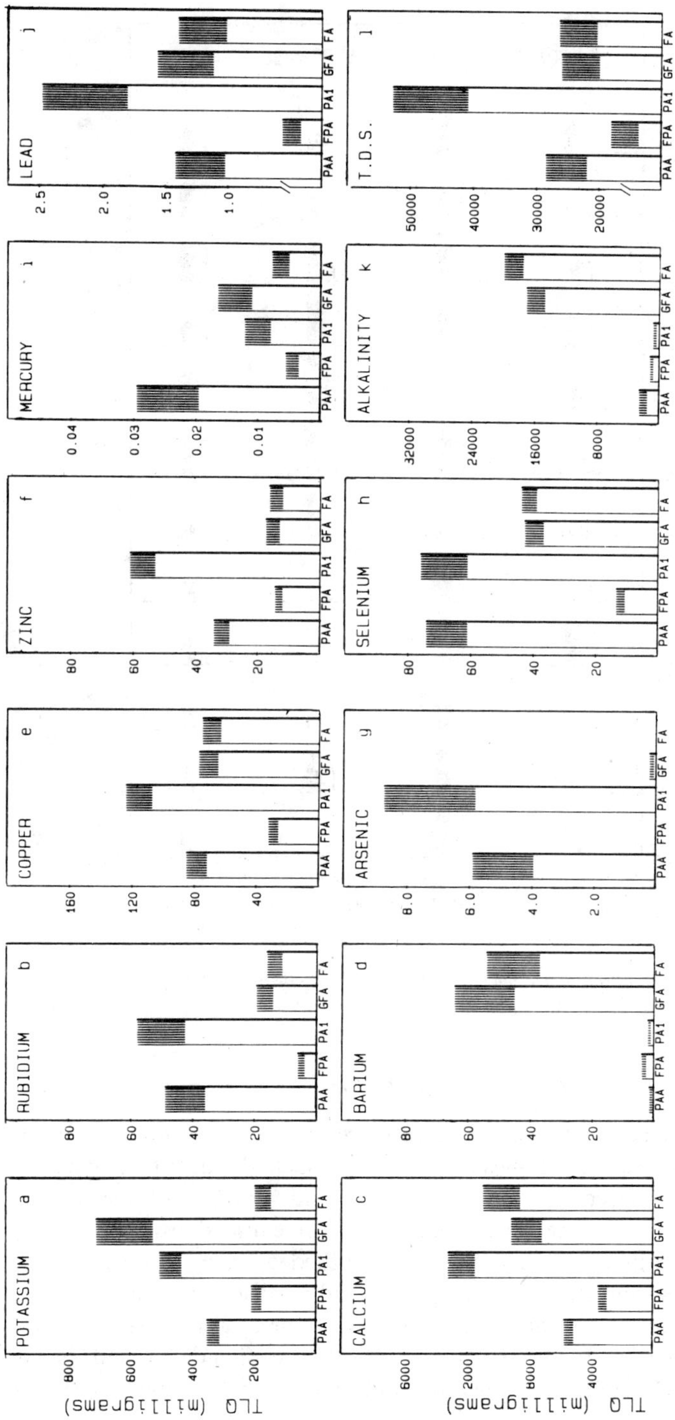

FIG. 6—*Total leached quantities of various elements in leachates from ash samples treated with phosphoric acid and from control samples. (The hatched areas represent probable error.)*

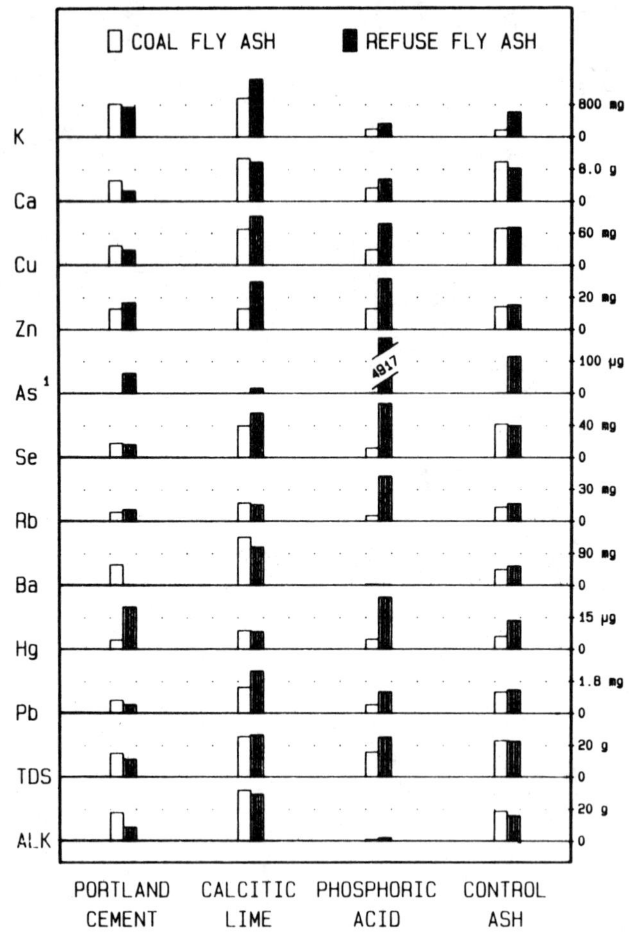

FIG. 7—*Comparison of average TLQs from coal-only fly ash and TLQs from RDF fly ash.*

The Effects of Compaction and Particle Size

In an attempt to change the surface area exposed to water passed through the ash samples, the samples that were compacted to about standard Proctor density and allowed to cure were crushed and sieved into three size groups. This procedure was repeated for each of the three test additives, making a total of nine different samples for placement in the leaching columns.

Table 5 lists the average sieve opening and particle surface area based on an assumed sphericity factor of 1.2. Total surface areas were calculated using the weight of fly ash added to each column, the specific gravity, and the surface areas for each size range. These computations yielded a ratio of surface areas

TABLE 5—*Particle size ranges and available surface area for compacted fly ash.*

Size Code	Average Sieve Opening, cm	Surface Area Per Particle, cm^2	Total Available Surface in Column, cm^2
1	2.22	18.62	622
2	1.11	4.67	1250
3	0.36	0.48	3770

in the columns close to 1:2:6 for the size ranges 1, 2, and 3, respectively (large to small). If metal ions appearing in the leachate originated from the surface only, a six-to-one difference in leachate concentrations might be expected.

This high ratio generally was not observed, as shown by the comparison of TLQs in Fig. 8. Typically, the increase in contaminant concentration as particle size decreased was less than twofold, and in some cases leachates from columns containing the larger grain sizes had the higher concentration of contaminants.

The number of times that the smaller particles yielded higher concentrations than the medium-sized particles, and those in turn that yielded higher values than the largest particles, was higher than would be expected from random chance, but the differences in TLQs were not statistically significant. This finding suggests that the metals were not fixed chemically in a manner that would prevent their leaching and that these metals could diffuse rapidly from solidified ash samples that were fractured into relatively large granular particles.

Conclusions

The results of this study have shown that chemical treatment of fly ashes using portland cement, lime, or phosphoric acid will not always reduce the leaching of metallic elements and can, in fact, produce an increase in the release of some elements. Portland cement reduced the total leached quantity of lead, selenium, copper, calcium, and total dissolved solids from 30 to 50% below the TLQs for untreated ash samples. TLQs for other elements increased slightly or there was no change.

Lime caused a significant reduction in mercury and arsenic but caused alkalinity and total dissolved solids to increase substantially. Barium, calcium, and potassium also increased when lime was used.

Phosphoric acid caused the leaching of a number of metals to increase—notably zinc, copper, selenium, lead, calcium, and arsenic. The total dissolved solids also increased. Concentrations of other metals as well as alkalinity were lower in the leachate from phosphoric acid-treated ash samples.

There was a general increase in the total level of contaminants in leachates from fly ashes produced when refuse-derived fuel was cofired with coal. However, each type of chemical treatment—with portland cement, lime, or phosphoric acid—showed increases in the leaching of different types of elements.

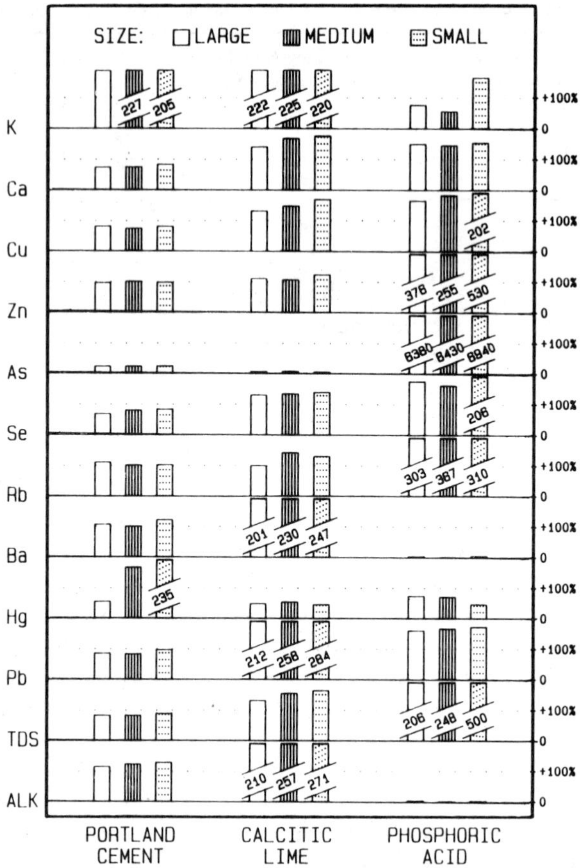

FIG. 8—*Comparison of TLQs in leachates from ash samples that were fixed, compacted, and crushed into large, medium, and small particle sizes.*

Compaction of chemically treated fly ash samples failed to reduce the leaching of metals to a significant extent. This suggests that the metals were not chemically fixed in a nonsoluble form and diffused from the granular matrix into the interstitial liquid.

Although insufficient data were collected to conclude definitely that chemical treatment is not an effective means of reducing the leaching of metals from fly ash to acceptable levels, the results of this study show that chemical treatment techniques should be evaluated more intensively before the practice is adopted widely.

Other chemical agents or processing techniques may be more effective than portland cement, lime, or phosphoric acid in eliminating leaching of undesirable elements. Use of controlled leaching tests, as described in this paper, should be beneficial in evaluating the efficacy of such agents.

Acknowledgments

This research work was supported in part by the Engineering Research Institute, Iowa State University, in part by the Ames Laboratory, Iowa State University, as part of U.S. Department of Energy Contract W-7405-ENG-82, and in part by the Department of Civil Engineering, the University of Arkansas at Fayetteville.

References

[*1*] "Solid Waste Research Needs for Emerging Coal Technologies," *Proceedings*, the American Society of Civil Engineers/Power Research Council-Electrical Power Research Institute Workshop, San Diego, Calif., American Society of Civil Engineers/Power Research Council, April 1979.

[*2*] Theis, T. L., Westrick, J. D., Hsu, C. L., and Marley, J. J., "Field Investigation of Trace Metals in Ground Water from Fly Ash Disposal," *Journal of the Water Pollution Control Federation*, Vol. 50, No. 11, November 1980, pp. 2457-2470.

[*3*] Kaakinen, J. W., Jorden, R. M., Lawasani, M. H., and West, R. E., "Trace Element Behavior in a Coal-Fired Power Plant," *Environmental Science and Technology*, Vol. 9, No. 9, September 1975, pp. 862-869.

[*4*] Young, H. W., "Stabilization of Ashes to Reduce Leaching of Heavy Metals," Master of Science Thesis, Iowa State University, Ames, Iowa, 1982.

[*5*] Rademaker, D. A. and Young, J. C., "Leachates from Solid Waste Recovery Operation," *American Society of Civil Engineers, Proceedings, Journal of the Energy Division*, Vol. 107, No. EY1, May 1981, pp. 17-29.

[*6*] Young, H. W., Boybay, M., Young, J. C., and Demirel, T., "X-Ray Fluorescence Spectrometry for Analysis of Fly Ash Column Leachates," in *Hazardous and Industrial Solid Waste Testing: Second Symposium, ASTM STP 805*, American Society for Testing and Materials, Philadelphia, 1983, pp. 214-228.

[*7*] *Standard Methods for the Examination of Water and Wastewater*, 15th ed., American Public Health Association, Washington D.C., 1982.

[*8*] Davidson, D. T., Handy, R. L., and Viskochil, R. K., "Effect on Density and Strength of Lime, Fly Ash Stabilized Soil," *Highway Research Board Bulletin*, Vol. 185, 1957, pp. 5-30.

[*9*] Mateos, M., "Effect of Trace Chemicals on Strength of Ottawa Sand-Lime-Fly Ash Mixtures," Master of Science Thesis, Iowa State University, Ames, Iowa, 1958.

Dale E. Baker,[1] *D. Scott Rasmussen,*[1] *and Janice Kotuby*[1]

Trace Metal Interactions Affecting Soil Loading Capacities for Cadmium

REFERENCE: Baker, D. E., Rasmussen, D. S., and Kotuby, J., **"Trace Metal Interactions Affecting Soil Loading Capacities for Cadmium,"** *Hazardous and Industrial Waste Management and Testing: Third Symposium, ASTM STP 851*, Larry P. Jackson, Alan R. Rohlik, and Richard A. Conway, Eds., American Society for Testing and Materials, Philadelphia, 1984, pp. 118–132.

ABSTRACT: The goal of this investigation was to establish a theoretically applicable model of, and develop an approach for, the direct measurement of cadmium loading capacities under laboratory conditions. Data from the literature were fitted to a model indicating a direct, linear relationship between the logarithm of the activity of Cd^{2+} in the soil solution and the logarithm of plant uptake of cadmium. These results support the model for passive absorption of cadmium by plants. It was postulated that the diffusion of Cd^{2+} from the soil-adsorbed phase to plant roots results from the electrochemical potential $\bar{\mu}_{Cd}$, gradient established between soil-adsorbed Cd^{2+} and that in solution, which is depleted as a result of passive absorption by plants.

Laboratory methods for relating the negative logarithm of the activity of Cd^{2+} in solution, pCd, to soil-adsorbed cadmium and plant-absorbed cadmium were developed, and test results for each are presented. It was concluded that this approach is theoretically valid and will provide routine laboratory methods for measuring soil loading capacities to protect the human food chain from undesirable concentrations of cadmium.

KEY WORDS: cadmium, soil test, cadmium loadings, food chain, sludges, passive absorption, electrochemical potential, diffusion, hazardous wastes, industrial wastes

Increasing use of cropland for the disposal of sludges containing variable concentrations of cadmium and other trace metals requires a greater understanding of soil–material interactions and the associated plant responses. With repeated application of sludges, the soil abundance of cadmium can increase 20 to 100-fold. Cadmium is the trace element that has the most potential for adverse effects on human health from a given quantity added to soils with sludges [*1–3*]. With no adverse effects on plant growth, plant uptake of cadmium may be sufficient to impose a long-range dietary intake by humans which is considered intolerable [*2*]. The National Research Council (NRC) [*4*] in 1980 set

[1]Professor of soil chemistry and graduate assistants, respectively, Pennsylvania State University, Department of Agronomy, University Park, Pa. 16802.

0.5 $\mu g/g$ as the maximum tolerable level of cadmium in rations for livestock and poultry. The tolerable level for cadmium set by the NRC was based on human food residue considerations and is lower than the concentrations found detrimental to animal health. The U.S. Environmental Protection Agency (EPA) Office of Research and Development determined that a daily intake of 200 μg cadmium per day could lead to adverse health effects in humans [*3,5*]. They concluded that an allowable increase of 30 μg cadmium per day over the existing intake by the median American would have no adverse health effects.

The problem in soil management is to determine annual and cumulative loadings with cadmium from sludges in relation to soil properties. Some increases in soil levels of cadmium are tolerable, but increases must not lead to excessive cadmium in livestock rations and ultimately in human diets. The Council for Agricultural Science and Technology (CAST) [*6*] concluded that cadmium in plants varies with (*a*) the species and cultivar grown, (*b*) environmental and management factors, (*c*) soil properties (pH is the most critical factor in controlling plant uptake of cadmium from a given loading), (*d*) the annual and cumulative amounts of cadmium applied to the soil, and (*e*) the plant part sampled.

CAST concluded that soil cation exchange capacity (CEC) does not adequately reflect the properties that control the availability to plants of cadmium in sludge-treated soils. Yet, the EPA has defined cumulative loading capacities for soils as a function of soil CEC [*3,5*]. For acid soils of the northeast with CECs of 5 or less, the cumulative loading capacity was set at 5 kg/ha. More recently, a regional committee from agricultural experiment stations in the northeast have proposed limits of 2 to 10 kg/ha, depending on soil texture and land use criteria [*7*].

The purpose of this investigation was to test the hypothesis that plant uptake of cadmium is a function of the chemical potential of cadmium, μ_{Cd}, in the soil solution.

$$\mu_{Cd} = \mu^{\circ}_{Cd} + RT \ln A_{Cd}/A^{\circ}_{Cd} \quad (1)$$

where μ°_{Cd} and A°_{Cd} relate to standard conditions for an ideal 1-*M* solution of Cd^{2+}, R is the molar gas constant, T is the temperature, and A_{Cd} refers to the solution activity of Cd^{2+}. Support for this hypothesis would indicate that the cumulative loading capacity of a soil for cadmium can be determined directly from the amount of cadmium adsorbed when the soil is in equilibrium with the desired solution pCd^{2+} (the negative logarithm of Cd^{2+} activity).

Theoretical Considerations

Passive Absorption of Cadmium by Plants

For passive absorption of an ion, plant accumulation occurs only to the accumulated concentration associated with the chemical potential, μ, of the ion in solution at the root surface [*8*]. From the electrochemical potential, $\bar{\mu}$, concept, the molar concentration of Cd^{2+} within the plant would reflect the chemical potential of the solution at the root surface plus the inner potential, ϕ, within the plant [*9*].

$$\bar{\mu}_i = \mu_i + z_i F\phi \quad (2)$$

For cadmium in the plant, $[M_{Cd}]_{in}$ compared with the concentration of Cd^{2+} in solution $[M_{Cd}]_{out}$, the electrochemical potential at 25°C takes the form

$$\bar{\mu}_{Cd} = \mu_{Cd} - 59 \log \frac{[M_{Cd}]_{in}}{[M_{Cd}]_{out}} \quad (3)$$

Since μ_{Cd} is a function of the logarithm of the activity of Cd^{2+} in solution, data indicating a linear relationship between the log soil solution activity of cadmium and the log of plant concentration of cadmium would support the hypothesis that Cd^{2+} uptake by plants involves passive diffusion in response to an electrochemical potential gradient from the soil solution to the plant part analyzed. The results of several investigations [*10–12*] indicate that uptake of Cd^{2+} is passive. The results of deVries and Tiller [*13*], showing the substantial effect of the glasshouse environment in increasing Cd^{2+} uptake by plants compared with plants grown in the field, suggest the importance of environmental factors.

Plant Absorption of Cd^{2+} from Soils

In most experimental work involving the growth of plants in soils, neither the ion activity nor the related μ for ions in solution are reported. The investigation of Baker and Low [*14*] indicated that plants respond to changes in $\bar{\mu}_{Na}$ in clay suspensions even when there is no change in concentration. For a given soil, changes in Cd^{2+} concentration would result in comparable changes in Cd^{2+} activity if the mechanism for specific adsorption of the ion remains constant. Therefore, a linear change in the log plant cadmium should be associated with a linear change in log solution cadmium, which in turn should be associated with a linear change in log Cd^{2+} adsorbed by the soil. Thus it is postulated that the diffusion of Cd^{2+} from the soil-adsorbed phase to the plant roots results from the $\bar{\mu}_{Cd}$ gradient established between soil-adsorbed Cd^{2+} and soil solution when the plant passively absorbs solution Cd^{2+}. This hypothesis was supported by the work of Browne, Wong, and Buhler [*15*]. Their results from many greenhouse studies showed a linear relationship between log soil cadmium extracted with diethylenetriamine pentaacetic acid (DTPA) and the log of plant cadmium. Since the chelator DTPA acts as a sink for Cd^{2+} diffusing from the adsorbed phase into solution, it is suggested that the results for DTPA soil tests

[*16,17*] reflect solution pCd in the soil assayed. The hypothesis for this investigation may be summarized from the following equation.

$$a \log Cd^{2+}_{soil} = pCd^{2+}_{solution} = b \log Cd^{2+}_{plant} \quad (4)$$

This equation suggests that Cd^{2+} is specifically adsorbed by soil and, within the tolerable range for pCd in soil solution, can be characterized by the Freundlich adsorption isotherm [*9,18*]. The coefficients a and b represent the buffer capacity of the soil and the inherent Cd^{2+} absorption capacity of the plant, respectively, in the specified environment.

Chemistry of Cd^{2+} in Soils and Interactions with Other Ions

In addition to its specific properties, the Cd^{2+} ion is similar to both Ca^{2+} and Zn^{2+}. All three ions form insoluble phosphates [*18*]; Cd^{2+} can form insoluble cadmium carbonate ($CdCO_3$) and calcium and cadmium carbonate solid solutions at pHs of 8.0 or greater; and Cd^{2+} forms acid insoluble sulfide precipitates typical of heavy metals whereas Ca^{2+} and Zn^{2+} do not. The toxicity of Cd^{2+} may be caused by its strong bond with sulfur, and hence cadmium can displace zinc and other essential metals from binding sites of certain enzymes [*19*].

Because of chemical element interactions in soils, the experimental results within the literature present a confusing picture. At high concentrations in soil-water systems, Cd^{2+} will enter into cation exchange reactions with Ca^{2+}. From these observations Cd^{2+} loading was related to the CEC. For a given Cd^{2+} loading, the mole fraction on the exchanger, which relates to Cd^{2+} activity, would decrease with an increase in CEC [*20*]. Baker [*21*] calculated that the mean activity of Ca^{2+} in a soil-water system for fertile, limed soil under field conditions should be 50 to 60 $\times 10^{-4}$ M for the soil solution. Therefore, to prevent cation exchange reactions involving Ca^{2+}, for laboratory studies all solution additions of Cd^{2+} to soils should be made in the presence of Ca^{2+} at the desired activity. Thus Ca^{2+} is not only essential for its role in membrane integrity in ion uptake studies [*22*], it is also essential to maintain the equilibria involving the electrochemical potentials of ions in soil-water systems.

The relative affinities for adsorption of trace metals by soil components have been summarized [*23*]. For productive agricultural soils below pH 7, the affinity for adsorption of trace metals should rank in the order lead $\geq$ copper $\geq$ nickel $\geq$ zinc $\geq$ cadmium when the soil solution activity for each chemical species is equal. From this relationship, it is apparent that Cd^{2+} will be relatively more available for plant uptake than the other trace metals contained in sludges. In addition, as the concentrations of the other metals increase, the specific adsorption of cadmium by soil should decrease.

In soil-plant relationship studies where the hypothesis summarized by Eq 4 is being tested, interaction effects from Ca^{2+} and other trace metals cannot be ignored, especially for Cd^{2+} and other ions that might be absorbed passively by plants.

Procedure

Reevaluation of Field Data

Results from data published in 1977 by Baker, Amacher, and Doty [*24*] were reevaluated to test the postulated relationships of Eq 4. Initially, the data had been used to relate concentrations of cadmium in plants to the negative logarithm of the activity of cadmium associated with DTPA [$pCd_{(DTPA)}$]. Essential information on experimental methods are presented with the results.

Laboratory Method for Measuring Cadmium Loading Capacities of Soils

A laboratory method is needed for direct measurement of the cadmium loading capacity of soils, which includes the following parameters:

1. The pH of the solution containing the cadmium to be added to the soil must be weakly buffered at the desired pH (usually pH 7 ± 0.1 would be representative of limed agricultural soils).
2. The activity of Cd^{2+} in the equilibrating solution should be in the range which is of concern from the point of view of plant uptake and human health considerations. A pCd of 6.0 was considered to represent a maximum solution level to be used with a pCd of 6.5 to represent a tolerable limit for cadmium in all soils.
3. The effects of other trace metals on cadmium adsorption should involve the competing effects of these metals at the respective activities considered applicable for soils. These activities have been summarized elsewhere [*17,21,25*].

The chemistry of the equilibrating solution was considered the more important factor in the development of a laboratory method for direct measurement of the cadmium loading capacity of soils. The approach being studied differed greatly from that proposed by the EPA [*5*]:

$$Cd_A = Cd_L - 2.2\,Cd_S \tag{5}$$

where Cd_A is the adjusted cumulative cadmium application limit (in kilograms per hectare), Cd_L is the cumulative cadmium application limit based on soil pH and CEC, Cd_S is the total background concentration of cadmium in soil (in micrograms per gram, dry weight); and 2.2 is the conversion for micrograms per gram to kilograms per hectare of furrow slice.

The concentrations of macrocations in solution were chosen to fix the ionic strength of the equilibrating solutions and at the same time provide a weak buffer at pH 7.0 ± 0.1 and prevent the soil binding of Cd^{2+} as an exchangeable cation. For the initial solution, a Ca^{2+} concentration of 60×10^{-4} *M* was chosen, and the pH of 7.0 was to be achieved using a combination of cadmium acetate [$Ca(CH_3COO)_2$] and calcium nitrate [$Ca(NO_3)_2$] in aerated, distilled, deionized water. It was found that 60×10^{-4} *M* $Ca(CH_3COO)_2$ provided the requirements stated previously. The selected levels for cadmium activities for repeated equilibrations were 10^{-8}, 10^{-7}, 10^{-6}, 10^{-5}, and 2×10^{-5} *M* Cd^{2+} as cadmium nitrate [$Cd(NO_3)_2$]. In order to produce the desired activities, the concentrations of cadmium were 1.68 times the desired activities to account for the ionic strength effect of $Ca(CH_3COO)_2$ on the activity.

Hagerstown silt loam was collected from field plots which had received no sludge in the field experiment [*24*]. The soil samples (2.5 g air-dry soil, <20 mesh) were equilibrated with 25 mL of solution for 2 h. Two replications were oven dried at 100°C after each equilibration, and two replications were held moist. The soil–solution equilibrations were repeated four times. The oven drying had no significant effect, so the data were handled as four replications. The respective equilibrating solutions contained enough cadmium to increase soil levels by 0.0188, 0.188, 1.88, 18.8, and 37.6 μg/g. The higher concentrations were applied to determine if the specific adsorption capacity of the soil for cadmium would be exhausted from repeated equilibrations. If all the added cadmium had been adsorbed, the background soil concentrations would have increased by 0.075, 0.75, 7.52, 75.2, and 150.4 μg/g. After the final equilibration, the samples were extracted with 1 *M* nitric acid (HNO_3) (two extractions with a 1:10 soil-to-solution ratio). The untreated soil (0 cadmium in solution) contained 0.54 μg cadmium per gram. The treated soil samples contained 0.54, 1.26, 7.96, 71.22, and 141.2 μg cadmium per gram, respectively, as the mean for the four replications. The coefficient of variation was −11% of the mean, indicating that for the higher concentrations the results were low by about 6%. Either a significant amount of the added cadmium was not adsorbed or the measured concentrations were in error [*26*]. The measured concentrations were taken as the amounts adsorbed by soils with each set of four equilibrations. From the results of the initial experiment, the solution with a pCd of 6.0 was chosen for repeated equilibration, with each equilibrium solution being subjected to a cadmium analysis. At a pCd of 6.0 and a soil-to-solution ratio of 1:10, the solution contained sufficient amounts of cadmium to bring the soil into equilibrium with a pCd of 6.5 after about five equilibrations.

The effect of time of equilibration on cadmium adsorption of soil was studied by comparing shaking times of 30 min and 2 h where the samples were shaken continuously on a horizontal, reciprocating shaker at 120 oscillations per minute. After shaking, the samples were centrifuged for 5 min at 2000 rpm. The centrifugate was analyzed for cadmium by use of an atomic absorption (AA) spectrometer equipped with a graphite furnace atomizer. Operation of the double-beam, dual-channel AA spectrometer and the graphite furnace have been described by Baker and Suhr [*26*].

Based upon ranges in trace metal solution activities reported by Baker et al

[7], the following concentrations of trace metals were included in the $Ca(CH_3COO)_2$ and $Cd(NO_3)_2$ solution to determine their effect on the adsorption of cadmium:

Metal	Source	Concentration	
		$\mu g/mL$	M
Zn	$Zn(NO_3)_2$	0.021	3.61×10^{-7}
Cu	$Cu(NO_3)_2$	6.35×10^{-6}	1.0×10^{-10}
Ni	$Ni(NO_3)_2$	1.86×10^{-7}	3.16×10^{-12}

In order to maintain a constant pH, enough calcium hydroxide [$Ca(OH)_2$] was substituted for $Ca(CH_3COO)_2$ to maintain a constant concentration of Ca^{2+} and pH 7.0 ± 0.1.

Plant Response to μ_{Cd} in Solution

The specific objective of this experiment was to test the hypothesis that μ_{Cd} in solution would be reflected linearly by the logarithm of plant concentrations of cadmium absorbed.

Doebler 59X hybrid corn seeds were selected for this experiment because, of the several hybrids tested, they were found to produce seedling radicles which were straight and uniform in length. These root characteristics are primary requisites for determining corn seedling uptake of ions using the method developed by Baker and Low [*14*].

The corn seeds were rinsed in distilled water and then placed in 3% hydrogen peroxide (H_2O_2) for 15 min with occasional stirring to sterilize the seed coat. The seeds were then air dried. The seeds were germinated within rolled, moistened paper towels, which were positioned with the narrow end of the seed pointed downward in 600-mL beakers. In this position the radicle for each seed grew straight downward to a length of 5 to 5.5 cm in a dark incubator held at 25°C for 72 h.

As a growing medium, the nutrient solution adapted from the Baker soil test solution [*17,21*] contained 50×10^{-4} M $Ca(NO_3)_2$, 10×10^{-4} M magnesium sulfate ($MgSO_4$), 2×10^{-4} M potassium chloride (KCl), and 0.5×10^{-4} M monobusic potassium phosphate (KH_2PO_4). Radioactive ^{109}Cd, with a specific activity of 2.24 mCi/mg, a Cd^{2+} concentration of 20.000 mCi/mL, and a purity of 99.000%, was utilized to tag the Cd^{2+} within each solution. The nutrient solution ^{109}Cd molar concentrations were 1.6×10^{-10} M, 8.0×10^{-10} M, 1.6×10^{-9} M, 8×10^{-9} M, 1.6×10^{-8} M, 8×10^{-8} M, 1.6×10^{-7} M, 8.0×10^{-7} M, and 1.6×10^{-6} M. Five replications were included for each variable concentration of cadmium. While the soil solution is expected to have 10^{-7} M Cd^{2+} or a pCd of 7.0 [*27*], the lower levels were included to ascertain a change in the mechanism of Cd^{2+} uptake at lower concentrations as suggested by Cataldo and Wildung [*28*].

The plant growth cells [*14*] consisted of round-bottomed, 15-mL plastic centrifuge tubes mounted in No. 2 rubber stoppers with 2.5 cm by 2.5 cm sections of PBC pipe mounted on top of the tube to act as a receptacle. The seed and seminal roots were mounted in moistened cotton to allow the corn seedling coleoptile to emerge while the radicle was immersed in the nutrient solution held within the tube. Each cell contained 15 mL of nutrient solution. The seedlings were exposed to the variable treatments for 24 h in a dark incubator at 25°C.

The seedlings were removed from the cell and the radicle was rinsed with 0.01*M* calcium chloride ($CaCl_2$) to remove any readily exchangeable cadmium. The endosperm portion of the seed was removed with a scalpel and the remaining parts—radicle, seminal roots, seed germ, and coleoptile—were folded and inserted in plastic tubes used for gamma counting. Two millilitres of concentrated nitric acid (HNO_3) and 1 mL of 30% H_2O_2 were added to each tube to hydrolyze the plant material and obtain a uniform distribution of the ^{109}Cd in a liquid phase. Each sample was counted for 20 min in a gamma spectrometer. The data were recorded as the log of the activity in nanocuries for each seedling.

Results and Discussion

Reevaluation of Field Data

Highly significant correlations were obtained by Browne, Wong, and Buhler [*15*] when they related the log plant cadmium to the log of soil cadmium removed by DTPA, as determined by Lindsay and Norvell [*16*] for greenhouse soils.

$$\log \mathrm{Cd}_{\mathrm{plant}} = a + b \log \mathrm{Cd}_{L\text{-DTPA}} \qquad (6)$$

For example, the data from Bingham et al [*29*] showed correlation coefficients, *r*, of 0.948 to 0.996 for seven crops grown under greenhouse conditions in soil treated with cadmium-enriched sewage sludge. These and data from other greenhouse experiments provide strong support for the hypothesis leading to Eq 4.

Soil extractants remove Cd^{2+} from solid solution-bound cadmium and other nonlabile sources not in equilibrium with Cd^{2+} in solution, and soils are expected to differ in their specific bonding of Cd^{2+}. These effects will cause each soil to give a different intercept and slope when log soil test cadmium is plotted against pCd ($-\log A_{Cd^{2+}}$ in solution). Environmental factors greatly influence plant uptake of cadmium, and plant species and cultivars differ in their absorption of Cd^{2+} [*6,24,28*]. Therefore, the intercept will change with experimental conditions, and the slope will change with plant growth when log plant cadmium is plotted against pCd for the soil solution. Because of the factors not accounted for by the thermodynamic considerations of Eq 4, the following statistical approach must be employed:

$$a + b \log \mathrm{Cd}_{\mathrm{soil}} = \mathrm{pCd}_{\mathrm{solution}} = c + d \log \mathrm{Cd}_{\mathrm{plant}} \tag{7}$$

Data for field plots treated with different rates of sludge containing cadmium which were used for the production of field corn and grain sorghum were reported previously [*24*]. The results by the Baker method of soil testing [*17*] used to predict the chemical potential of cadmium in solution ($\mathrm{pCd}_{\mathrm{DTPA}}$) were regressed on the results from different soil extractants. The extractants included the Baker–DTPA method [*17*], the Lindsay–DTPA method [*16*], and the 0.1 *N* HCl method [*30*]. The following regression coefficients and r^2 values were obtained.

$$\mathrm{pCd}_{\mathrm{DTPA}} = 12.26 - 1.11 \log \mathrm{Cd}_{B\text{-DTPA}} \qquad r^2 = 0.95 \tag{8}$$

$$\mathrm{pCd}_{\mathrm{DTPA}} = 12.21 - 1.15 \log \mathrm{Cd}_{L\text{-DTPA}} \qquad r^2 = 0.89 \tag{9}$$

$$\mathrm{pCd}_{\mathrm{DTPA}} = 12.44 - 1.12 \log \mathrm{Cd}_{\mathrm{HCl}} \qquad r^2 = 0.89 \tag{10}$$

The uptake of cadmium by corn leaves and sorghum leaves was also correlated with $\mathrm{pCd}_{\mathrm{DTPA}}$.

$$\mathrm{pCd}_{\mathrm{DTPA}} = 12.08 - 1.57 \log \mathrm{Cd}_{\mathrm{corn}} \qquad r^2 = 0.88 \tag{11}$$

$$\mathrm{pCd}_{\mathrm{DTPA}} = 11.38 - 1.49 \log \mathrm{Cd}_{\mathrm{sorghum}} \qquad r^2 = 0.73 \tag{12}$$

These data indicated that the cumulative loading rate for Cd^{2+} in soils could be determined under laboratory conditions; also, it was concluded that the soil level of cadmium associated with a $\mathrm{pCd}_{\mathrm{DTPA}}$ of 12.0 should not be exceeded [*24*]. Concentrations of cadmium in corn leaves and soil test cadmium values for 1974 and 1975 are plotted in Fig. 1. As expected from Eqs 8 and 11, the relationships were found to be excellent and represented a great improvement over the relationship reported previously [*24*].

Laboratory Method for Measuring Cadmium Loading Capacities of Soils

In order to test the feasibility of developing a laboratory procedure for measuring soil cadmium adsorption as a function of pCd in solution, samples of Hagerstown silt loam were collected from check plots of the field experiment discussed earlier. The samples were thoroughly mixed and sieved through a 20-mesh screen. Samples of the soil were treated and equilibrated four times with solutions containing Cd^{2+} activities of 10^{-8}, 10^{-7}, 10^{-6}, 10^{-5}, and 2 ×

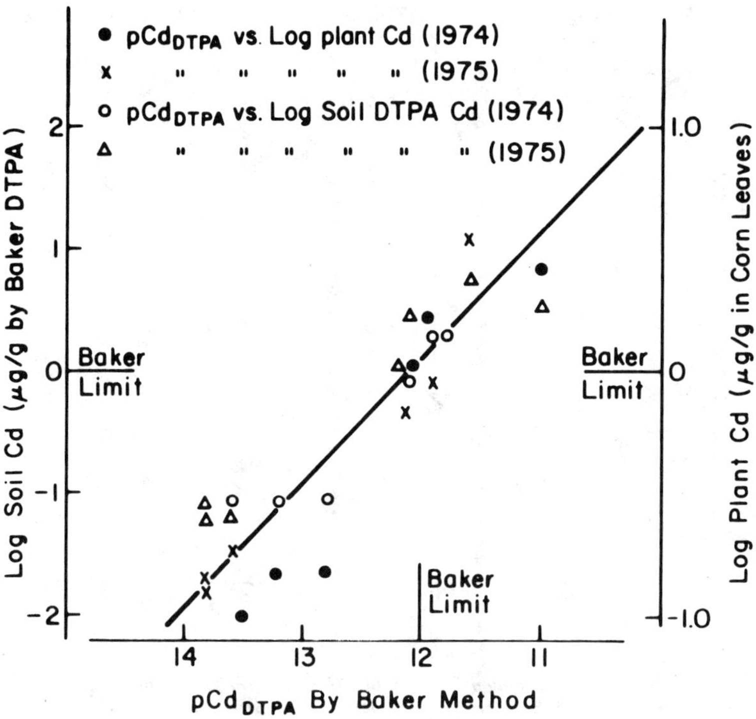

FIG. 1—*Interrelationships among* pCd_{DTPA} [16], *the logarithm of soil cadmium extracted with DTPA by the Baker method* [16], *and the logarithm of cadmium concentrations in corn leaves taken at silking time from plants grown in 1974 and 1975 in field plots receiving different levels of cadmium from sewage sludge* [23].

10^{-5} *M* Cd^{2+} as $Cd(NO_3)_2$ in 60 × 10^{-4} *M* $Ca(C_2H_3O_2)$ at pH 7.0. The linear regression coefficient, *r*, relating log $Cd_{B\text{-DTPA}}$ and pCd_{DTPA} was 0.999. An equal regression coefficient was obtained when pCd_{DTPA} was regressed with log cadmium extracted with 1 *N* HNO_3. These results and the regression line for the field data are shown in Fig. 2. Although the slopes of the two regression lines are nearly identical, the intercept values are different. The different intercept values may have resulted from the use of cadmium nitrate ($CdNO_3$) in the laboratory while sludge cadmium was applied in the field plots, or other trace metals in the sludge could have reduced cadmium adsorption in the field, or both. Analysis for cadmium in the equilibrium solutions with activity coefficients calculated by the extended form of the Debye-Hückel equation indicated that a pCd_{DTPA} of 12 for this soil by the Baker method [*17*] was equivalent to a pCd in solution of 6.5. Amacher [*25*] reported recently that a pCd of 12.5 for a soil sample by the Baker method corresponded to a pCd value of 7.03 for a saturation extract with activity calculated by GEOCHEM [*31*]. The pCd_{DTPA}-

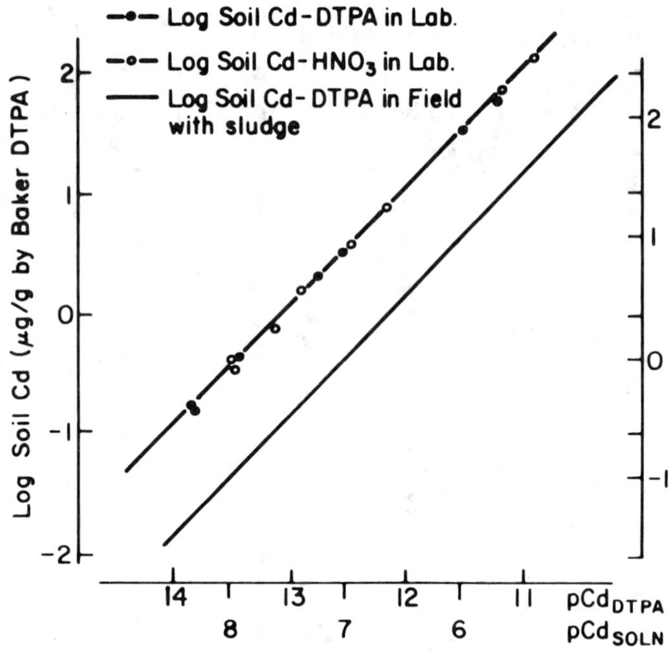

FIG. 2—*Interrelationships among* pCd_{DTPA} [16], $pCd_{solution}$, *the logarithm of soil cadmium extracted with DTPA* [16], *and the logarithm of cadmium extracted with 1* M HNO_3 *in the laboratory compared with the relationships shown in Fig. 1.*

$pCd_{solution}$ comparisons of Fig. 2 indicate a constant difference of 5.5 for $pCd_{solution}$ values in the range of 5.5 to 8.0.

For use in routine laboratory determinations of cadmium loading capacities of soils to a desired pCd in solution, the Hagerstown soil was equilibrated with the $Ca(CH_3COO)_2$ at $60 \times 10^{-4} M$, pH 7.0 ± 0.1, which had a pCd of 6.0. The variable for the experiment was the shaking time. A short shaking time of 30 min is preferred to the 2-h period used previously. The results for the experiment are summarized as follows:

$$\text{Time} = 30\,\text{min}; N = 10; \text{pCd} = 6.60 - 0.313 \log \text{Cd}_{soil} \qquad r = 0.960$$

$$\text{Time} = 2\,\text{h}; N = 6; \text{pCd} = 6.71 - 0.299 \log \text{Cd}_{soil} \qquad r = 0.929$$

For the 2-h equilibration, the first increment of added cadmium did not measurably increase the solution concentration of cadmium over the value for the untreated soil. Therefore, the correlation was lower than for the 30-min shaking time. The greater intercept value suggests that more cadmium was adsorbed with time by the soil. It was concluded that the method will require a 2-h shaking time, which must be carefully controlled if laboratories are to obtain comparable results.

To determine the effects of low, biologically important concentrations of nickel, copper, and zinc in solution on the adsorption of cadmium, the $Cd(CH_3COO)_2$ plus $Cd(NO_3)_2$ solution was fortified with these elements for comparison with the solution used previously. The results are summarized as follows:

$$N = 8; \text{no metals added}; \text{pCd} = 6.49 - 0.943 \log \text{Cd}_{\text{soil}} \qquad r = 0.943$$

$$N = 8; \text{copper, zinc, nickel added}; \text{pCd} = 6.36 - 0.765 \log \text{Cd}_{\text{soil}} \qquad r = 0.968$$

The difference between the results for the solution without metals and those of the previous experiment was greater than the difference caused by the presence of metals. Therefore, the small decrease in cadmium adsorption resulting from the presence of other metals must receive further study. The plan is to improve the precision of the method and apply it to selected modal soils representative of soils of the northeast.

Plant Response to μ_{Cd} in Solution

With the selected standard state associated with Eq 1, the A_{Cd} is equal to the concentration of Cd^{2+}, C_{Cd}, times the activity coefficient, γ_{Cd}. The value for γ_{Cd} for all nine of the solutions used to measure cadmium accumulation by corn seedlings was calculated to be 0.59 at the ionic strength of 0.0177 calculated for the nutrient solution. The results for the experiment led to the following linear relationship:

$$\text{pCd}_{\text{solution}} = 7.669 - 1.020 \log \eta\text{CT}_{\text{seedlings}} \qquad r^2 = 0.99$$

As was reported by Baker and Low [*14*] and Shuman, Baker, and Thomas [*32*], the uptake of elements by corn seedlings is very precise when the growth cell technique is employed. It was concluded that the uptake of Cd^{2+} is passive for corn seedlings even at very low levels of Cd^{2+} in the nutrient solution.

The plant growth cell technique provides an excellent approach for precisely measuring the laboratory equivalent of the relationship found for the Baker soil test (pCd_{DTPA}) and plant accumulation of cadmium under field conditions (Eqs 11 and 12). The growth cell technique offers an excellent approach for testing the validity of relating soil-adsorbed cadmium to pCd in solution where the results for the nutrient solution are compared with those obtained with soil-water suspensions. For soils previously treated with sludges, cadmium species other than Cd^{2+} may have to be measured to obtain accurate values for $\text{pCd}_{\text{solution}}$ [*18*].

Summary and Conclusions

Existing data summarized by Browne, Wong, and Buhler [*15*] for uptake of cadmium from soils treated with sludges under greenhouse conditions and the results of several investigators indicated passive uptake of Cd^{2+} by plants. These results and those for the Baker method of soil testing [*17,21*] led to the postulated relationships presented and defined in Eq 7

$$a + b \log \mathrm{Cd}_{soil} = \mathrm{pCd}_{solution} = c + d \log \mathrm{Cd}_{plant}$$

An evaluation of field data relating the Baker soil test results for pCd^{2+}_{DTPA} ($-\log A_{Cd}$ calculated from the chemistry of DTPA) indicated Eq 7 provided excellent linear relationship for both soil-adsorbed and plant-absorbed cadmium. The linear regression equations provided r^2 values of 0.89 to 0.95 for pCd^{2+}_{DTPA} related to plant-adsorbed cadmium.

A laboratory method to relate pCd_{DTPA} and $\mathrm{pCd}^{2+}_{solution}$ to cadmium adsorbed by soil was developed. While the data from the laboratory were not identical to those for the field, the equations of best fit were identical except for the intercept values, and the correlation coefficients were very high.

Preliminary data indicate that, with a 2-h shaking time and careful control of pCd, pH, and the concentration of Ca^{2+}, a laboratory equilibration procedure may be used to obtain a direct measurement of the cadmium loading capacity of different soils as a function of pCd in solution. The effect of other trace metals on the cadmium loading capacity requires further study involving different soils.

The growth cell technique of Baker and Low [*14*] was found to be very precise for relating pCd in solution to uptake of cadmium by corn seedlings. The technique is considered excellent for evaluating the validity of the laboratory-measured cadmium loading capacity for different soils.

It was concluded that the approach and methods developed in this investigation have substantial potential for providing a routine laboratory method to directly measure the cadmium loading capacity of soils. The approach could be extended to include copper, zinc, and nickel as well as cadmium.

References

[*1*] Baker, D. E. and Chesnin, L., "Chemical Monitoring of Soils for Environmental Quality and Animal and Human Health," *Advances in Agronomy*, Vol. 27, 1975, pp. 305–374.

[*2*] Baker, D. E., Amacher, M. C., and Leach, R. M., "Sewage Sludge as a Source of Cadmium in Soil-Plant-Animal Systems," *Environmental Health Perspectives*, Vol. 28, 1979, pp. 45–49.

[*3*] "Criteria for Classification of Solid Waste Disposal Facilities and Practices," U.S. Environmental Protection Agency, *Federal Register*, Vol. 44, No. 179, 1979, pp. 53438–53464.

[*4*] *Mineral Tolerance of Domestic Animals*, National Research Council, National Academy of Sciences, Washington, DC, 1980.

[*5*] *Background Document: Cumulative Cadmium Application Rates*, Docket 4004, U.S. Environmental Protection Agency, Washington, DC, 1979.

[6] "Effects of Sewage Sludge on the Cadmium and Zinc Content of Crops," No. 83, Council for Agriculture Science and Technology, Ames, Iowa, 1980.
[7] Baker, D. E., "Criteria and Recommendations for Land Application of Sludges in the Northeast," Bulletin 851, Pennsylvania State University, College of Agriculture, University Park, Pa., 1983, in press.
[8] Levitt, Jacob, *Introduction to Plant Physiology*, Mosby, St. Louis, 1969.
[9] Adamson, A. W., *Physical Chemistry of Surfaces*, 3rd ed., Wiley, New York, 1976, pp. 230-234.
[10] Cutler, J. M. and Rains, D. W., "Characterization of Cadmium Uptake by Plant Tissue," *Plant Physiology*, Vol. 54, 1974, pp. 67-71.
[11] Petit, C. M., Ringoet, A., and Myttenaere, C., "Simulation of Cadmium Uptake in Relation to the Cadmium Content of Plants," *Plant Physiology*, Vol. 62, 1978, pp. 554-557.
[12] Sharma, R. P., "Soil-Plant-Animal Distribution of Cadmium in the Environment," in *Cadmium in the Environment, Part 1: Ecological Cycling*, J. O. Nriagu, Ed., Wiley, New York, 1980, pp. 587-605.
[13] deVries, M. P. C. and Tiller, K. G., "Sewage Sludge as a Soil Amendment, with Special Reference to Cd, Cu, Mn, Ni, Pb, and Zn—Comparison of Results from Experiments Conducted Inside and Outside a Greenhouse," *Environmental Pollution*, Vol. 16, 1978, pp. 231-240.
[14] Baker, D. E. and Low, P. F., "Effect of the Sol-Gel Transformation in Clay-Water Systems on Biological Activity: II. Sodium Uptake by Corn Seedlings," *Soil Science Society of America, Proceedings*, Vol. 34, 1970, pp. 49-56.
[15] Browne, C. L., Wong, Y. M., and Buhler, D. R., "A Predictive Model for the Accumulation of Cadmium by Container-Grown Plants," *Journal of Environmental Quality*, Vol. 13, No. 2, April/June 1984, pp. 184-188.
[16] Lindsay, W. L. and Norvell, W. A., "Development of a DTPA Soil Test for Zinc, Iron, Manganese, and Copper," *Soil Science Society of America Journal*, Vol. 42, 1978, pp. 421-428.
[17] Baker, D. E. and Amacher, M. C., "The Development and Interpretation of a Diagnostic Soil-Testing Program," Bulletin No. 826, Pennsylvania State University, Agricultural Experiment Station, University Park, Pa., June 1981.
[18] Pickering, W. F., "Cadmium Retention by Clays and Other Soil and Sediment Components" in *Cadmium in the Environment, Part 1: Ecological Cycling*, J. O. Nriagu, Ed., Wiley, New York, 1980, pp. 365-397.
[19] Baes, C. F., Jr. and Mesmer, R. E., *The Hydrolysis of Cations*, Wiley, New York, 1976.
[20] Sposito, G., *The Thermodynamics of Soil Solutions*, Oxford Clarendon Press, New York, 1981.
[21] Baker, D. E., "A New Approach to Soil Testing: II. Ionic Equilibria Involving H, K, Ca, Mg, Mn, Fe, Cu, Zn, Na, P, and S," *Soil Science Society of America, Proceedings*, Vol. 37, 1973, pp. 537-541.
[22] Epstein, E., *Mineral Nutrition of Plants: Principles and Perspectives*, Wiley, New York, 1972.
[23] Rose, A. W., Hawkes, H. E., and Webb, J. S., *Geochemistry in Mineral Exploration*, 2nd ed., Academic Press, New York, 1979, p. 202.
[24] Baker, D. E., Amacher, M. C., and Doty, W. T., "Monitoring Sewage Sludges, Soils and Crops for Zinc and Cadmium," *Land as a Waste Management Alternative*, R. C. Loehr, Ed., Ann Arbor Science, Ann Arbor, Mich., 1977, pp. 261-281.
[25] Amacher, M. C., "Determination of Ionic Activities in Soil Solutions and Suspensions: Principle Limitations," *Soil Science Society of America, Journal*, Vol. 48, pp. 519-524.
[26] Baker, D. E. and Suhr, N. H., "Elemental Analysis by Atomic Absorption and Flame Emission Spectrometry," *Methods of Soil Analysis*, 2nd ed., Soil Science Society of America, Madison, Wis., 1983, pp. 13-27.
[27] Lindsay, W. L., *Chemical Equilibria in Soils*, Wiley, New York, 1979.
[28] Cataldo, D. A. and Wildung, R. E., "Soil and Plant Factors Influencing the Accumulation of Heavy Metals by Plants," *Environmental Health Perspectives*, Vol. 27, 1978, pp. 149-159.
[29] Bingham, F. T., Page, A. L., Mahler, R. J., and Ganje, T. J., "Growth and Cadmium Accumulation of Plants Grown on Soil Treated with Cadmium-Enriched Sewage Sludge," *Journal of Environmental Quality*, Vol. 4, 1975, pp. 207-211.

[*30*] Jones, J. B., *Handbook on Reference Methods for Soil Testing*; Council on Soil Testing and Plant Analysis, Athens, Ga., 1981.

[*31*] Sposito, G. and Mattigod, S. V., *GEOCHEM: A Computer Program for the Calculation of Chemical Equilibria in Soil Solutions and Other Natural Water Systems*, Kearney Foundation of Soil Science, University of California, Riverside, Calif., 1980.

[*32*] Shuman, L. M., Baker, D. E., and Thomas, W. I., "Zinc Accumulation Characteristics of Corn Hybrids," Bulletin No. 811, Pennsylvania State University, Agricultural Experiment Station, University Park, Pa., 1976.

Amelioration of Wastes in the Disposal Environment

György Mucsy,[1] *Erzsébet Kránicz-Pap,*[2] *and György Urbányi*[1]

Disposal of Oil-Containing Sludges on Farmlands

REFERENCE: Mucsy, G., Kránicz-Pap, E., and Urbány, G., "**Disposal of Oil-Containing Sludges on Farmlands,**" *Hazardous and Industrial Waste Management and Testing: Third Symposium, ASTM STP 851,* Larry P. Jackson, Alan R. Rohlik, and Richard A. Conway, Eds., American Society for Testing and Materials, Philadelphia, 1984, pp. 135–151.

ABSTRACT: The response of soils and plants to oil-containing sludges was studied by pot, small plot, and field experiments. The objective was to develop a simple method of farmland disposal without detrimental effects on the environment.

Dewatered sludge from a petroleum storage plant, in combination with nitrogen fertilizer, was found most suited to farmland disposal. The sludge was applied to sandy soil at rates from 3 to 8 g oil per kilogram of soil, and during a single growing season (six months) 60 to 90% of the oil was decomposed. No oil was detected below a 60-cm depth. Some physical properties of the soil, particularly the texture and water-retaining capacity, were improved. The plants were analyzed for protein, phosphate, and potassium oxide and found to have suffered no detriment and no loss in crop yield. The method of disposal by conventional agrotechnical operations proved environmentally safe and economically superior to other alternatives.

KEY WORDS: petroleum refineries, liquid hydrocarbons, oil-containing sludges, fertilizers, agriculture, soil texture, groundwater, water budget, biological activity, bacteria, fungi, yeast mould, nitrogen-binding bacteria, actinomyces, azobacter, oil-decomposing bacteria, fertility, indicator plant, maize, hybar, sunflower, quality analysis, environmental protection, hazardous wastes, industrial wastes

Petroleum production and processing have multiplied over the past decades, in the wake of which environmental pollution has assumed global proportions. Production, refining, transport, and use of petroleum are invariably sources of pollution. In these processes liquid, crude, or refined hydrocarbons are the main pollutants which have aroused the interest of professionals and the general public alike. Attention has been concentrated, especially in recent years,

[1] Civil engineer and senior research engineer, and chemist-biologist and research associate, respectively, Research Centre for Water Resources Development, Budapest, Hungary.

[2] Civil engineer, agricultural engineer, and senior research engineer, Keszthely University of Agriculture, Department of Agronomy, Mosonmagyaróvár, Hungary.

on the sludges containing oils of different origin, which present a steadily growing hazard to the environment.

The collection, treatment, and disposal of oil-containing sludges has raised a growing problem in Hungary, inhabited by 10 million people at a population density of 113 per square kilometre. The importance of the problem is illustrated by the fact that petroleum consumption has quadrupled over the past two decades and has by now attained the 10 million tonnes per year level. A slight decline has been observed within the last two years.

A number of methods have been developed for treating the oil-containing sludges, of which incineration, use for landfilling, and farmland disposal represent the feasible alternatives [*1*,*2*]. The volume of sludge, the method of final disposal, the level of technology in a particular country, and the funds available are the considerations involved in any choice.

In Hungary the oil-containing sludges have until recently been dewatered on drying beds, and then used for landfilling, or discharged directly into sludge reservoirs. No complete solution has been found thus far to the problem of treatment and disposal of oil-containing sludges.

In 1972, the National Water Authority (OVH) assigned the Research Centre for Water Resources Development (VITUKI) the task of developing a rational method for the disposal of oil-containing sludges, which can be realized by simple means available in Hungary, without detriment to the environment.

A review of the literature has revealed that the problems associated with land disposal of oil-containing sludges have already received considerable attention. According to Dotson et al [*3*] and Fehér [*4*], the topsoil is extremely rich in microorganisms, a number of which are bacteria and fungi capable of decomposing several kinds of hydrocarbon. Spreading liquid oily sludges in a 15-cm thick layer on farmlands is reported as routine practice at oil refineries in Texas [*3*]. After draining to separate the oil and water, the sludge is mixed with the soil three to five times each month. Hydrocarbons are decomposed most effectively on warm, dry soils, as confirmed by Beerstecher [*5*]. Further, Sabanov [*6*], Nagiev and Halilov [*7*], and Tahmazov [*8*] registered increased crop yields of 9 to 18% in maize and wheat irrigated with petroleum derivatives, occasionally in combination with the application of fertilizer. Mamedow [*9*] reported on very favorable changes in the physical soil properties after applications of oily sludges, in that the density of, and evaporation losses from, the soil decreased, while its permeability and water-retaining capacity increased.

Encouraged by the foregoing, attention has been concentrated on farmland disposal, all the more so because poor fertility soils could be expected to be improved by the oily sludge applied.

The experiments have been performed by VITUKI in cooperation with the Mosonmagyarovar Department of Agronomy, Keszthely University of Agriculture.

The objective of these experiments was to study the disposal of various oily sludges on farmlands with due regard to the environmental requirements. A further objective was to develop a harmless method, which can be introduced in Hungary without additional agrotechnical measures and expenditure. Preliminary experiments have shown dewatered sludges to be most readily applicable for large-scale land disposal.

Pot Experiments

On the basis of petri dish tests, pot experiments were performed in 1973 using six different kinds of sludge and maize as the indicator plant. The origin and characteristics of the sludges used have been compiled in Table 1. Fertilizer was added as required by the plants. The duration of the experiments was three months.

The main results of the pot experiments are as follows:

1. The oily sludge caused no substantial change in the soil pH.
2. All the oily sludges caused an increase in the total nitrogen and phosphorus content of the soil.
3. The potassium content was not influenced perceptibly by the addition of oily sludge alone, but when combined with fertilizers it decreased considerably.

Microbiological tests and analyses have shown the oily sludge to promote biological activity in the soil, as demonstrated in the higher counts of germs, yeast mold, and nitrogen-combining bacteria by the end of the growing season. These data are presented in Table 2.

The sprouting ratio was not affected perceptibly by the oily sludges, except for the bentonitic and No. 4 sludges, to which the plants responded favorably. Early development was accelerated by adding fertilizer to all the sludges, especially the aluminum sludge and Sludges 2 and 4. The effect has been improved further by the combined supplementary addition of straw and fertilizer, except for Sludge 2, with which the sprouting percentage decreased.

The height of the indicator plant, silage maize, was not increased by the application of oily sludge alone, but when fertilizer was added to the oily sludge-soil mixture, the plants grew 3 to 35% higher than the reference plants.

The green mass of the plants was increased considerably, by 24 to 32%, over that of the reference in the soils mixed with aluminum and bentonitic sludge. The improvement upon the addition of fertilizer was 1 to 2% only.

The dry content yield of maize was not influenced by any of the oily sludges applied.

The raw protein content of the plants was increased by the aluminum-, bentonitic, and No. 2 sludge, of the oily sludges applied alone. Sludges 1, 2, and 3 added in combination with straw and fertilizer also resulted in higher protein contents of the plants.

Except for those to which the aluminum, bentonitic, and No. 3 sludges were applied, the phosphorus content of the plants decreased in response to sludge application. With supplementary fertilizer application, the phosphorus con-

TABLE 1—*Data from pot experiments.*[a]

Item	Sludge Type					
	A[b]	B	1	2	3	4
Water content, %	94.5	93.2	33.7	39.4	36.4	3.7
Oil as percentage of dry substance	17.8	15.2	4.3	4.1	50.0	6.0
Oil content of dry sludge, mg/g	176.8	151.9	43.0	42.0	500.0	60.0
Oil content in soil before the growing season, mg/g	0.25	0.79	2.20	0.58	0.79	2.20
Oil content in soil after the growing season, mg/g	0.20	0.81	0.45	0.25	0.61	0.83
Oil removal efficiency, %	19.2	...	79.8	57.0	22.8	64.2

[a] Extract content of oily sludges and soils: Soxlet extraction by carbon tetrachloride; determination by gravimetry according to "Comecon Uniform Hydroanalytical Methods" and Hungarian Standard Specifications MSZ 260/22.74.

[b] Key to abbreviations:

A = oily aluminum hydroxide sludge from pilot experiment.
B = oily sludge with activated bentonite from pilot experiment.
1 = sludge from drying bed of storm sewer at petroleum refinery KOF.
2 = sludge from drying bed of industrial sewer at KOF.
3 = bottom sludge from storage lagoon at KOF.
4 = mixed sludge from sludge storage at KOF.

TABLE 2—*Microbiological data from soil samples mixed with different sludges.*

Treatment	In 1 g of Soil			
	Total Bacterium Count	Actonomices Count	Yeast Fungus Count	Nitrogen-Binding Bacteria Count
References	$3\ 200 \times 10^3$	4×10^3	11×10^3	1.5×10^3
Sludge A	$4\ 600 \times 10^3$	50×10^3	15×10^3	3.5×10^3
Sludge B	$5\ 000 \times 10^3$	30×10^3	10×10^3	5.5×10^3
Sludge 1	$17\ 500 \times 10^3$	70×10^3	8×10^3	8.0×10^3
Sludge 2	$16\ 000 \times 10^3$	55×10^3	13×10^3	7.0×10^3
Sludge 3	$14\ 000 \times 10^3$	90×10^3	11×10^3	42.5×10^3
Sludge 4	$31\ 000 \times 10^3$	160×10^3	6×10^3	14.0×10^3
Sludge 3 + fertilizer	$65\ 000 \times 10^3$	120×10^3	43×10^3	44.0×10^3
Sludge 3 + fertilizer + straw	$32\ 000 \times 10^3$	120×10^3	40×10^3	20.0×10^3

tent decreased appreciably but increased when the oily sludge was applied in combination with fertilizer and straw.

Of the oily sludges used alone, the bentonitic and Nos. 1 and 2 sludges proved most beneficial to the potassium content. Except for Sludges 3 and 4, all the sludges applied in combination with fertilizer produced higher potassium contents. When used in combination with straw and fertilizer, the potassium content was not increased further.

Small Plot Experiments

On the basis of the results of the pot experiments, Sludges 1, 2, and 4 were used in the small plots (15 m^2) on calciferous alluvial soil under field conditions, with each combination applied simultaneously on five plots. The main data of these experiments are shown in Table 3. It can be seen from these data that the oil-decomposition efficiencies observable on the plots with supplementary fertilizer agreed well with the results of the pot experiments. The results of soil tests are presented in Table 4, indicating that the soil parameters tested were little affected by the oily sludges. The microbiological effects of sludge application, that is, the welcome increase of bacteria counts by the end of the growing season, can be perceived from Table 5. The crop yields of the indicator plant, maize KSC-360, are presented in Table 6, showing the adverse effects of oily sludges alone, and the definite increase attained in combination with fertilizer. To obtain information on the downward migration of oil in the soil, samples were taken for analysis from the 0 to 30-cm, 31 to 60-cm, and 61 to 90-cm layers. The results, shown in Table 7, indicate that the oil was decomposed with good efficiency below the topsoil, too, and that no oil penetrated deeper than 60 cm.

TABLE 3—*Data from small plot experiments.*

Item	Sludge Type[a]		
	1	2	3
Water content, %	73.9	34.0	34.4
Oil as percentage of dry substance	7.4	22.9	73.9
Sludge application rate, tonne/ha	12	12	12
Oil application rate, kg/m^2	0.088	0.273	0.776
Fertilizer, g/kg soil	0.24	0.76	2.16
Effective N, kg/ha	110	110	110
Effective P, kg/ha	35	35	35
Effective K, kg/ha	78	78	78
Oil decomposition efficiency, %	79.5	68.5	86.4
Growing season, months	6	6	6

[a] For origin of sludge, see Table 1.

Plot Experiments

The 80 m^2 plots were situated in the area of a state farm, on sandy adobe soil. The oil-containing sludges originated from the mechanical oil traps of Refineries 1 and 2 of petroleum company KIV. The sludge from the oil traps was deposited on dumps and retrieved therefrom for the experiments. Maize KSC-360 combined with hybar Sudan grass served as the indicator plant.

The main data of the sludges and the experiments have been compiled in Table 8. Four different kinds of treatment have been repeated four times in these experiments.

During the three months of the growing season, the oil removal efficiencies were comparable to those in the pot and small plot experiments. These values refer to the plots to which fertilizer was also applied.

The results of the soil analyses are presented in Table 9. As can be seen there, the pH of the soil was hardly affected by the kinds of treatment applied.

The nitrogen level in the soil decreased slightly in relation to the original medium-to-poor value. With supplementary fertilizer application the phosphorus and potassium levels developed favorably. From the results of the microbiological analyses compiled in Table 10, it can be seen that the bacteria counts in the soil are increased by the application of oily sludges alone, but even higher counts are attainable when these are combined with fertilizer.

Examination of the plant height data in Table 11 again reveals the favorable effect of fertilizer added to the sludge. A similar trend is observable also in the crop yields, which were higher when oily sludge was applied in combination with fertilizer. The results reveal also that oily sludge applied alone has no detrimental effects.

The results of the analyses for protein, phosphate (P_2O_5), and potassium oxide (K_2O) compiled in Table 12 demonstrate that the protein and potassium

TABLE 4—*Soil analyses.*

Treatment	pH, KCl	Total N[c]	P_2O_5, mg/g[d]	K_2O, mg/g[e]	$CaCO_3$, %[f]
Reference[a]	7.16	0.156	0.84	0.30	1.28
[b]	6.73	0.071	0.90	0.20	0.77
Sludge 1[b]	6.71	0.090	0.85	0.19	0.74
Sludge 2[b]	6.72	0.089	0.80	0.26	0.78
Sludge 4[b]	6.70	0.059	0.97	0.23	0.86
Sludge 1 + fertilizer[b]	6.70	0.055	0.89	0.23	0.71
Sludge 2 + fertilizer[b]	6.68	0.076	0.84	0.22	0.72
Sludge 4 + fertilizer[b]	6.70	0.084	0.88	0.20	0.90
SD_5, %[g]	...	...	0.098	...	...

[a] Before experiment.
[b] After experiment.
[c] Kjeldahl.
[d] From soil extract by ammonium lactate ($NH_4C_2H_5O_3$), with ammonium molibdate [$(NH_4)_6MO_7O_{24} \times 4H_2O$] reagent and determination by photometric method.
[e] By atom absorption.
[f] With Scheibler calcimeter.
[g] Significant difference of 5%.

TABLE 5—*Counts of typical oil decomposing bacteria in soils treated with different oily sludges.*

Treatment	Bacteria Count in 1 g Soil
Reference	70×10^3
Sludge 1	260×10^3
Sludge 2	290×10^3
Sludge 3	280×10^3
Sludge 4 + fertilizer	540×10^3
Sludge 2 + fertilizer	410×10^3
Sludge 4 + fertilizer	360×10^3

TABLE 6—*Crop yields from plots.*[a]

Treatment	Grain Crop		Percentage Ratio, %
	kg/Plot	tonne/ha	
Reference	15.80	10.5	100.0
Sludge 1	12.76	8.5	80.8
Sludge 2	13.40	8.9	84.8
Sludge 4	14.90	9.9	94.3
Sludge 1 + fertilizer	17.80	11.9	112.7
Sludge 1 + fertilizer	19.12	13.7	121.0
Sludge 4 + fertilizer	20.80	13.9	131.7

[a] No control plot with fertilizer alone was used.

TABLE 7—*Distribution and decomposition of oil content in the soil layers.*

Type of Sludge[a]	Depth, cm	Oil Decomposition		
		Beginning of Growth, g/kg[b]	End of Growth, g/kg[c]	%
Sludge 4				
	0 to 30	2.67	0.75	71.9
	31 to 60	0.30	0.02	93.3
	61 to 90	0.00	0.00	...
Sludge KIV 1				
	0 to 30	5.45	1.56	71.4
	31 to 60	0.90	0.25	72.2
	61 to 90	0.00	0.00	...
Sludge KIV 2				
	0 to 30	2.98	1.14	61.7
	31 to 60	0.23	0.07	69.6
	61 to 90	0.00	0.00	...

[a] Sludge types:
Sludge 4, mixed sludge from KOF sludge dump.
KIV 1, from Refinery 1 of petroleum company.
KIV 2, from Refinery 2 of petroleum company.

[b] At the beginning of the growth period and after 8 mm of rain.

[c] At the end of the six-month growing season.

TABLE 8—*Main data on sludges and plot experiments.*

Item	Sludge Type	
	KIV 1[a]	KIV 2[b]
Sludge water content, %	41.2	59.3
Oil as percentage of dry substance	4.1	6.6
Sludge application rate, tonne/ha	125	125
Oil application rate, kg/m^2	0.30	0.34
Oil application rate, g/kg soil	0.84	0.93
Fertilizer application		
Effective N, kg/ha	175	175
Effective P, kg/ha	20	20
Effective K, kg/ha	45	45
Oil decomposition efficiency, %	64.5	57.8
Growing season, months	3	3

[a] From refinery 1 of petroleum company.
[b] From refinery 2 of petroleum company.

TABLE 9—*Results of soil analyses.*

Treatment	pH, KCl	Total nitrogen, %	P_2O_5, mg/g	K_2O, mg/g	$CaCO_3$, %
Reference[a]	6.98	0.069	0.12	0.13	0.24
[b]	6.98	0.108	0.19	0.13	0.17
Fertilizer[b]	7.16	0.072	0.24	0.14	0.39
Sludge[b]	7.35	0.069	0.31	0.11	0.54
Sludge + fertilizer[b]	6.82	0.081	0.43	0.25	0.20

[a] At the beginning of the growing season.
[b] At the end of the growing season.

contents of the plants increased slightly when oily sludge was applied together with fertilizer.

Field Experiments

Drawing on the results of the pot, small plot, and normal plot experiments, field application was started in 1980 with the aim of improving sandy soil in the area of a farming cooperative. An abandoned sand pit was selected as a soil of low agricultural quality, with the hope of improving its fertility.

The groundwater table was situated between 4 and 6 m below the terrain and never rose to the 2.5-m deep bottom of the observation wells sunk for the duration of the experiment. The field was 4.5 ha (0.045 km^2) large. The soil characteristics are shown in Table 13.

The oily sludge was withdrawn from a shallow secondary treatment lagoon, overgrown with reeds, to which the oily plant effluents and storm runoff from a

TABLE 10—*Results of microbiological analyses.*

Treatment	In 1 g of Soil				
	Total Bacterium Count	Azobacter	Oil Decomposing Bacteria	Actonomices	Other Microscopic Fungi
Reference	$23\,000 \times 10^6$	5×10^6	20×10^6	201×10^6	25×10^6
Fertilizer	$22\,000 \times 10^6$	15×10^6	31×10^6	216×10^6	21×10^6
Oil sludge	$110\,000 \times 10^6$	83×10^6	64×10^6	148×10^6	27×10^6
Oil sludge + fertilizer	$53\,000 \times 10^6$	183×10^6	83×10^6	150×10^6	17×10^6

TABLE 11—*Crop yields of maize plus hybar.*

	Height					
	Maize		Hybar		Green Weight	
Treatment	cm	%	cm	%	kg	%
Reference	152.5	100	120.0	100	121.1	100
Fertilizer	202.5	133	120.0	100	130.4	108
Oily sludge	165.0	108	107.5	90	121.1	100
Oily sludge + fertilizer	180.0	118	137.5	115	128.7	114

TABLE 12—*Quality of experimental plants.*

Treatment	Protein, %	P_2O_5, mg/g	K_2O, mg/g
Stalk			
Reference	6.79	5.30	232.5
Fertilizer	7.43	5.30	234.0
Sludge	5.11	5.30	230.0
Sludge + fertilizer	7.49	4.45	272.0
Foliage			
Reference	13.16	7.73	182.5
Fertilizer	10.74	5.10	194.8
Sludge	10.73	6.55	169.3
Sludge + fertilizer	13.58	7.15	195.0

TABLE 13*a*—*Analytical data on sandy soil prior to its disposal.*

pH KCl	7.63	
Plasticity	30%	(liquid limit)
Total salts	...	(by electrical conductivity)
$CaCO_3$	3.12%	
Humus	0.72%	
Total N	0.042%	
$NO_3^- + NO_2^-$	9.6 mg/kg	(with photometric method)
P_2O_5	136.0 mg/kg	
K_2O	105.0 mg/kg	

petroleum storage plant are discharged after primary mechanical treatment. The data of the sludge are presented in Table 14.

The area was subdivided into plots on which sludge was spread in amounts of 0, 130, 186, and 260 tonnes in a random pattern. These amounts were determined so as not to surpass the 3 to 8-g oil per kilogram of soil value, the latter calculated for the top 20-cm-thick layer. The sizes of the plots and the actual application rates have been compiled in Table 15. Before the experiment was

TABLE 13*b*—*Metal and heavy metal content in sandy soil prior to its disposal.*

Metals and Heavy Metals	Total Content, mg/kg[a]	Accessible Content, mg/kg[b]
Mg	...	168.0
Na	...	27.0
Fe	17 000.0	...
Mn	380.0	34.6
Pb	22.0	...
Cr	28.0	...
Cu	11.0	0.4
Zn	80.0	0.4
Cd	1.0	...
Zn/Cd	80	...

[a] After treatment of the dried soil with concentrated nitric acid, determination made by atom absorption.

[b] By treatment with Complexon III solution.

TABLE 14—*Analytical data on the oily sludge used.*

Constituent	Measurement Data
Water content, %	26.3
Total inorganics in dry substance, %	90.2
Total organics in dry substance, %	5.3
Total extract content in dry substance, %	4.5
Sludge density, tonne/m^3	1.55
Total metal content, mg/kg	
Fe	29 500.0
Mn	640.0
Pb	56.0
Cr	50.0
Cu	34.0
Zn	220.0
Cd	2.0
Zn/Cd	110

TABLE 15—*Plot sizes and oily sludge application rate.*

Plot No.	Area, ha	Oily Sludge		Total Extract, mg/g Soil
		tonne/Plot	tonne/ha	
1	0.421	...	...	...
2	0.414	130	315	3.96
3	0.386	130	338	4.25
4	0.455	260	572	7.20
5	0.414	260	629	7.92
6	0.383	130	340	4.28
7	0.439	...	...	...
8	0.386	260	674	8.50
9	0.423	186	440	5.52
10	0.439	...	...	...

started, potassium fertilizer was applied to all plots at the rate of 160 kg of potassium per hectare.

The sludge was worked into the soil by conventional agrotechnical methods down to 20 cm depth, after which sunflower seeds were sown. The normal farming routine was followed thereafter.

The analyses during the growing season revealed the total extract content to have decreased steeply in the early vegetation period; one month after disposal and working in, 77 to 79% were already decomposed. No substantial decrease was detected over the remaining part of the growing season.

The moisture content of the soil in the fields over which oil was spread was higher by 17.7, 38.8, and 18.0% at the application rates of 130, 186, and 260 tonnes, respectively—thus, 20.8% higher on the average than in the reference fields.

In the soils of farmlands on which oil-containing sludges are disposed, an increase in metals content must be anticipated.

The metals content of oily sludge is given in Table 14. These concentrations are substantially lower than the limits considered acceptable by Tietjen [*10*]. The amounts of heavy metals introduced into the soil with the sludge are presented in Table 16, demonstrating that the original heavy metals content of the soil was not increased significantly by the sludge. The variation in the amount of some readily accessible metals present in the soil is seen in Table 17.

By the middle of the growing season, the indicator plants were taller in the experimental fields than in the reference fields. The crop yields at harvesting ranged from 2140 to 2560 kg per hectare in these fields. Appraising the impact of sludge application, it is concluded that, on the average, crop yields were not diminished (Table 18).

Other Considerations

In selecting the experimental conditions, materials, and technique described in the foregoing, the governing consideration was to obtain results and experiences which could be used as directly and as early as possible.

This is why sludges from petroleum refineries and storage facilities were adopted, these—especially the latter—being of primary interest for disposal on and valorization of farmlands.

The sandy adobe and sandy soils selected for the experiments are representative of very large areas in Hungary.

The oil application rates were selected to ensure that a large part of the oil content should be decomposed within a single growing season, without requiring additional operations.

Because of the need to apply oily sludges of different qualities and volumes to farmland, controlled conditions are considered essential, regardless of the type of soil and the crop to be grown. Wherever oily sludges are disposed of on

TABLE 16—*Total metals introduced with the sludge.*

Metal	Tolerable Level[a]	Plot No.										Originally in Soil
		1[b]	2	3	4	5	6	7[b]	8	9	10[b]	
Fe, g/kg	...	...	2.6	2.8	4.7	5.2	2.8	...	5.5	3.6	...	17.0
Mn, g/kg	...	...	0.056	0.060	0.10	0.11	0.060	...	0.12	0.08	...	0.38
Pb, mg/kg	100	...	3.6	3.9	6.6	7.2	3.9	...	7.7	5.0	...	22.0
Cr, mg/kg	100	...	3.2	3.5	5.9	6.4	3.5	...	6.9	4.5	...	28.0
Cu, mg/kg	100	...	2.2	2.3	4.0	4.4	2.4	...	4.7	3.1	...	11.0
Zn, mg/kg	300	...	14.2	15.2	25.8	28.3	15.3	...	30.4	19.8	...	80.0
Cd, mg/kg	5	...	0.12	0.14	0.23	0.26	0.14	...	0.28	0.02	...	1.0
Zn/Cd, mg/kg	...	...	118.3	108.6	112.2	108.8	109.6	...	108.6	990.0	...	80

[a] According to Tietjen [*10*].
[b] Reference field.

TABLE 17—*The changes of accessible metal content of the soil.*[a]

Plot No.	Manganese, mg/kg			Copper, mg/kg			Zinc, mg/kg		
	April	June	September	April	June	September	April	June	September
1	34.6	24.2	47.2	0.4	0.6	0.6	0.4	0.9	0.8
2	. . .	37.3	20.5	. . .	0.8	0.5	. . .	0.9	1.5
3	. . .	14.3	16.5	. . .	1.2	0.7	. . .	1.5	1.6
4	. . .	27.0	14.6	. . .	0.7	0.7	. . .	1.0	0.8
5	. . .	30.0	46.9	. . .	1.4	1.7	. . .	1.9	2.3
6	. . .	12.2	24.0	. . .	0.7	1.0	. . .	0.7	0.9
7	34.6	18.5	46.9	0.4	0.5	1.7	0.4	1.3	2.3
8	. . .	58.5	66.4	. . .	1.3	1.5	. . .	1.9	1.5
9	. . .	50.2	69.4	. . .	1.0	1.8	. . .	0.7	1.3
10	34.6	39.0	31.5	0.4	0.7	0.7	0.4	0.7	0.8

[a] The accessible metals content was determined by the Complexon III solution method.

TABLE 18—*Typical average data on indicator sunflower crop for different treatments.*

Treatment	Height, cm	Crop Yield, tonne/ha
Reference	96	2.31
130 tonne/plot	115	2.46
186 tonne/plot	115	2.14
260 tonne/plot	105	2.56

farmlands, long-term investigations should be carried out into their aftereffects and the changes in quality caused by them.

Conclusions

From the results of the experiments performed thus far, the following conclusions have been made:

1. Oily sludges can be disposed of on farmlands without detrimental effects on the soil.
2. Microorganisms are invariably present in the soil, which are mobilized by the oil applied, prolificate, and decompose the oil. No inocculation of the soil with bacteria is necessary.
3. The microorganisms living in the soil increase in number and are stimulated in their biological activity by oily sludge.
4. During a single growing season 60 to 90% of the oil applied at the 3 to 8 g/kg soil recommended on the basis of these experiments is decomposed. These data apply to six-months-long growing seasons.
5. Some physical properties, particularly the water-retaining capacity of the soil, are improved by oily sludge.
6. No adverse change was observable in the quality of the plants grown on soils treated with oily sludge.
7. Oily sludge must be applied to the soil in combination with a fertilizer.
8. The groundwater table should not rise to within 2 m of the terrain.
9. Industrial crops—which do not serve for direct human consumption, such as corn for distillation, sunflower, hemp, and rape—should be grown on soils treated with oily sludge.
10. The oily sludge and the soil should be checked for heavy metal content before and following application.
11. The main effect of oily sludges disposed of on farmlands is to improve the soil.
12. The method of oily sludge disposal described is, except for special local conditions, superior economically to most other solutions.
13. Oil-containing sludges can be disposed of on farmlands without detri-

ment to the environment. The substances which are hazardous in concentrated form can thus be made harmless.

Acknowledgment

Appreciation for their financial and moral support is expressed to the National Water Authority, the Ministry of Food and Agriculture, Pesticide and Agrochemical Centre, the Mosonmagyarovar Department of the University of Agriculture, the Research Centre for Water Resources Development VITUKI, the farming cooperatives and state farms participating in the work, and, further, to all the agencies that provided valuable assistance in these efforts.

References

[*1*] *Manual on Disposal of Refinery Wastes, Vol. VI, Solid Wastes.* 1st ed., American Petroleum Institute, Division of Refining, Washington, D.C., 1963.

[*2*] Adams, Carl E., Jr., "The Economics of Handling Refinery Sludges," *Proceedings*, Second Forum on Management of Petroleum Refinery Wastewater, EPA-600/2-78-058 ad., University of Tulsa, Tulsa, Okla., March 1978, pp. 503–511.

[*3*] Dotson, C. K., Dean, R. B., Cook, W. B., and Kenner, B. A., "Land Spreading, a Conserving and Nonpollution Method of Disposing of Oily Wastes," *Advances in Water Pollution Research, Proceedings of the 5th International Conference Held in San Francisco and Hawaii, 1970*, S. H. Jenkins, Ed., Pergamon Press, New York, 1971, p. II-36.

[*4*] Fehér, D., *Soil Biology* (in Hungarian), Akademiai Kiado, Budapest, 1954.

[*5*] Beerstecher, E., *Petroleum Microbiology*, Elsevier Press, New York, 1970.

[*6*] Sabanov, M. Z., "Effect of Petroleum-Based Stimulant on Maize Yield in Deteriorated Soil," (in Russian), *Naphtyanie udobrenia i stimulatory. Izdatielstuo Akademii Nauki Aserbaidshanskoi Sovietskoi Socialisticheskoi Respubliki, Baku*, 1963, pp. 219–222.

[*7*] Nagiev, G. M. and Halilov, G. R., "Effect of Petroleum-Based Organic Compound on Maize Yield," (in Russian), *Naphtyanie udobrenia i stimulatory. Izdatielstuo Akademii Nauki Aserbaidshanskoi Sovietskoi Socialisticheskoi Respubliki, Baku*, 1963, pp. 353–356.

[*8*] Tahmazov, G. H., "Effect of Petroleum-Based Stimulant on Autumn Wheat Yield," (in Russian), *Naphtyanie udobrenia i stimulatory. Izdatielstuo Akademii Nauki Aserbaidshanskoi Sovietskoi Socialisticheskoi Respubliki, Baku*, 1963, pp. 375–377.

[*9*] Mamedov, R. G., "Effect of Petroleum-Based Organic Fertilizer on the Agrophysical Properties of Soils," (in Russian), *Izd. Sowhoz XXII Partsiesda Astarinskova Raiona Aserbaidshanskoi Soviet Socialisticheskoi Republiki*, 1971, pp. 31–98.

[*10*] Tietjen, C., "Limits of Use of Liquid Sludges in Cultivation of Plants," *Vizugyi Muszaki Gazdasagi Tajekoztato*, VIZDOK, Budapest, Vol. 90, 1977, p. 107.

Christopher Barber,[1] *Peter J. Maris,*[1] *Susan C. Bull,*[2] *and Ralph G. Johnson*[1]

Codisposal of Oil Wastes with Domestic Solid Wastes in Landfills: Leaching and Persistence of Oil

REFERENCE: Barber, C., Maris, P. J., Bull, S. C., and Johnson, R. G., **"Codisposal of Oil Wastes with Domestic Solid Wastes in Landfills: Leaching and Persistence of Oil,"** *Hazardous and Industrial Waste Management and Testing: Third Symposium, ASTM STP 851*, Larry P. Jackson, Alan R. Rohlik, and Richard A. Conway, Eds., American Society for Testing and Materials, Philadelphia, 1984, pp. 152-170.

ABSTRACT: Controlled landfill offers a possible alternative for the disposal of oily wastes, and research has been carried out to assess the impact of these practices using laboratory simulations and a site study.

These studies showed that oil emulsions (cutting oil and mousse) were rapidly broken down to free oil and an associated aqueous phase in the landfill. The oil was readily sorbed by domestic waste solids, whereas the aqueous phase was flushed out with leachate. Concentrations of oil in leachate were, in most cases, similar to those found in leachate from domestic wastes only (within the range of 5 to 10 mg/L). It was probable that under anaerobic conditions in the fill, oil was not significantly degraded by microorganisms, small losses of oil being mainly due to leaching.

It is concluded that water pollution (surface or groundwater) from the disposal of small volumes of oily wastes with domestic wastes will not be significantly greater than pollution by leachates from domestic wastes only. It is possible that, at small landfills, disposal of excessive quantities of aqueous oil emulsion wastes could exceed the sorptive capacity of the domestic wastes and give rise to leaching of oil.

The persistence of oil in a landfill for long periods of time is an additional problem, which has consequences for the long-term development of land "reclaimed" with wastes containing oils.

KEY WORDS: oil emulsion wastes, domestic (municipal) wastes, codisposal simulation, site study, leaching, persistence, groundwater, hazardous wastes, industrial wastes

[1]Head, senior scientist, and scientist, respectively, Landfill Section, Groundwater Group, Water Research Centre, Stevenage Laboratory, Elder Way, Stevenage, Herts, United Kingdom SG1 1TH.

[2]Scientist, Aquifer Pollution Section, Groundwater Group, Water Research Centre, Environment, Medinenham, Marlow, Buckinghamshire, United Kingdom SL7 2HO.

In the United Kingdom, as in most developed countries, the use of oil and petroleum products is widespread both in industry and for domestic purposes. The wide usage of mineral oils inevitably gives rise to a considerable quantity and variety of mineral oil wastes from the distribution, storage, and processing of oils and from the use of petroleum products. Separation (for example, from aqueous effluent) and reuse of spent oils and product wastes has increased recently in response to rising costs of crude oil and with the establishment of environmental legislation to control disposals of wastes. Incineration of oil wastes is also widely practiced, where inflammable oils can reduce fuel costs [*1*]. This is often the preferred method of waste disposal, although controlled landfill, that is, disposal to land (in landfills), offers a possible alternative route for some wastes; indeed, in the past this method was used for a variety of petroleum residues. At present, for certain types of oily wastes, which are unsuitable for recovery of oil or for disposal by incineration, landfill offers a viable alternative. Examples of wastes for which disposal to landfill might be appropriate include dilute aqueous emulsions (cutting and lubricating oils from metal finishing industries) and oil mousse and contaminated sand arising from marine oil spills. This paper summarizes the results of laboratory and field studies carried out to assess the behavior of these wastes when codeposited with domestic (municipal) solid wastes in landfills. This work follows from previous investigations of several landfill sites which had received oily wastes and found considerable attenuation of oil by sorption within the landfill [*2*].

Method of Study

In order to study the behavior of wastes under controlled experimental conditions and also provide an indication of how representative these experiments were of the situation in the field, both small-scale experimental work (simulated landfill disposal) and a site investigation have been carried out. These are detailed in the following section.

Experimental Details

Simulated Landfills

Small-scale simulations of the codisposal of domestic and oily wastes were set up to investigate the effects of varying the moisture regimen and addition of a nutrient (nitrate) on the leaching and degradation of oil. Experimental units are shown in Fig. 1, and experimental details are given in Table 1.

Dry-pulverized (milled) domestic wastes (average particle size 5 to 7 cm) were used in all the experiments. Two types of oily waste emulsions were also used; these were a "cutting-oil" waste (used Mobil Oil Solvac Double One), containing 180 g oil per litre, and an oil mousse. The latter was obtained from

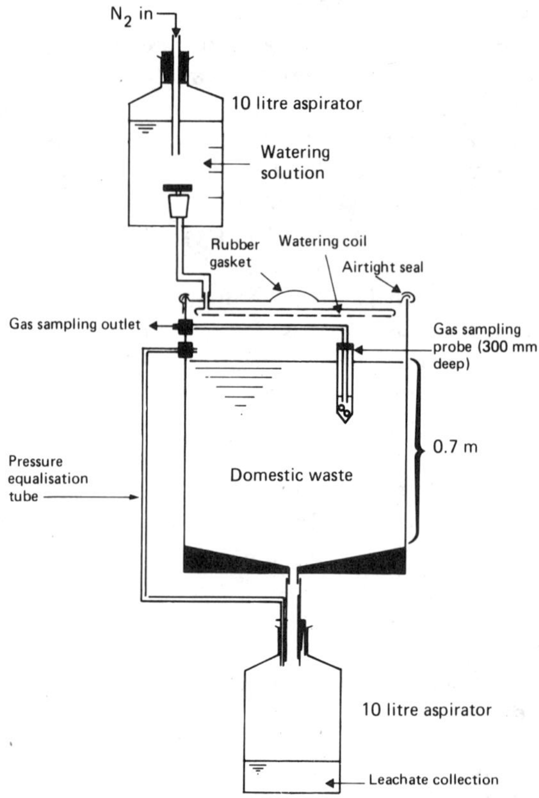

FIG. 1—*Experimental unit showing irrigation, gas sampling, and leachate collection systems.*

Angle Bay, Pembrokeshire, following a marine oil spill from the bulk carrier *Cristos Bitas*, which was carrying Iranian crude oil. The collected mousse contained 76% moisture (seawater), 24% weathered crude oil, and possibly some lubricating oil or other hydrocarbons.

All the units were irrigated with either distilled water, to simulate infiltration from rainfall, or nitrate solution, to assess whether increased biochemical degradation could be induced by addition of this nutrient (Table 1). A tritium tracer (2.74 μCi/L) was added to the irrigant of these units on commencement of the experimental work to determine flow characteristics within the wastes. The leachates were collected at monthly intervals, volumes were measured, and samples were taken for analysis. Gas was extracted from probes installed within each unit (Fig. 1) at approximately three-month intervals, using a 10-mL gas-tight syringe, and analyzed for hydrogen, oxygen, nitrogen, methane, and carbon dioxide by a gas chromatographic method described in the Analytical section.

TABLE 1—*Experimental conditions and mass balance of oil in laboratory studies.*

Unit	Experimental Detail (all units irrigated at 500 mm/year)	Initial Condition		Final Condition	
		Domestic Waste Dry Weight, kg	Weight of Oil Added, g	Residual Oil, g[a]	Oil leached, g
		OIL EMULSION			
OE1	fresh refuse only—control	51.6	...	205	0.58
OE2	fresh refuse + oil emulsion	66.5	2696	3300	0.77
OE3	fresh refuse at field capacity + oil emulsion	56.1	2696	3000	1.01 (4.76)[b]
OE4	aged refuse at field capacity + oil emulsion	105.5	2696	[a]	0.61 (3.51)[b]
OE5	fresh refuse, irrigated with nitrate solution (10 mg/L) + oil emulsion	51.7	2696	2700	0.88
		OIL MOUSSE			
BO1	fresh refuse + oil layer	81.9	3504	4300	0.29
BO2	fresh refuse + oil dispersed	107.2	3424	5200	0.24
BO3	fresh refuse at field capacity + oil layer	83.3	3561	2900	0.31
BO4	fresh refuse at field capacity + oil dispersed	82.1	3544	4400	0.24
BO5	fresh refuse, irrigated with nitrate solution (10 mg/L) + oil dispersed	64.2	3811	[c]	2.62
BO6	fresh refuse only—control	102.4	...	475	0.25

[a] Values are approximate because of difficulties in subsampling and in estimating the weight of residual domestic wastes. Approximate values are given by (depth sampled/total depth) × (total dry weight of refuse added) × (grams oil/dry kilogram waste samples). This includes the oil originally present in domestic wastes.

[b] Figure in parentheses includes oil emulsion eluted immediately after the addition of oil emulsion waste.

[c] No adequate control was available; significant amounts of oil were present.

Site Study—Bergh Apton, Norfolk

On 3 May 1978, a marine oil spill occurred from the bulk carrier *Eleni V*, which subsequently affected beaches in Norfolk and Suffolk in the United Kingdom. Some of the oil-contaminated beach material was buried *in situ*. A further 4 $\times$ 10^4 tonnes of beached material, containing on average 3.5% oil, was transported to an inland area and stored for six months in heaps before being codeposited with domestic waste in a sandpit being used for the disposal of solid domestic waste at Bergh Apton, Norfolk. The oily sand/shingle was deposited in a 2-m-thick layer "sandwiched" between 2-m layers of crude domestic solid wastes, these being deposited in the base of the site. The sand pit was located in Norwich Crag deposits (sand and gravel, with some clays) overlying Upper Chalk, a soft microporous limestone. The boreholes were drilled using percussion techniques through the wastes into the underlying strata in the spring of 1980. Solid core samples were recovered for analysis at 1-m intervals from the Norwich Crag and Upper Chalk rocks beneath the wastes and from the wastes themselves during the drilling.

Interstitial (pore) water was extracted for analysis from core samples from Borehole BA1 within a few days of recovery to determine the extent of pollution of the unsaturated zone by aqueous landfill leachate.

Devices for recovery of the water samples from unsaturated sands and from saturated sands and chalk were installed in the boreholes, shown along with the geological succession in Fig. 2.

Analytical

Liquid (leachate) samples from the experimental units and from the field site were analyzed for chemical oxygen demand (COD), five-day biochemical oxygen demand (BOD_5), volatile acids, chloride, ammonia, and a range of metals using methods developed at the Water Research Centre, United Kingdom [*3–8*]. The concentration of total mineral oil was determined by infrared (IR) spectrophotometry, litre samples of leachate and groundwater being extracted into carbon tetrachloride (CCl_4) (250 mL) using a liquid/liquid technique [*9*]. Solid samples of drill core and waste (approximately 80 g per sample) were extracted with carbon tetrachloride using a Soxhlet apparatus. The carbon tetrachloride extracts were concentrated where necessary by distilling them down to a small volume over a water bath [*9*]. The extracts were then subjected to a cleanup procedure by passing 1 or 2 mL of extract through an anhydrous sodium sulfate/florisil column to remove fats, grease, and biological liquids before extracting the mineral oil.

Quantitative determination of oil was obtained from absorbance at a frequency of 2930 cm^{-1}, the IR spectrophotometer being calibrated for each type of oil used. Qualitative characterizations of oil extracts were performed on a modified Pye Unicam Series 104 chromatograph fitted with an on-column injector, operated under the following conditions:

Type of injection:	on column
Injection volume:	2 μL
Column:	wall coated open tubular (WCOT), 50-m fused silica
Column coating:	OV1
Oven program:	25°C for 5 min initially, programmed at 4°C/min to 250°C and held
Carrier gas:	hydrogen
Carrier gas flow rate:	1 to 2 mL/min
Detection:	flame ionization detector (FID)

Gas generated in the experimental units was analyzed for hydrogen, oxygen, nitrogen, methane, and carbon dioxide using a Perkin Elmer Sigma 2B gas chromatograph fitted with a thermal conductivity detector (TCD) and connected to a Sigma 15 chromatography data station. Two 3-mm outside diameter stainless steel columns were connected in series with a ten-port gas sampling valve providing backflush of one of the columns. One column was packed with 80/100 mesh molecular sieve 13X for the determination of hydrogen, oxygen, nitrogen, and methane, and the other with 80/100 mesh Poropak N, for the determination of carbon dioxide. The switching system previously outlined combines the benefits of both columns to achieve a comprehensive analysis from one sample injection. The instrument was operated under the following conditions:

Injection temperature:	40°C
Oven temperature:	40°C
Detector temperature:	50°C
Detector filament current:	150 mA
Carrier gas:	helium, 30 mL/min
Sample volume:	0.1 mL

Results

Experimental Work

The volumes of liquid obtained initially from each of the experimental units were higher for those that were brought to "field capacity" than for the units that were unwetted before irrigation was begun, as expected. In general, liquid added balanced liquid output over monthly periods once significant leachate production occurred, indicating free gravity drainage through the units, although not necessarily in the fully saturated mode. The discharge curves for the tritium tracer added as a "slug" to each unit showed that the flow through each unit was not homogeneous, particularly for units to which oil emulsion waste had been added. In these, the tritium discharge curves were strongly asymptotic, whereas in the units containing dispersed beached

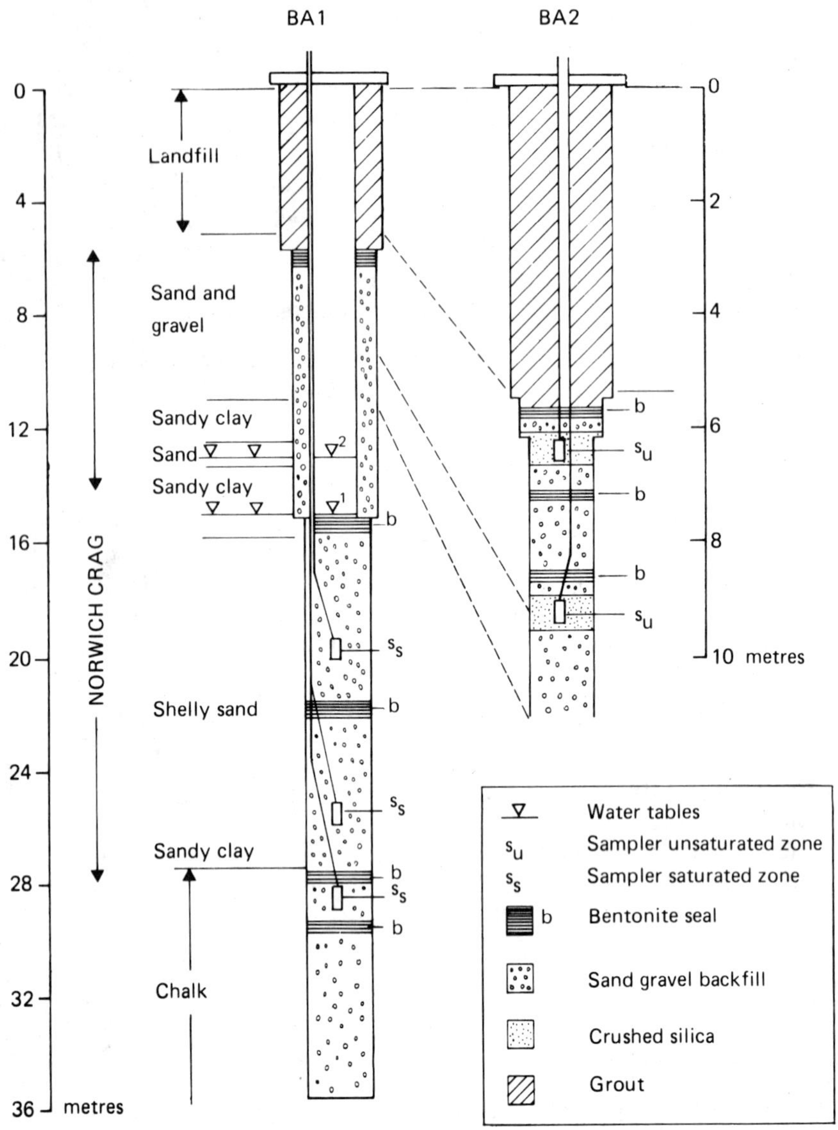

FIG. 2—*Geological succession and borehole construction, Bergh Apton.*

oil, the discharge curves were much closer to the expected Gaussian distribution. Less liquid also was obtained from the controls, which contained no oil additions and were not wetted before irrigation commenced.

Determinations of the composition of the gas within the wastes showed high concentrations of methane (60% by volume) within two months after waste emplacement in units containing oil mousse. Similar conditions were reached after six months in the control units.

The quality of aqueous leachates from the experiments varied considerably, although the major organic and inorganic constituents found were typical of leachates from domestic wastes, and the decrease in concentration with time for all the leachate followed a similar pattern. The leachate from one experiment (BO5) containing dispersed oil mousse and domestic wastes obtained only a few days prior to emplacement did show unusually high concentrations of organic carbon, ammonia, metals, and chloride and low pH values during the first twelve months of operation. These leachates were much stronger than those from the other experiments which contained domestic wastes stored for a few months before emplacement.

The main difference between aqueous leachates from the units containing oil mousse and from the controls was in the concentration of chloride. The mass of chloride leached from these has been calculated and an estimate made of the amount of chloride originally held within the mousse (that is, in salt water) on emplacement (Table 2). It is apparent from these figures that, allowing for variations in moisture and chloride content of mousse, a considerable portion of this brine had been leached from the units where oil was dispersed. In those units where oil mousse was emplaced in one layer, more chloride was leached than in the controls, although 50% or more of the seawater brine still remained within these units (compare the last two columns in Table 2). Elution of the brine from all the units occurred within the first twelve months of operation, that is, in the first few bed volumes of leachate.

The amount of oil leached from all the units that contained oily wastes was a negligible portion of the amount of oil added (Table 1). Taking into account the difficulties in sampling and analysis of solids on completion of the investi-

TABLE 2—*Mass balance of chloride leached from experimental units containing oil mousse and domestic solid wastes.*

	Oil Mousse				
Unit	Deposition	Total Weight Added, kg[a]	Chloride Leached, g	Excess Leached Chloride Above Control, g	Estimated Chloride in Mousse, g[b]
BO1	layer	15.025	290	160	218
BO2	dispersed	14.684	374	204	213
BO3	layer	15.270	192	60	222
BO4	dispersed	15.199	338	208	221
BO5	dispersed	16.342	631	[c]	238
BO6	control	0	164	0	0
BO7	control	0	135	0	0

[a]Seawater content of the mousse is 76.68%.
[b]Estimated from the total seawater content of the added mousse, assuming this contains 19 g/L chloride.
[c]No adequate domestic waste control was available.

gation, it would also appear that most of the oil which was added remained within the units (that is, had not degraded or leached to any large extent—Table 1). The oil mousse wastes were found in approximately the same positions as they had been emplaced within the domestic wastes. Oil from the oil emulsions remained within the upper 20 to 30 cm of each of the units that received aqueous oil emulsion.

Only small amounts of oil were leached from these units in concentrations similar to those from the controls. Concentrations in leachate (measured as total mineral oil) were 5 to 10 mg/L in units containing oil emulsion waste during the first twelve months of operation. This compared with concentrations of less than 5 mg/L in the controls and in later leachates from all the units (Fig. 3). Generally, lower concentrations of oil were found in leachates from the units containing oil mousse, with the exception of Unit BO5, in which a peak of 96 mg oil per litre was found (Table 3).

The nature of the oil mousse, residual oil, and leached oil from Units BO1 through BO7 were investigated by gas chromatography of extracted oil (in CCl_4). The results are shown in Table 4 and in Figs. 4 through 6. There are a number of differences between the chromatograms of the original oil in the mousse and residual oil in layers and dispersed within Units BO2, BO4, and BO5 on completion of the experiments (Fig. 4, Table 4). These differences indicate a loss of normal alkanes in relation to unresolved hydrocarbons, and a decrease in the ratio of normal alkanes to branched alkanes (pristane and phytane—Table 4). These changes were most noticeable in the units in which oil was dispersed rather than emplaced as a layer. Chromatograms for these oils differ considerably from that for oil extracted from the control Unit BO6 (Fig. 4).

In the units in which oil was added in a layer, some movement of oil within the underlying domestic wastes is indicated, although the nature of oil changes with depth (Fig. 5). These changes with depth show a gradual loss of lighter oil fraction (to approximately C_{20}), leaving heavier *n*-alkanes and unresolved hydrocarbons. The oil in leachate obtained during the first six months of operation was relatively enriched in lighter normal alkanes to C_{17}, which predominate over heavier hydrocarbons. The later leachates did not show this predominance (Fig. 6).

Site Study

Aqueous leachate from the emplaced wastes have migrated into the underlying sands of the Norfolk Crag deposits, with the leachate "front" being at about a 12.5-m depth (Table 5). Although contaminants were dispersed in pore water below this depth, only low levels of contamination occurred in groundwater associated with a perched water table at about 18-m depth and in groundwater in the crag and chalk below 21-m depth (Table 5).

Sampling of leachate, pore water, and groundwater from *in situ* devices in

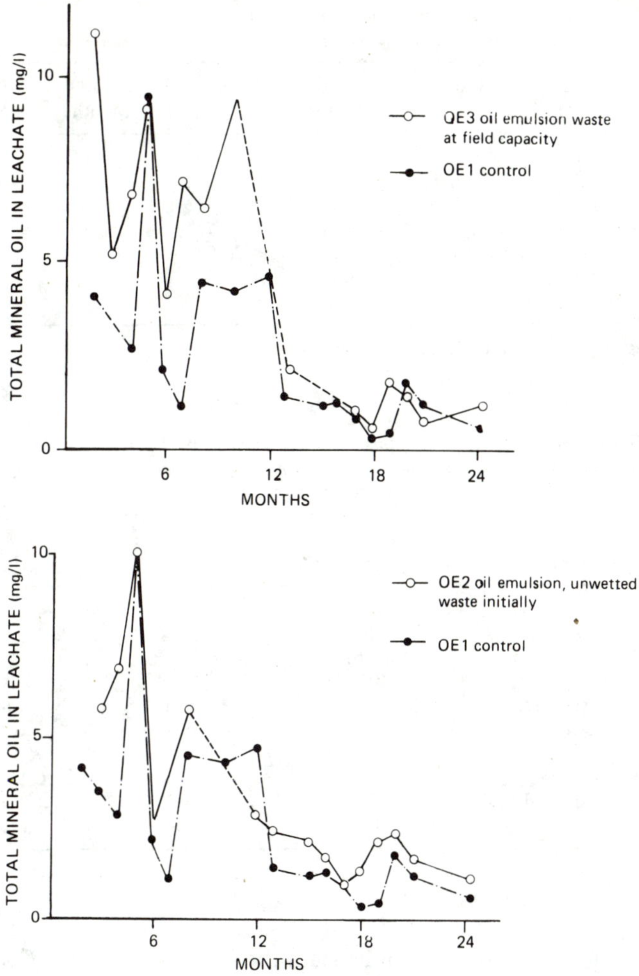

FIG. 3—*Variation with time in concentration of mineral oil: experimental units with added oil emulsion and control.*

both the unsaturated (to 18-m depth) and saturated zones (below 18-m depth) have shown little change over an 18-month period (Table 6). It appears from these data that total organic carbon (TOC), chloride, sulfate, sodium, magnesium, and iron are most mobile, while potassium and ammoniacal-nitrogen have migrated least in the unsaturated zone.

The concentrations of oil in the core samples taken from below the landfill were low but showed maxima at 12.5 m (within the pollution front) and at 18 m (associated with perched area of saturation within the crag deposits). The highest concentrations of oil in liquid samples from the unsaturated zone

TABLE 3—*Variation in the concentration of mineral oil in leachate from experimental units containing domestic wastes and oil mousse (Units 1 to 5) and controls (Units 6 and 7).*[a]

Unit	1 to 2 Months	3 to 4 Months	5 to 6 Months	7 to 8 Months	9 to 11 Months	12 to 14 Months
1	1.1	0.8	3.94	2.26	2.3	1.4
2	0.7	0.6	2.45	3.0	2.7	1.1
3	1.3	0.7	2.89	4.2	1.7	1.3
4	1.0	0.7	3.10	1.7	1.9	1.2
5	...	10.7	96.46	23.9	9.0	1.3
6	2.2	nd	3.58	2.3	1.6	1.3
7	2.6	nd	3.85	2.3	1.7	1.5

[a]All values are in milligrams per litre (nd = not detectable).

TABLE 4—*Details of chromatograms of oil extracted from residual material in units containing oil mousse and related control showing ratio of* n-*alkanes (C_{17},C_{18}) to isoprenoids pristane and phytane.*

Unit	Oil Emplacement	Depth, cm	C_{17}/Pristane	C_{18}/Phytane
...	original mousse	...	2.1	1.3
BO1	layer	20 to 30	1.1	0.8
		30 to 40	0.8	1.0
		40 to 50	1.6	1.0
BO2	dispersed	0 to 70	1.1 to 1.3	0.9 to 1.4
BO3	layer	10 to 20	1.3	0.6
		20 to 30	1.2	0.5
BO4	dispersed	0 to 70	0.2 to 0.4	0.1 to 0.4
BO5	dispersed	0 to 70	1.7 to 1.9	1.2 to 1.9
BO6	none	0 to 70	unresolved	unresolved

(17.8 mg/L) were found 1 to 2 m beneath the landfill (Table 6). Concentrations of oil in the saturated zone were highest in groundwater from the perched water table (up to 9.4 mg/L), although groundwater at a depth of 20 m below this contained 4.7 mg oil per litre. The latter value was in excess of that normally associated with uncontaminated groundwater (less than 1 mg/L TOC) and suggests low-level contamination of groundwater by oil hydrocarbons in the crag aquifer.

Characterization of the oil extracted from the samples collected from *in situ* devices in the unsaturated zone and from groundwater at 18-m depth showed consistent differences (Fig. 7). In the unsaturated zone at a depth of 9 m, chromatograms of oils were characterized by a preponderance of heavy *n*-alkanes in relation to lighter (to C_{18}) alkanes. At the water table, the reverse was the case. Here heavier *n*-alkanes (C_{20} to C_{35}) were barely resolvable above a large bulge of unresolved hydrocarbons. Also lighter *n*-alkanes (to C_{17}) were much more important than at higher levels in the unsaturated zone.

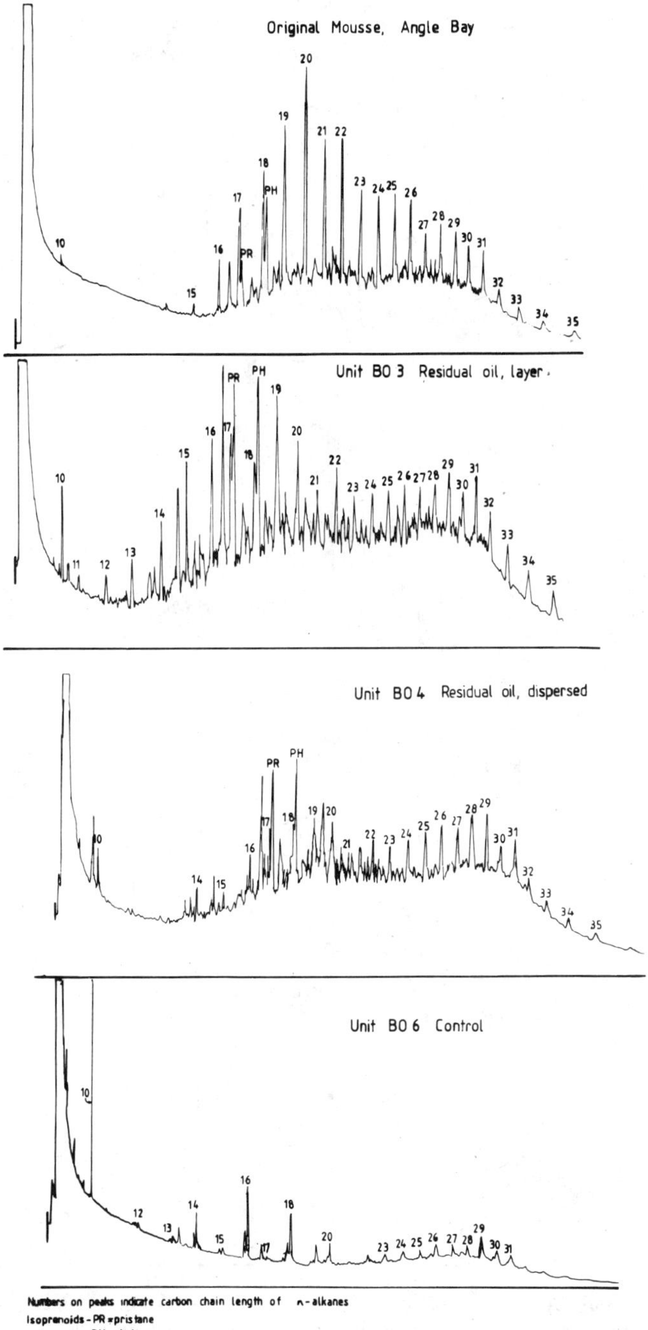

FIG. 4—*Comparison of chromatograms of original oil mousse and residual oil in experimental units to which mousse has been added.*

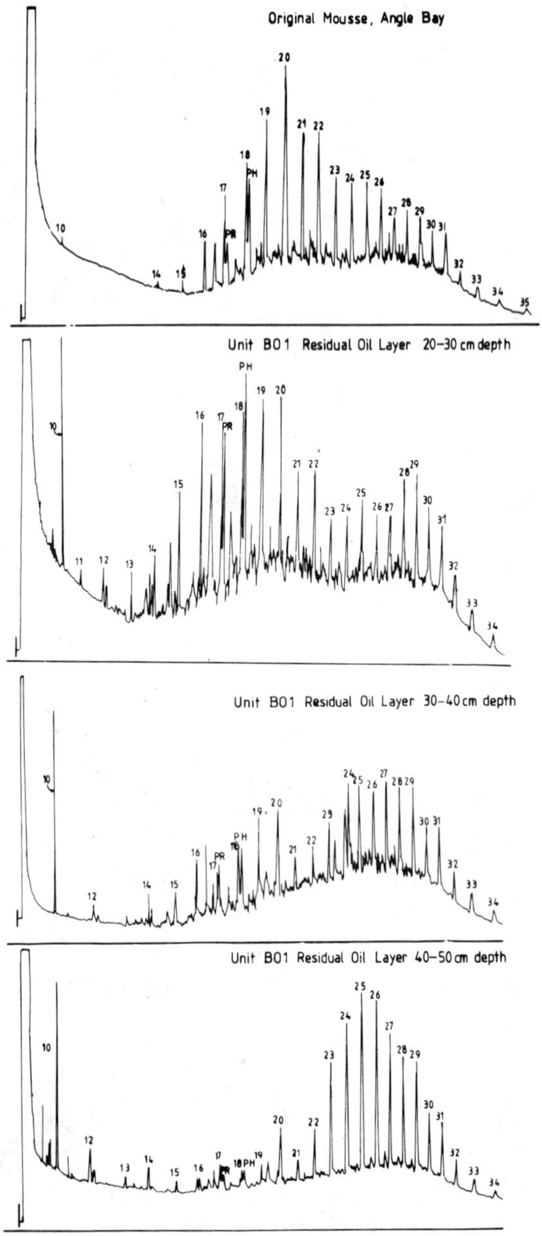

FIG. 5—*Comparison of chromatograms of original oil mousse with residual oil within Unit BO1 in which the mousse was emplaced in a layer at a depth of 20 to 30 cm.*

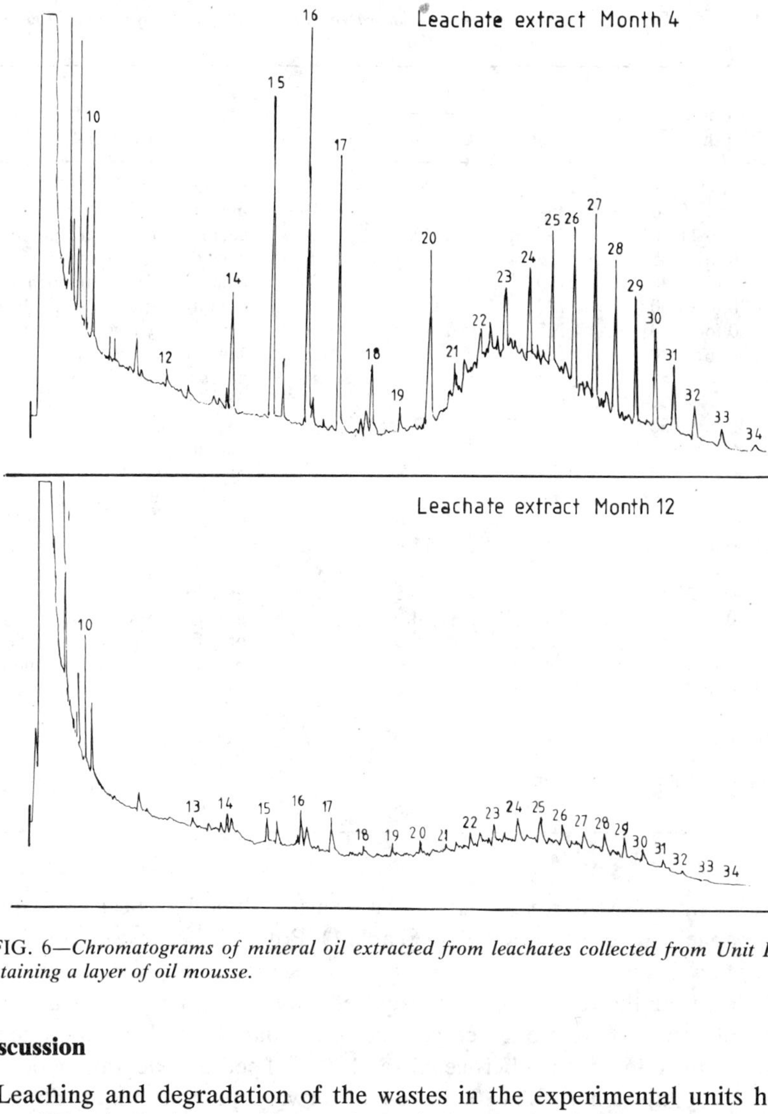

FIG. 6—*Chromatograms of mineral oil extracted from leachates collected from Unit BO1 containing a layer of oil mousse.*

Discussion

Leaching and degradation of the wastes in the experimental units have been affected to only a small extent by variations in the chemical conditions of the wastes (addition of oil to domestic wastes, irrigation with nitrate). Earlier generation of methane in units containing oil was noted in the Units BO1 through BO5 compared with the control. However, increasing the moisture content of domestic wastes (produced by addition of mousse to these units) is known to enhance methane production [*10*]. This, rather than the addition of oil, may have given rise to the early generation of methane. The different moisture regimens, however, did affect the initial behavior of the oil emulsion wastes in Units OE2 through OE5. Here, unwetted domestic wastes retained

TABLE 5—*Results of analyses of interstitial water and of drill core samples from Bergh Apton, Borehole BA1, May 1980.*

Sample Depth, m	Interstitial Water		Oil mg/kg	Material
	Cl^-, mg/L	TOC, mg/L		
9.0 to 9.25	1406	5600	26	medium sand + pebbles
10.15 to 10.40	2464	8100	10	medium sand + few pebbles
11.25 to 11.50	2297	8200	34	coarse sand + many pebbles
12.30 to 12.55	2218	7800	97	coarse sand + many pebbles
13.10 to 13.40	2180	4560	36	coarse sand + many pebbles
14.20 to 14.50	797	435	23	medium sand + few pebbles
15.60 to 16.60	105	40	29	clay + very few pebbles
16.90 to 17.20	124	...	29	sandy clay + very few pebbles
17.65 to 17.80	...	...	16	coarse sand + clay + many pebbles
18.30 to 18.50	92	17	91	coarse sand + clay + many pebbles
19.50 to 20.50		13.8	17	black sandy clay + no pebbles, oily smell
20.50 to 21.50	73	9.0	...	black sandy clay + no pebbles
22.50 to 23.50	46	16.3	...	dark fine sand
24.50 to 25.50	39	4.9	...	dark fine sand
26.50 to 27.50	39	5.5	...	grey fine sand
28.50 to 29.50	41	4.0	...	grey fine sand
30.50 to 31.50	39	5.5	...	grey fine sand and marl pieces
32.50 to 33.50	29	8.0	...	mollusc shell pieces + fine sand
34.50 to 35.50	26	8.0	...	fine sand + some mollusc shell pieces
36.50 to 37.50	26	5.5	...	fine sand + some mollusc shell pieces
38.50 to 39.50	25	4.0	...	fine sand + white chalky clay
40.5	25	6.5	...	white chalk

the oil emulsion to a greater extent than wastes which had been brought to field capacity before addition of oil (Table 1). Probably because of inhomogeneous flow in these units, the volume of aqueous emulsion retained was somewhat less than the volume of water required to reach field capacity (in unwetted units) and to elute one bed volume of leachate [*11*]. The experimental units retained loadings of between 39 and 50 g oil per dry kilogram of domestic waste, equivalent to 0.2 to 0.3 m^3 emulsion waste per tonne of dry domestic wastes. Increasing the loadings above these values (that is, overloading with liquid wastes) would inevitably have given rise to elution of much greater quantities of oil immediately following the addition of oil waste. In contrast, all oil mousse was initially retained in experimental Units BO1 through BO5, probably because of the high viscosity of this waste.

The main long-term effects of the addition of oil to domestic wastes in the experimental studies were (1) to slightly increase the leaching of oil (measured as total mineral oil) and (2) to increase the leaching of chloride in the experimental units containing oil mousse.

TABLE 6—*Summary of analysis of pore water and groundwater recovered from* in situ *samples, April 1981–October 1982, Bergh Apton site, Norfolk.*[a]

Borehole/Zone	Depth	TOC	Total Mineral Oil	Ammoniacal Nitrogen	Chloride	Sulfate	Potassium	Sodium	Iron
BA2/unsaturated zone	9 m	104 to 199	5.8 to 17.8	232 to 655	627 to 1060	33 to 102	480 to 867	618 to 2600	1.4 to 14
BA2/unsaturated zone	13 m	971 to 2190	5.2	13 to 205	2091 to 4291	53 to 818	52 to 160	3190 to 9990	6 to 33
BA1/perched saturated zone	18 m	1.4 to 6.3	4.7 to 9.4	0.1 to 2.4	51 to 66	72 to 82	2.6 to 3.7	17 to 38	2.2 to 43
BA1/saturated zone (crag)	37 m	...	4.74	0.6	23	45	1.2	17	34
BA1/saturated zone (chalk)	41 m	...	...	0.04	21	36	2.6	13	2.2

[a] All results are in milligrams per litre.

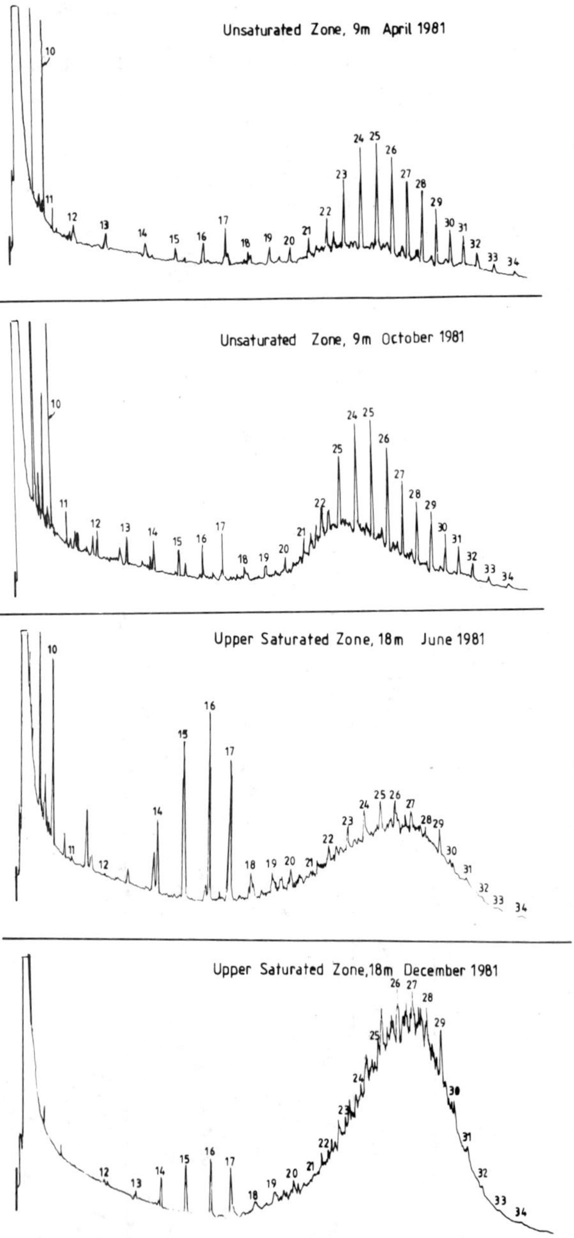

FIG. 7—*Chromatograms of mineral oil extracted from the liquid samples recovered from the boreholes at Bergh Apton.*

Generally, concentration of oil in leachates from the units containing oil wastes were 5 to 10 mg/L or less (Fig. 3, Table 3). Negligible amounts of total added oil were leached. The highest concentration in leachate (96 mg/L) was found in Unit BO5 containing dispersed mousse and very fresh domestic wastes (the latter being different from other units which contained slightly weathered domestic waste). Unit BO5 also produced a very strong leachate (COD up to 88 800 mg/L), which may indicate that with leachates of high organic strength, solubilization of oil was enhanced. This requires further investigation.

Investigation of the composition of leached oil indicates that lighter hydrocarbons (normal alkanes and aromatic compounds) predominate, while heavier alkanes and unresolved hydrocarbons remain within the wastes.

The leaching of brine from the dispersed oil mousse emulsion is also interesting. This and the fact that a minimal amount of oil was eluted from the units which received the aqueous oil emulsion waste indicate that emulsions are rapidly broken down within domestic waste/oil waste mix. This process releases free oil which is sorbed and aqueous fluid which is leached or eluted from the wastes. The layered oil mousse would appear to have broken down more slowly, releasing smaller quantities of chloride in the leachate.

The results of the experimental work compare well with those of the site investigation reported here and in other studies [*1*]. Minimal leaching of oil was found at the Bergh Apton site containing domestic waste and oil-contaminated beach sand and small amounts of mousse from the *Eleni V* spill. There were also some indications of greater leaching of chloride from these wastes than from domestic wastes alone (Table 6). Characterization of the composition of oil in leachate (pore water) and groundwater beneath the Bergh Apton site show similarities to oils leached from the experimental units. Leachate 1 to 2 m below the landfill some two years after emplacement of the waste at the landfill site contained relatively small amounts of lower *n*-alkanes (Fig. 7). These were similar to the older leachates found in the experimental units (Fig. 6). In the samples recovered from the water table (18-m depth) beneath the site, lighter *n*-alkanes are more abundant; these oils are similar to leachates from the experimental units produced in the first six months of operation. These results from both the experimental and field studies indicate the more extensive leaching of lighter hydrocarbon fractions from oil/domestic wastes and their greater mobility both within the landfill and in the underlying sand aquifer.

Although the leaching of oil is likely to give rise to only low-level organic pollution of groundwater (minimal when compared to pollution from other organic constituents of leachate), a possible problem is the persistence of high concentrations of residual oil within a landfill. Despite some fractions of these wastes (lighter normal alkanes) being removed with time, a large amount of added oil remains within the landfill, consisting of higher normal alkanes and unresolved hydrocarbons. It is probable that further leaching and microbial degradation of oil (if, indeed, this occurs) will not reduce these concentrations while anaerobic conditions prevail.

The persistence of oil wastes in landfills for long periods of time would consequently need to be taken into account in the long-term development of land reclaimed with wastes containing oils.

Acknowledgments

This work was carried out as part of a larger research program on the disposal of wastes in landfills funded by the Department of the Environment. We wish to acknowledge help and encouragement given by colleagues at the Water Research Centre, particularly C. P. Young, N. C. Blakey and Dr. M. Fleet. The paper is published with the permission of the director, Water Research Centre, Environment.

References

[*1*] "Mineral Oil Wastes. A Technical Memorandum on Arisings, Treatment and Disposal Including a Code of Practice," Waste Management Paper No. 7, Department of the Environment, Her Majesty's Stationery Office, London, 1976.

[*2*] "Co-operative Programme of Research on the Disposal of Hazardous Wastes in Landfill Sites," Final Report of the Policy Review Committee, Department of the Environment, Her Majesty's Stationery Office, 1978.

[*3*] Petts, K. W., "The Determination of the Biochemical Oxygen Demand in 5 Days at 20°C," Laboratory Procedure No. 2, Water Research Centre, Herts, England, 1980.

[*4*] Musselwhite, C., The Determination of Chemical Oxygen Demand Using Micro-Digestion in Sealed Tubes and Block Thermostat Followed by Micro-Titration with Ferrous Ammonium Sulphate," Laboratory Record, LR906, Water Research Centre, Herts, England, 1978.

[*5*] Petts, K. W., "Determination of Nitrogen Compounds by Technicon Autoanalyser, AA II," Technical Memorandum, TM 108, Water Research Centre, Herts, England, 1975.

[*6*] Petts, K. W., "Determination of Organic Carbon in Liquid Samples by Instrumental Analysis (Bechman Carbon Analyser, 915B)," Laboratory Procedure No. 22, Water Research Centre, Herts, England, 1982.

[*7*] Petts, K. W. and Welch, Victoria C., "An Automated Method for the Determination of Chloride in Sewage, Sewage Effluent and River Water," Laboratory Report No. 62-S, Water Research Centre, Herts, England, 1981.

[*8*] *Determination of Volatile Fatty Acids in Sewage Sludge*, Her Majesty's Stationery Office, London, 1979.

[*9*] Mosey, F., "A Procedure for the Determination of Mineral Oils in Aqueous Samples," Laboratory Report No. 946, Water Research Centre, Herts, England, 1978.

[*10*] Klink, R. E. and Ham, R. K., "Effects of Moisture Movement on Methane Production in Solid Waste Landfill Samples," *Resources and Conservation*, Vol. 8, 1982, p. 29-41.

[*11*] Blakey, N. C., "Absorptive Capacity of Refuse," *Proceedings*, Symposium on Landfill Leachate, 19 May 1982, Harwell Laboratory, Oxon, United Kingdom, 1982.

Marjory B. Rinaldo-Lee,[1] *James A. Hagarman,*[2] *and Andrew F. Diefendorf*[2]

Siting of a Metals Industry Landfill on Abandoned Soda Ash Waste Beds

REFERENCE: Rinaldo-Lee, M. B., Hagarman, J. A., and Diefendorf, A. F., **"Siting of a Metals Industry Landfill on Abandoned Soda Ash Waste Beds,"** *Hazardous and Industrial Waste Management and Testing: Third Symposium, ASTM STP 851,* Larry P. Jackson, Alan R. Rohlik, and Richard A. Conway, Eds., American Society for Testing and Materials, Philadelphia, 1984, pp. 171–192.

ABSTRACT: A recent application by a steel-manufacturing plant to obtain a permit for an industrial landfill on abandoned soda ash waste beds near the city of Syracuse, New York, resulted in an extensive hydrogeologic and geochemical investigation. This investigation was initiated because of (1) previous disposal of waste by the metal manufacturer at this site and (2) the unique location of the landfill on top of preexisting waste beds on the shores of Onondaga Lake. The waste beds cover approximately 1 473 000 m^2 to a depth of 20 m and consist of hydraulically placed hydroxides, salts, and residues, all by-products of soda ash production. The metals-manufacturing wastes have been placed as a thin (0.5 to 2 m) layer over an area of 69 000 m^2 on the soda ash waste beds during the past ten years. The primary leachate contaminant of concern has been hexavalent chromium.

The groundwater monitoring program and investigations included installation of 76 monitoring points including wells, piezometers, lysimeters, and seepage galleries. The hydrogeologic investigation included 30 permeability tests, over 600 water quality samples, and development of a three-dimensional, seven-layer groundwater model of the site. A significant groundwater mound was identified within the waste beds. Groundwater flows predominantly downward through the soda ash wastes into thick underlying alluviolacustrine and glaciolacustrine deposits and ultimately into Onondaga Lake. The results of groundwater monitoring over a one-year period indicate no detectable chromium from the metal-waste leachate escaping through the soda ash wastes. Retention of hexavalent chromium within the underlying highly alkaline soda ash wastes by adsorption, reduction, and precipitation suggests a viable means for *in situ* treatment of several metals-manufacturing waste products.

KEY WORDS: groundwater sampling, chromium, waste disposal, hazardous waste disposal, groundwater, leachate, hazardous wastes, industrial wastes

[1]Senior hydrogeologist Thomsen Associates, Groton, N.Y. 13073.

[2]Project chemist and manager, Geotechnical Studies, respectively, Calocerinos and Spina, Syracuse, NY 13088.

Obtaining a permit for an existing landfill is a situation confronting many industries because of recent state and federal solid waste regulations. Facilities which had been used for many years to dispose of a variety of wastes, including some now classified as hazardous under new federal regulations, are being upgraded to meet the new regulations. This investigation was initiated to obtain a permit for an existing disposal site for Crucible, Inc., a specialty steel mill located near Syracuse, New York.

Although only nonhazardous wastes are currently disposed of in the landfill, previous disposal of materials classified as hazardous wastes at the landfill and the unique location of the landfill on top of preexisting soda ash waste beds on the shores of Onondaga Lake resulted in an extensive hydrogeologic and geochemical investigation. Several of the metals-manufacturing wastes placed on the landfill contained high concentrations of soluble hexavalent chromium which pose a potential threat to the groundwater and surface water. However, the hexavalent chromium leachate has been immobilized within the underlying soda ash process wastes by adsorption, reduction, and precipitation. Thus, a very effective *in situ* treatment process has protected groundwater and surface water from contamination by chromium released by the steel mill wastes. The purpose of this paper is to present the results of the hydrogeologic and geochemical investigation conducted at the site.

Procedure

A hydrogeologic investigation was initiated to examine the extent and hydrogeologic characteristics of the unconsolidated deposits beneath the steel mill wastes, the groundwater flow regime in these deposits, and the extent of contamination from the steel mill wastes. Because of the unique site setting, this investigation was more extensive than it would have been for a similar waste type in another setting.

Seventy-six monitoring points were installed in the vicinity of the steel mill landfill, including 41 monitoring wells, 25 piezometers, 5 lysimeters, and 5 seepage galleries.

Groundwater levels were measured over a 16-month period to define the groundwater flow regime. Hydraulic conductivity of the underlying deposits was examined using slug tests on 16 wells following the methodology of Bouwer and Rice [*1*] and by performing five laboratory permeability tests on undisturbed samples [*2*]. Physical characteristics of the unconsolidated deposits were determined by grain-size analyses on 15 soil samples and 1 soda ash waste sample in accordance with ASTM Particle-Size Analysis of Soils [D 422-63 (1972)]. A three-dimensional groundwater flow model was used to estimate flow patterns and to investigate the route of leachate migration generated by the steel mill wastes [*3,4*].

The chemistry of the steel mill waste was investigated by determining the chemical composition of the various wastes [*5*], extraction procedure (EP) toxicity testing [*6*], and laboratory column leach testing. The chemistry of the interaction of steel mill waste leachate and the underlying soda ash wastes was also examined by adsorption isotherm measurement and column leach testing. Fifteen sets of water samples were withdrawn from selected wells, two sets from lysimeters, and two sets from seeps and seepage galleries during 1981 and 1982. These samples were analyzed to determine the extent of metals migration [*5, 7*].

Results and Discussion

Site Setting

The steel mill landfill is on the southwestern shore of Onondaga Lake near Syracuse, New York (Fig. 1). The steel mill wastes have been placed over the past ten years as a thin (0.5 to 2.0 m) cap on 69 000 m^2 of an existing soda ash wastebed. The soda ash wastes were hydraulically deposited between the 1880s and 1950s in large impoundments eventually covering 1 473 000 m^2 to a maximum depth of 20 m [*8*]. After they were filled with the slurried waste, the impoundments were allowed to dewater. The dewatered waste material was used to form new dikes on top of the old waste beds, resulting in a terraced configuration. To allow for expansion of the waste beds during the early part of the century, Ninemile Creek was diverted to its present location. The former creek bed is buried beneath the waste beds directly below the steel mill landfill.

Geology

The steel mill wastes are underlaid by at least 42 m of unconsolidated deposits, consisting of four geologic units: soda ash waste, swamp and lacustrine deposits, glaciolacustrine deposits, and ablation till. The interrelationship of these units is shown in Figs. 2 and 3.

The soda ash wastes consist primarily of silt-sized particles of insoluble oxides and silicates, alkaline hydroxides, and soluble chlorides and sulfates of sodium, calcium, and magnesium. Hydraulic placement has resulted in interbedded and interfingered layers of waste ranging from thin to massive layers 20 cm or more in thickness. Frequent partings and laminae, as well as 2 to 5-cm layers of gray waste consisting of mixed fly ash and soda ash waste, are found throughout the beds.

Although secondary permeability and channelization of flow within the waste was an area of concern because of evidence of vertical jointing, laboratory and field tests indicated that the joints do not cross individual stratigraphic boundaries. Laboratory permeability tests on undisturbed soda ash waste samples were used to estimate vertical hydraulic conductivity of the waste [*2*]. Field slug tests were used to estimate horizontal hydraulic conductivity of the waste beds [*1*]. The mean horizontal hydraulic conductivity was 6×10^{-5} cm/s, and

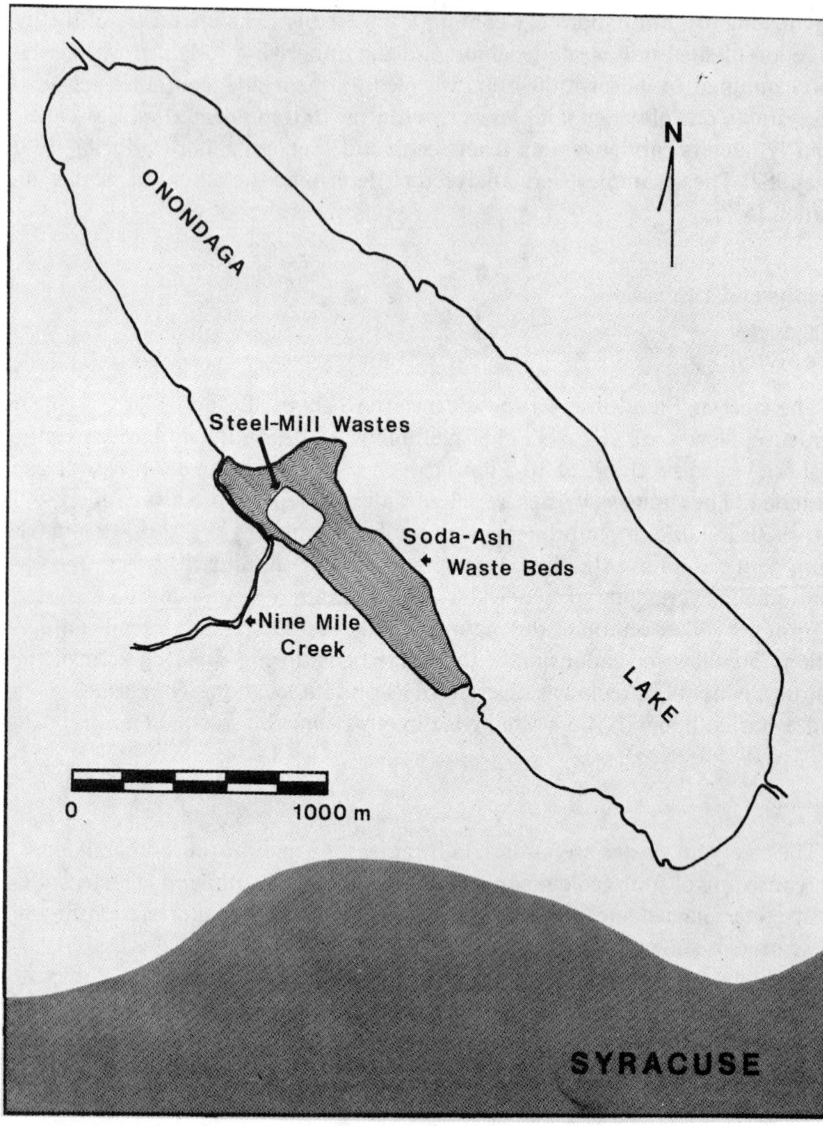

FIG. 1—*Location map of the steel mill landfill.*

the mean vertical hydraulic conductivity was 8×10^{-6} cm/s. Thus, the horizontal permeability is about seven times the vertical permeability. Since the waste was deposited in layers, the horizontal hydraulic conductivity would be expected to be higher than the vertical permeability.

Swamp and lacustrine deposits found immediately beneath the soda ash wastes are 2.5 to 5.0 m thick and consist of peat, marl, silt, fine sand, and clay.

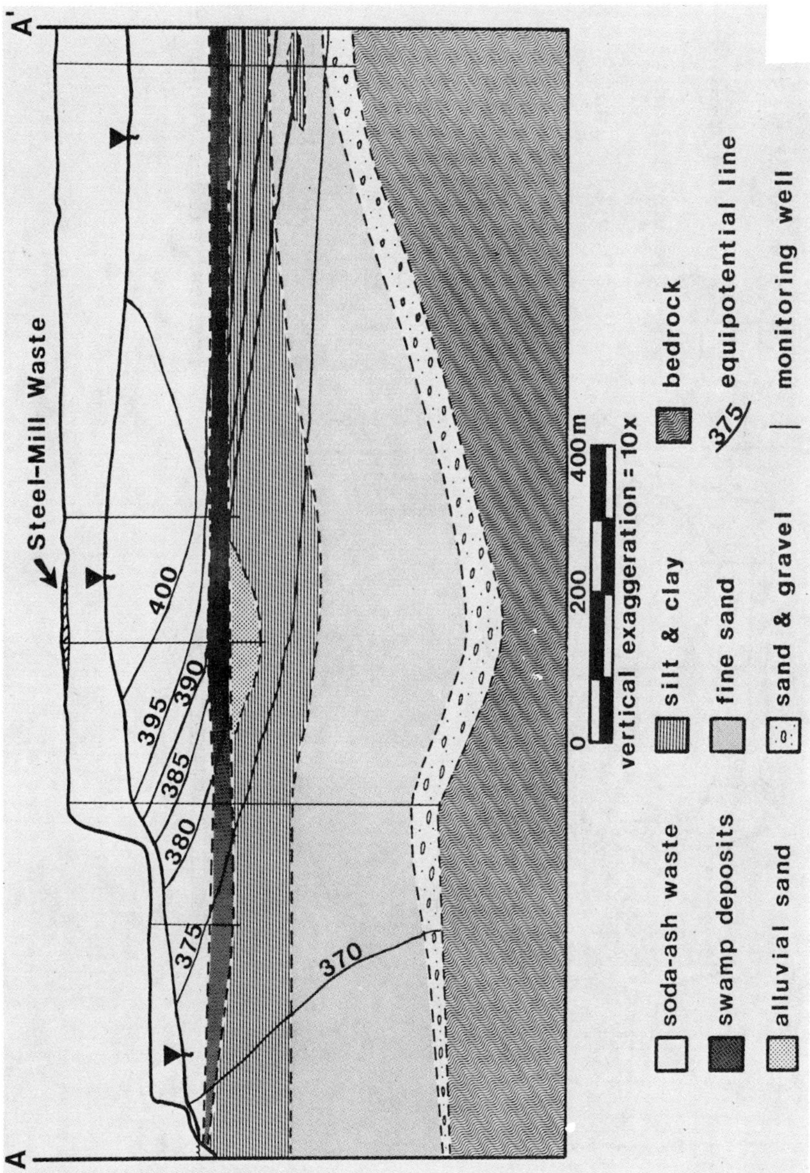

FIG. 2—*Geologic cross section A-A' parallel to the lake shore.*

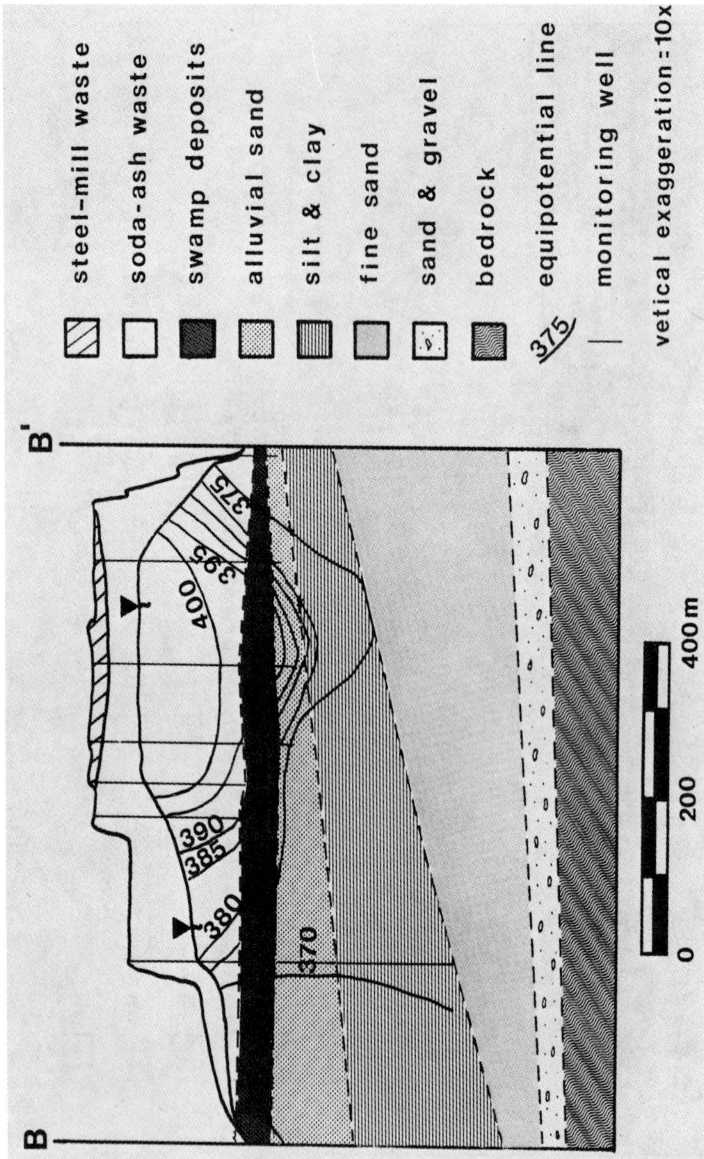

FIG. 3—*Geologic cross section B–B′ perpendicular to the lake shore.*

The low vertical hydraulic conductivity of this layer (7×10^{-7} cm/s) retards the downward flow of groundwater.

Alluvial silty sand deposits of postglacial Ninemile Creek are bound beneath the swamp and lacustrine deposits. The alluvial deposits range from 2.5 to 6.0 m in thickness and increase in thickness in a deltaic configuration toward the northeast where the postglacial stream channel entered the lake.

A glaciolacustrine deposit consisting of alternating layers of silt, clay, and sandy silt lies beneath the stream channel deposits and directly beneath the swamp deposits in areas outside of the stream channel. Toward the lake trough and in the Ninemile Creek channel, the silt and clay layer increases in thickness to a range of 8 to 13 m.

The glaciolacustrine silt and clay unit is less permeable than the overlying alluvial sands. Field tests indicate that the horizontal hydraulic conductivity is on the order of 4×10^{-5} cm/s, while a laboratory permeability test indicates that this unit has a very low vertical hydraulic conductivity (5×10^{-8} cm/s). The low vertical hydraulic conductivity of this unit further restricts the downward movement of groundwater.

The glaciolacustrine unit coarsens with depth. Beneath the upper silt and clay layers is a more permeable layer of silty sand. This sequence of thick silt and clay overlying more granular silty sand is found throughout the area. The top of the silty sands is 33 to 46 m below the steel mill wastes. This layer also increases in thickness toward the lake trough and toward the center of the buried Ninemile Creek channel. The sandy silt ranges from 15 to 21 m in thickness beneath the site. The horizontal hydraulic conductivity of this unit is estimated at 10^{-4} cm/s.

Beneath the thick sequence of glaciolacustrine deposits is a thin deposit of ablation till. This unit consists of sand and gravel 4 to 5 m thick. The hydraulic conductivity of the ablation till measured in field tests was 4×10^{-3} cm/s [*1*].

Groundwater Flow

A large groundwater mound is found in the soda ash waste beds beneath the steel mill wastes (Fig. 4). This mound is due to (1) the groundwater contours generally following the topographic contours of the soda ash waste beds, (2) the low permeability of the soda ash waste, and (3) greater infiltration through the steel mill wastes than in adjacent areas of soda ash waste. Groundwater from the mound flows toward Onondaga Lake and Ninemile Creek. A groundwater divide exists south of the site, and all the water from the site ultimately discharges into Onondaga Lake. Groundwater flow beneath the site is complex because of the subsurface deposits. The swamp and lacustrine deposits retard downward flow, resulting in high vertical gradients in this unit. Groundwater from the site discharges into both the lake and stream through the more permeable underlying units. Although some groundwater from the site discharges into Ninemile Creek directly from the soda ash waste, the majority of ground-

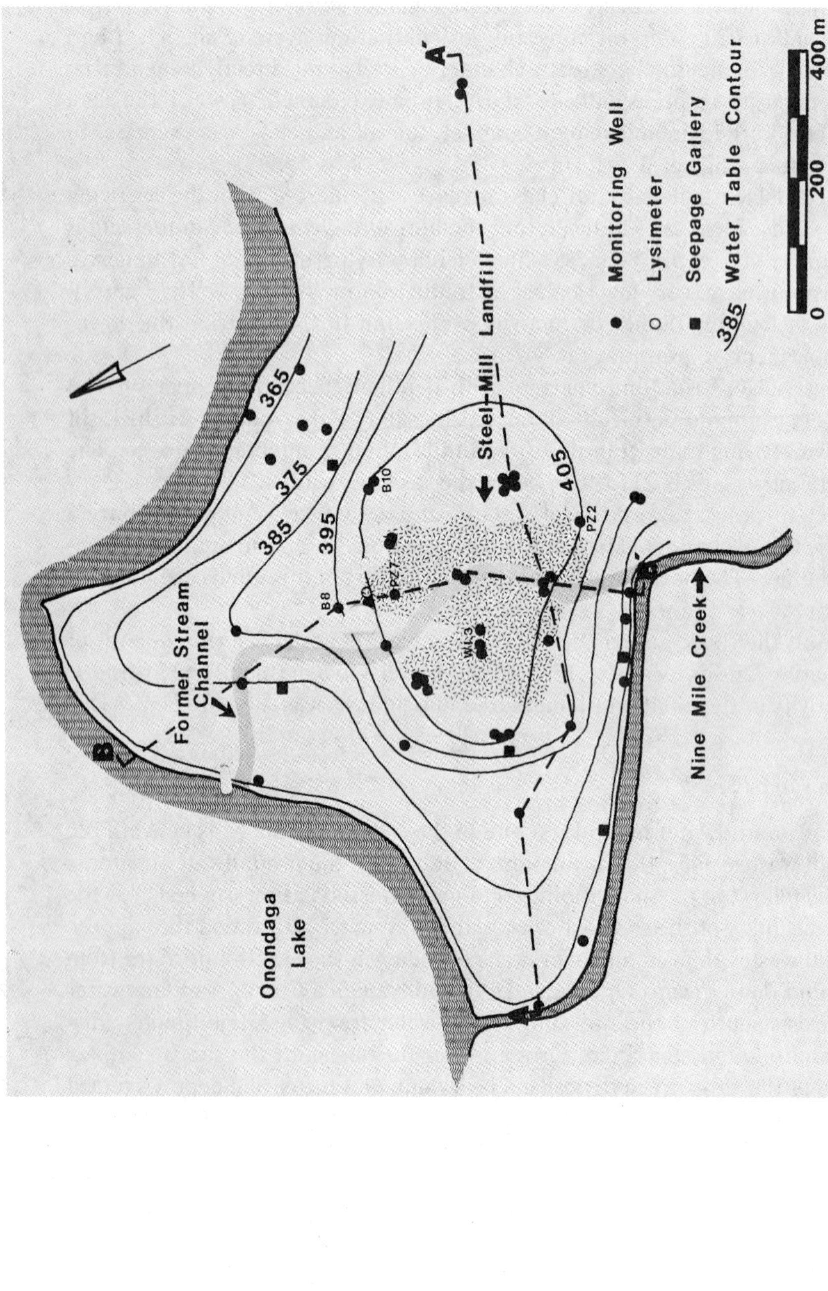

FIG. 4—*Water table map.*

water flow from the site moves through the swamp and lacustrine deposits and underlying silt and sand before it discharges into surface waters. Thus, leachate from the site will follow a long flow path through soda ash waste, swamp and lacustrine deposits, and silts and sands before entering surface water.

Seeps along the slopes of the soda ash waste beds were identified during the spring melt. The seeps are connected to the water table. The water table beneath the top of the soda ash waste exhibits large head changes during the year (up to 7.5 m) and responds quickly to recharge events. These fluctuations in head are due to the fine-grained nature of the soda ash waste. Since the waste is at or close to saturation, a small amount of recharge can raise the water table elevation significantly. Moisture content measurements [5] on Boring B-8 (Fig. 4) showed an average moisture content of 131% (standard deviation = 36) for 12 samples from the unsaturated zone and an average moisture content of 142% (standard deviation = 24) for 17 samples from the saturated zone. The small difference between saturated and unsaturated moisture contents is reflected in the response of the water table to recharge water. During spring melt there is a large recharge event resulting in a rapid rise in the water table. Seepage out the side of the waste beds occurs from this rapid rise in the water table. During these periods, leachate from the steel mill waste will reach surface water from seeps along the soda ash waste beds. However, the seeps are at least 74 m from the edge of the steel mill waste, and groundwater is 6.5 m below the ground surface. Thus, leachate from the site would have to travel a minimum of 82 m through the soda ash waste before it discharged out the side of the waste beds.

Vertical and horizontal hydraulic gradients were computed from the equipotential lines shown on the geologic cross sections (Figs. 2 and 3). Using representative gradients and hydraulic conductivity of the various units, the average linear velocity of groundwater flow beneath the site was estimated using Darcy's law, $\bar{v} = kI/n$, where $\bar{v}$ is the average linear velocity of flow, k is the hydraulic conductivity, I is the hydraulic gradient, and n is the effective porosity. Table 1 shows that the rates of groundwater flow in both the soda ash waste

TABLE 1—*Groundwater flow rates beneath steel mill wastes.*

Geologic Unit	Direction of Flow	K, cm/s	I, m/m	n	$\bar{v}$, m/year
Soda ash	vertical	8×10^{-6}	0.4	0.1	10
	horizontal	6×10^{-5}	0.05	0.1	9
Swamp/lacustrine	vertical	7×10^{-7}	0.8	0.25	1
deposits	horizontal	2×10^{-5}	0.1	0.25	3
Alluvial silty sand	vertical	2×10^{-5}	1.0	0.3	21
	horizontal	9×10^{-4}	0.1	0.3	95
Silt, clay, and sandy silt	vertical	5×10^{-8}	0.2	0.1	0.03
	horizontal	4×10^{-5}	0.01	0.1	1

and swamp and lacustrine deposits are very similar in the horizontal and vertical directions. The decreased permeability in the vertical direction is compensated by an increased gradient. However, the horizontal flow rate through the alluvial silty sands is greater than the vertical flow rate. Groundwater flow through the silt, clay, and sandy silt is much slower than through the alluvial silty sands.

These results indicate that groundwater from the site flows rapidly through the soda ash waste, and then slowly through the swamp and lacustrine deposits. In the areas of the old stream channel groundwater flow is rapid toward the lake, but in areas of silt, clay, and sandy silt, groundwater will move more slowly toward the lake.

In order to examine the groundwater flow regime in the vicinity of the steel mill waste landfill, a three-dimensional groundwater flow model was developed for the area. The purpose of the model was (1) to investigate the ultimate fate of leachate generated by the steel mill wastes and (2) to provide a means of verifying our field data and estimated groundwater flow patterns. The groundwater flow model was developed by the U.S. Geological Survey [*3,4*]. The model uses a finite-difference approximation to solve the partial differential equations of groundwater flow. A steady-state simulation was used to examine groundwater flow patterns.

The area around the Crucible landfill was divided into a series of blocks for the model (Fig. 5). A variable grid spacing was used to concentrate nodes in the area of Crucible waste. The boundary of the area used in the model was Onondaga Lake on the north, east, and northwest; Ninemile Creek on the southwest; and a groundwater divide on the south. Ninemile Creek and Onondaga Lake were assumed to be constant-head boundaries, and the groundwater divide was assumed to be a no-flow boundary. To simulate three-dimensional flow, the unconsolidated deposits were divided into seven layers. Transmissivity and infiltration values were changed by trial and error to produce the results shown on the two cross sections (Figs. 6 and 7).

The final values used for transmissivity were very close to the original estimates obtained from field data (Table 2). Estimated values for hydraulic conductivity for the upper layers were based on a large number of samples. Very few hydraulic conductivity tests were run on samples from the lower deposits. The final values used for the model show closer agreement for the upper layers. However, the hydraulic conductivity values used for the model for even the lower layers show very good agreement with the estimated values.

Infiltration for the area was estimated using the U.S. Environmental Protection Agency (EPA) model [*9*]. The EPA model predicted 22 cm/year of infiltration through the soda ash waste beds and 33 cm/year infiltration through the steel mill wastes. The final value for infiltration through the soda ash waste beds in the model (22 cm/year) is identical to the EPA estimate of 22 cm/year. However, a much higher infiltration was needed than predicted by the EPA model in areas of steel mill waste and the parking lot to produce the large

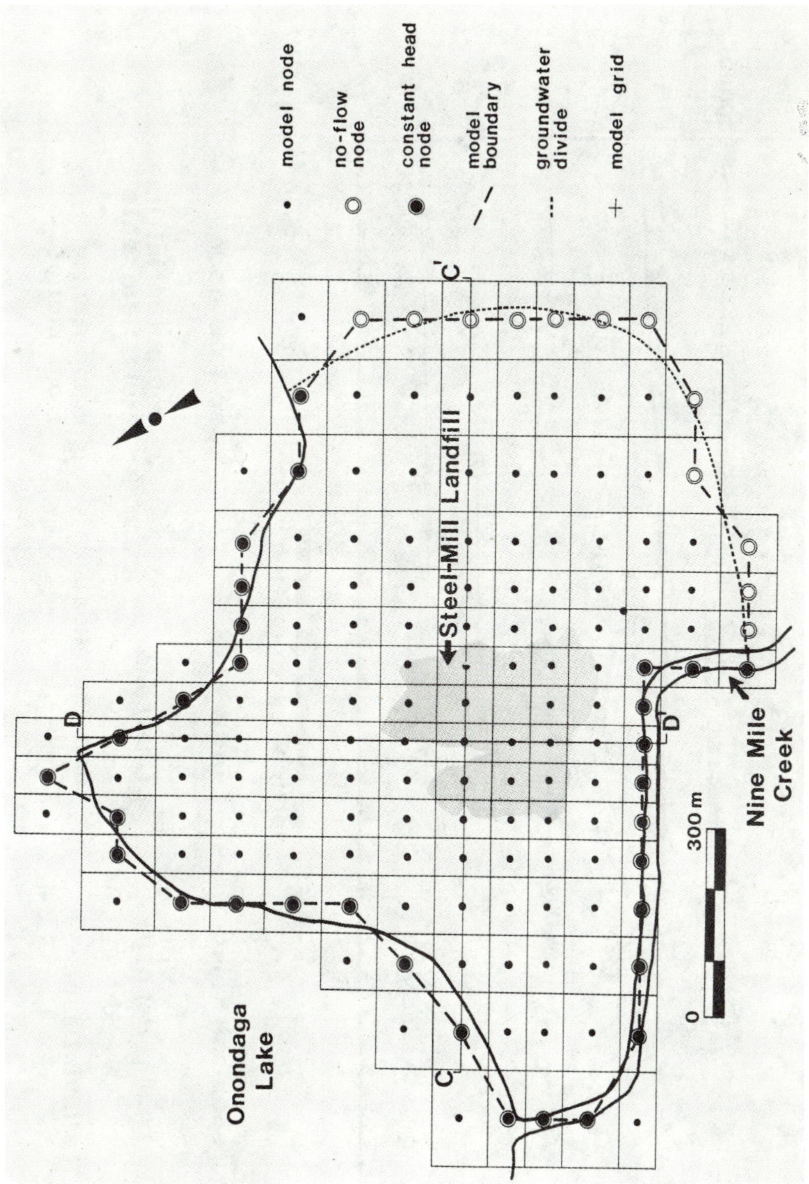

FIG. 5—*Groundwater model of area around the Crucible landfill.*

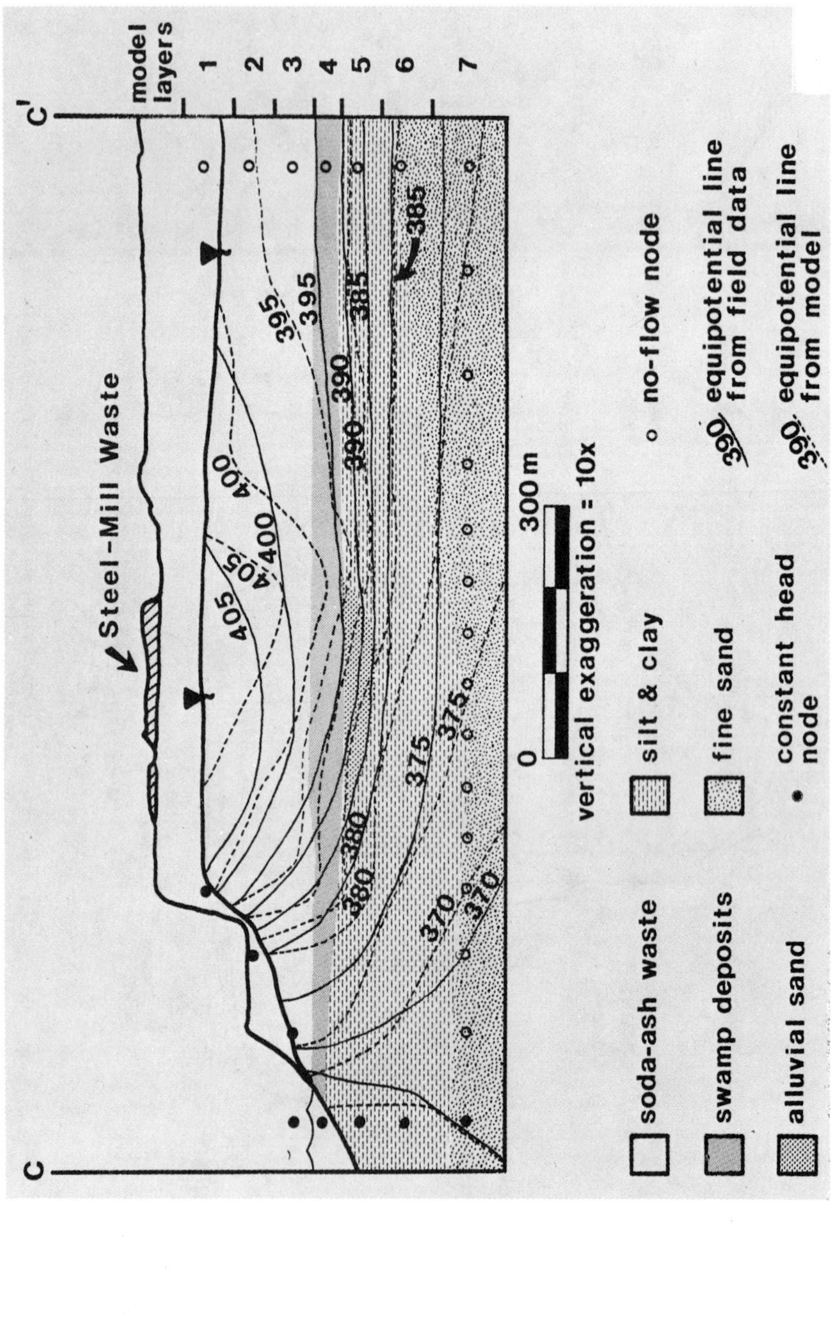

FIG. 6—*Model cross section C–C′ parallel to the lake shore.*

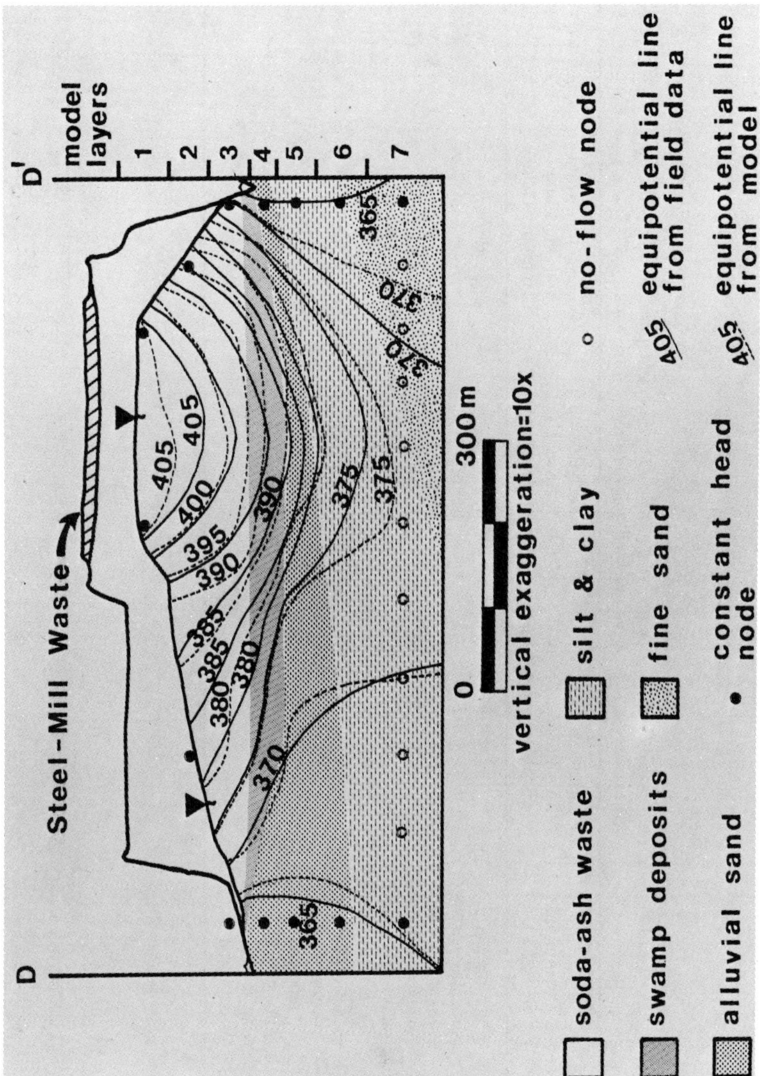

FIG. 7—*Model cross section D-D′ perpendicular to the lake shore.*

TABLE 2—*Comparison of hydraulic conductivity estimated from field and laboratory data and hydraulic conductivity values used in groundwater flow model.*

Model Layer	Predominant Geologic Unit	Hydraulic Conductivity Input to Model from Field Data, cm/s	Final Hydraulic Conductivity Used in Model, cm/s	K_H/K_V Estimated from Field and Laboratory Data	Final K_H/K_V Used in Model
1	soda ash waste	6×10^{-5}	6×10^{-5}	7	6
2	soda ash waste	6×10^{-5}	6×10^{-5}	7	6
3	soda ash waste	6×10^{-5}	6×10^{-5}	7	6
4	swamp/lacustrine	2×10^{-5}	2×10^{-5}	30	15
5	alluvial silty sand	9×10^{-4}	2×10^{-3}	50	50
6	silt, clay, and sandy silt (upper glaciolacustrine deposits)	4×10^{-5}	1×10^{-4}	800	100
7	Silty sand (lower glaciolacustrine deposits)	4×10^{-4}	1×10^{-3}	100	100

groundwater mound found beneath the site. The EPA model predicts 33 cm/year of infiltration through steel mill wastes whereas 58 cm/year was used for the groundwater model. The discrepancy is probably due to (1) using the water table elevations measured during the spring when recharge is high and (2) the EPA model's assumption that 33 cm of water could run off, whereas runoff from the steel mill is negligible.

The results of the model agree closely with the estimated equipotential lines drawn from measured water level (Figs. 6 and 7). These results verify conclusions from field data that (1) the majority of water infiltrating through the steel mill wastes flows vertically downward through the underlying soda ash waste and swamp and lacustrine deposits before it moves laterally toward the lake and (2) a small percentage of water passing through the steel mill wastes moves vertically downward through the soda ash wastes before it discharges into the creek.

Geochemistry

Wastes from the steel mill have been placed on the soda ash waste beds for the past ten years. The wastes from the steel mill include wastewater treatment plant sludge, air pollution dust, grinding dust, waste caustic solids, acid pickling solids, boiler house ash, slag, coolant swarf, mill scale, and construction and refractory debris. Air pollution dust, waste caustic solids, and acid pickling solids are classified as hazardous by EP toxicity testing [*6*]. Disposal of hazardous wastes at the landfill was discontinued in 1982. Both air pollution dust and waste caustic solids are classified as hazardous because of leaching of soluble hexavalent chromium. Leachate produced from the EP toxicity test on these wastes contained between 10 and 180 mg/L total chromium, substantially higher than the regulatory limit of 5 mg/L. Leach studies by both column and successive batch procedures give a total leachable hexavalent chromium content of approximately 430 mg/kg for air pollution dust. Approximately 1×10^6 kg of air pollution dust were placed in the landfill. Because of the small volumes of landfilled waste caustic solids in relation to air pollution dusts and the ease of locating air pollution dust deposits in the landfill, further testing centered on the interaction of hexavalent chromium leached from air pollution dust with soda ash waste.

Both laboratory column testing and field observations in landfill test pits dug through air pollution dust deposits show that hexavalent chromium leaches from the waste within the first year after landfilling. Field evidence of the effectiveness of the soda ash waste in attenuating chromium was seen in the test pits dug through deposits of air pollution dust, 2.5 to 3 m into the underlying soda ash waste. Four test pits excavated through air pollution dust showed a surface brown layer in the soda ash wastes underlain by a yellow stain of various intensities and depths. Samples from the yellow-stained layers were centrifuged at 3000 rpm for 10 min, and the supernatant was analyzed for hexavalent chromium. Results from the analysis showed up to 800 mg/L hexavalent chromium. Comparison of the depth of the stain and age of the air pollution dust shows that over a period of three years the yellow hexavalent chromium stain spreads to a depth of 2 m with no further spread over three more years.

A series of adsorption isotherms were measured for hexavalent chromium onto soda ash waste to further define the nature of observations in test pits. Four samples of soda ash waste (20 g wet) were mixed with solutions of potassium dichromate varying in concentration from 10 to 1000 mg/L. The first soda ash sample was material collected from the surface of the landfill remote from the steel mill wastes. This material is termed coarse-edge soda ash waste (CESW-2) and has been exposed to the environment for at least 40 years. A second sample was collected from the four test pits dug through deposits of air pollution dust at the edge of the steel mill wastes. The test sample (BWSW) was a composite of samples taken beneath the yellow layers observed at these locations at an approximate depth of 2 m. The third and fourth soda ash waste samples (B10S8 and B10S2, Fig. 4) were taken from split-spoon samples from a boring remote from the steel mill waste at depths of 8 and 3 m, respectively. Moisture content of the soda ash waste samples was determined by standard methods [*5*]. Weighed soda ash waste samples were mixed with 100 mL of potassium dichromate ($K_2Cr_2O_7$) solution. The mixtures were shaken and allowed to stand overnight, and the equilibrium hexavalent chromium concentrations were determined by standard wet chemical methods [*5*]. The results of these experiments are shown in Fig. 8, where the logarithm of equilibrium hexavalent chromium concentration is plotted against the logarithm of adsorbed hexavalent chromium. Adsorbed hexavalent chromium (S) was calculated as follows:

$$S(\text{mg/kg}) = (C_i - C_{\text{eq}})V_T/1000X \tag{1}$$

where

C_i = initial hexavalent chromium concentration, mg/L,
C_{eq} = equilibrium hexavalent chromium concentration, mg/L,
V_T = solution volume, mL, and
X = dry mass of soda ash waste, kg.

The linear plots shown in Fig. 8 indicate that these results may be fit by the Freundlich isotherm given by

$$\log S = b \log C_{\text{eq}} + \log K_d \tag{2}$$

or

$$S = K_d C_{\text{eq}}^b$$

where K_d is termed the distribution coefficient (in millilitres per gram) [*10*].

The interaction of hexavalent chromium with soils, natural sediments,

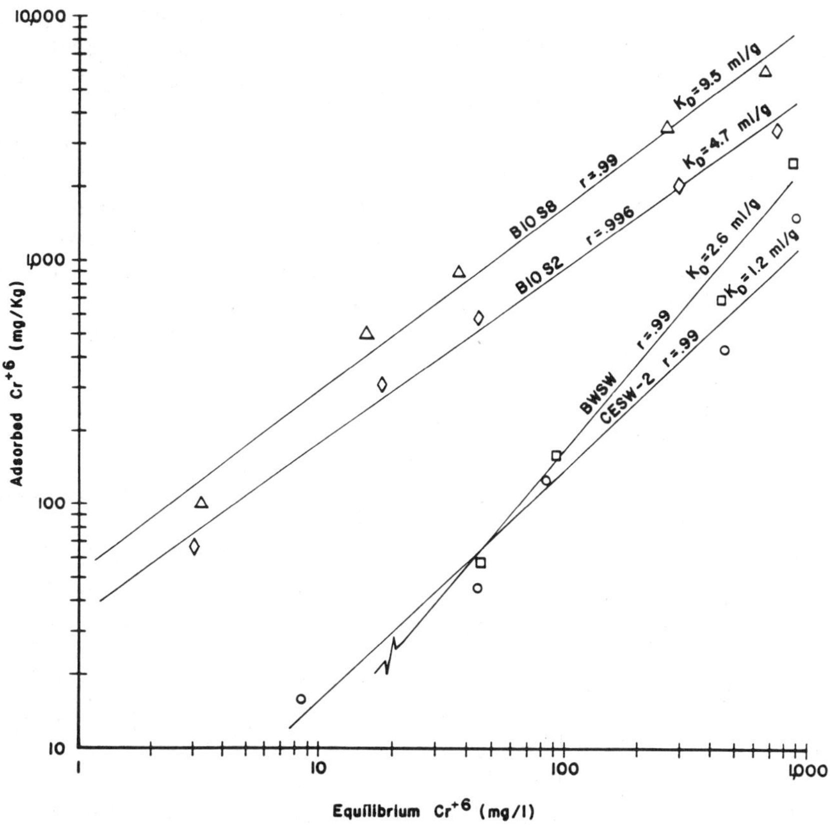

FIG. 8—*Adsorption of hexavalent chromium on soda ash waste.*

model substrates, crushed limestone, clay minerals and activated carbon has been studied previously [*11–15*]. These studies have determined that both adsorption and reduction and precipitation of hexavalent chromium to trivalent chromium occur depending on the nature of the solid phase. The maximum adsorption level observed for hexavalent chromium on montmorillonite clay was 0.400 mg/g at pH 3.0 [*11*]. The four samples of soda ash waste produced adsorption maxima from 1.51 mg/g (CEWS-2) to 5.93 mg/g (B10S8). The pH range for these adsorptions varied from approximately pH 7 for CESW-2 to pH 12 for B10S2 and B10S8. The adsorptive capacity of undisturbed soda ash waste is related to the solution pH with high pH samples (B10S2 and B10S8) producing the highest K_d values shown in Fig. 8. The lower solution pH in CESW-2 relates to leaching of calcium hydroxide [$Ca(OH)_2$] content in this sample. Typically, undisturbed surface soda ash samples show reduced leaching pH in relation to deeper samples. Soda ash samples from test pits beneath air pollution dust deposits (BWSW) also show reduced leach solution pH (8 to

11) and a lower K_d value (Fig. 8). The BWSW sample also shows the highest slope (1.2) of the four soda ash samples tested. Testing of BWSW at equilibrium hexavalent chromium concentrations of 10 mg/L showed negligible adsorption (discontinuity for BWSW isotherm shown in Fig. 8). Centrifuge testing of BWSW showed traces of hexavalent chromium (2 mg/L) in two of the four samples from beneath yellow deposits. Comparison of the adsorptive capacity of BWSW and B10S2, both taken from similar depths in the soda ash waste, shows that the waste from beneath the contaminant enclave (BWSW) has reduced adsorptive capacity for hexavalent chromium, especially in the lower equilibrium chromium concentration range.

Observations in tests pits and laboratory isotherm measurements are consistent with adsorption of hexavalent chromium leached from air pollution dust onto soda ash waste. The stability of test pit observations over five years, the magnitude of measured K_d values, the linearity of isotherms, and the finite amount of hexavalent chromium leached from air pollution dust are also consistent with formation of a "contaminant enclave" at the surface of the soda ash waste.

The effectiveness of the soda ash waste in adsorbing hexavalent chromium was also seen in results from water sampling. No chromium above the detection limit of 0.01 mg/L was found in any of the water samples from monitoring wells, seeps, or seepage galleries. Monitoring wells below the steel mill waste sampled groundwater at three elevations within the soda ash waste and in each of the underlying deposits. Water samples from seeps and seepage galleries monitored water quality from seeps out the side of the soda ash waste beds. All the results show no chromium contamination of groundwater.

The lysimeter results confirm the rapid attenuation of chromium in the upper few meters of the soda ash waste beds. The lysimeters were installed at 2-m intervals below an area containing most recently landfilled air pollution dust. The uppermost lysimeter (at a depth of 2 m) had hexavalent chromium concentrations to 1300 mg/L whereas the next lysimeter (at a depth of 4 m) had only trace concentrations of hexavalent chromium. No detectable hexavalent chromium was found in either lysimeter at the 6-m or 8-m depth.

To examine whether hexavalent chromium was escaping the "contaminant enclave" into the deeper layers of soda ash waste, 195 split-spoon samples from borings were analyzed for chromium. Fourteen of the samples of soda ash waste beneath steel mill wastes were found to contain low concentrations of trivalent chromium (9 to 133 ppm). These samples occur at the surface of the soda ash waste, near the water table and deeper in the soda ash waste. The chromium in these samples is in the solid phase and was not detected in either pore water from the samples or groundwater from the sample location. None of the samples analyzed from borings remote from the steel mill waste contained chromium.

Soda ash waste samples from borings were also analyzed for iron content.

Comparison of the results from iron and chromium analyses on soda ash waste samples shows that chromium-containing samples have significantly higher iron content than nonchromium samples. A statistical Student's *t* comparison of chromium-containing samples to nonchromium samples in the same borings and to all other nonchromium borings indicates the probability that the high iron content in chromium-containing layers arises from chance is exceedingly small (less than 0.0005). These data indicate that chromium and iron found in samples are derived from the metals waste rather than the soda ash waste. Since the pH in these samples is in the 11 to 12 range, it is hypothesized that hexavalent chromium has reacted with ferrous iron in these layers, precipitating ferric and chromic hydroxides.

The interaction of soda ash wastes and air pollution dust leachate was further studied in column experiments. Four 2.5-mm diameter Plexiglas columns were filled to 0.5 m with previously described CESW and connected in series to form a 2-m column. A leachate solution was generated by mixing air pollution dust with tap water to produce a water-phase hexavalent chromium concentration of 140 mg/L. The soda ash column was leached to breakthrough with a yellow coloration developing throughout. One section of the column was drained and the solids analyzed at natural moisture content. A second section was washed with tap water until the hexavalent chromium concentration in the eluent dropped below 0.1 mg/L. After they were drained, the solids in this column were also analyzed at natural moisture content. The solids from the first drained section showed a total chromium concentration of 590 mg/kg and a hexavalent chromium concentration of 350 mg/kg. The water-washed second section showed a total chromium concentration of 180 mg/kg and a hexavalent chromium concentration of 8 mg/kg. These results are consistent with the adsorption observed in isotherm measurement and with reduction of hexavalent chromium to trivalent chromium. The interaction of trivalent chromium with CESW was investigated by passing a solution of 1000 mg/L trivalent chromium through a 15-cm column of CESW. A noticeable blue-gray coloration, similar to the color of the trivalent chromium solution, developed immediately in the CESW at the top of the column. The eluent from the bottom of the column remained clear. When the blue-gray section of the CESW had extended 5 cm, the column was drained and the blue-gray solids were analyzed at natural moisture content. The total chromium content of this layer was 68 000 mg/kg. A cementation of the CESW in the blue-gray layer was noticed, indicating that trivalent chromium precipitates are not merely being filtered by soda ash wastes but are attaching to at least some of the constituents of the soda ash wastes.

In order to determine whether the iron and chromium layers found deep in the soda ash wastes could remobilize, five of the chromium-containing samples from various levels were subjected to leaching tests utilizing the standard EP toxicity test leaching procedure [*6*]. Even after the maximum amount of

TABLE 3—*Resolubilization of chromium and iron in soda ash waste.*

Boring Location	Sample Depth Below Ground Surface, m	Solids Analysis (Natural Moisture Content), mg/kg			EP Toxicity Results, pH 8.5 to 10.0 mg/L			EP Toxicity Results, pH 7.0, mg/L		
		Iron	Chromium (Total)[a]	Chromium (Hex)[a]	Iron	Chromium (Total)[b]	Chromium (Hex)[b]	Iron	Chromium (Total)[b]	Chromium (Hex)[b]
W1.3	6.5 to 7	2550	BDL	BDL	0.09	BDL	BDL	0.20	BDL	BDL
PZ2	15 to 16	3900	9	BDL	0.10	BDL	BDL	0.28	BDL	BDL
PZ2	18 to 18.5	6000	9	BDL	0.22	BDL	BDL	0.32	BDL	BDL
PZ7	7.5 to 8	3850	60	BDL	0.12	BDL	BDL	0.20	BDL	BDL
PZ7	14 to 14.5	3330	44	BDL	0.24	BDL	BDL	0.30	BDL	BDL

[a]BDL = below detection limit of 0.5 mg/L.
[b]BDL = below detection limit of 0.01 mg/L.

acid allowed under the test procedure was added, the pH of the mixture remained in the 8.5 to 10.0-pH range. Further tests were performed utilizing additional acetic acid to reduce the mixture to pH 7.0 during the 24-h stirring.

Table 3 summarizes the test results for total iron and chromium concentration in the solid sample of soda ash waste and the iron and chromium concentrations found in the EP toxicity test leachate. The data indicate that the chromium is immobilized in the soda ash wastes and that low concentration of iron can be resolubilized at pH 7.

Reduction and precipitation of hexavalent chromium to trivalent chromium in deeper layers in the waste beds provides another attenuation mechanism for chromium in the waste beds. This secondary treatment process, following the primary *in situ* treatment process of hexavalent chromium adsorption at the surface of the waste beds, effectively retains chromium within the soda ash wastes.

Conclusions

The location of a hazardous waste site on top of preexisting soda ash process wastes has provided an unusual opportunity to examine the feasibility of *in situ* treatment of metals manufacturing wastes. The hazardous wastes placed in the landfill release high concentrations of soluble hexavalent chromium. Although the landfill is adjacent to Onondaga Lake and leachate from the steel mill wastes flows into the lake, the landfill has had no adverse affects on groundwater or surface water quality because of rapid attenuation of hexavalent chromium in the underlying soda ash waste beds. Attenuation of the majority of hexavalent chromium released from the wastes by adsorption in the upper few meters of the soda ash waste beds is followed by reduction and precipitation of the low levels of hexavalent chromium escaping the "contaminant enclaves" to insoluble trivalent chromium hydroxides deeper in the waste beds. This results in a very effective *in situ* treatment process.

The results from this study show that hazardous waste sites should not always be confined to traditionally approved hydrogeologic settings. Unconventional locations (such as the soda ash waste beds) for landfilling certain waste types may be more effective in attenuating leachate than a conventional site. Adsorption, reduction, and precipitation within the highly alkaline soda ash waste suggest a viable means for disposal and *in situ* treatment of several metals-processing waste products.

References

[1] Bouwer, H. and Rice, R. C., *Water Resources Research*, Vol. 12, No. 3, 1976, pp. 423–428.

[2] U.S. Army Corps of Engineers, *Laboratory Soils Testing*, Engineering Manual EM 1110-2-1906, Appendix 7, Section 7, 1970.

[3] Trescott, P. C., "Documentation of Finite-Difference Model for Simulation of Three-Dimensional Ground-Water Flow," USGS Open File Report 75-438, U.S. Geological Survey, Boulder, Colo., 1976.
[4] Trescott, P. C. and Larson, S. P., "Supplement to Open File Report 75-438," USGS Open File 76-591, U.S. Geological Survey, Boulder, Colo., 1976.
[5] *Methods of Chemical Analysis of Water and Wastes*, EPA Manual 400, U.S. Environmental Protection Agency, Washington, D.C., 1979.
[6] "Code of Federal Regulations," *Federal Register*, Title 40, Part 261, Appendix II, July 1983.
[7] American Public Health Association, *Standard Methods for the Examination of Water and Wastewater*, 15th Ed., Washington, D.C., 1980.
[8] Grove, C. S., Boecher, F. W., Biemiller, C., and Li, W. H., "The Rehabilitation of the Solvay Process Waste Beds," Bureau of Solid Waste Management, Environmental Control Administration, Department of Health, Education and Welfare, Cincinnati, Ohio, December 1969.
[9] *Hydrologic Simulation on Solid Waste Disposal Sites*, SW-868, U.S. Environmental Protection Agency, Washington, D.C., 1980.
[10] Freeze, R. A. and Cherry, J. A., *Groundwater*, Prentice-Hall, Englewood Cliffs, N.J., 1979, pp. 403–408.
[11] Griffen, R. A., Au, A. K., and Frost, R. R., "Effect of pH on Adsorption of Chromium from Landfill-Leachate by Clay Materials," *Journal of Environmental Science Health*, Vol. A12, No. 8, 1977, pp. 431–449.
[12] Antiole, J. and Fuller, W. H., "Effect of Crushed Limestone Barriers on Chromium Attenuation in Soils," *Journal of Environmental Quality*, Vol. 8, No. 4, 1979, pp. 503–510.
[13] Mayer, L. M. and Schick, L. L., "Removal of Hexavalent Chromium from Estuarine Waters by Model Substrates and Natural Sediments," *Environmental Science and Technology*, Vol. 15, No. 12, 1981, pp. 1482–1484.
[14] Bartlett, R. J. and Kimball, J. M., "Behavior of Chromium in Soils: II. Hexavalent Forms," *Journal of Environmental Quality*, Vol. 5, 1976, pp. 383–386.
[15] Wu, M. H., Hsu, D. V., and Huang, C. P., "Regeneration of Activated Carbon for the Adsorption of Chromium" *Proceedings*, 32nd Industrial Waste Conference, Purdue University, 1976, Lafayette, Ind., pp. 409–419.

Peter J. Young,[1] *Grant Baldwin,*[1] *and David C. Wilson*[1]

Attenuation of Heavy Metals Within Municipal Waste Landfill Sites

REFERENCE: Young, P. J., Baldwin, G., and Wilson, D. C., **"Attenuation of Heavy Metals Within Municipal Waste Landfill Sites,"** *Hazardous and Industrial Waste Management and Testing: Third Symposium, ASTM STP 851*, Larry P. Jackson, Alan R. Rohlik, and Richard A. Conway, Eds., American Society for Testing and Materials, Philadelphia, 1984, pp. 193–212.

ABSTRACT: The Waste Research Unit has developed several laboratory tests to investigate the attenuation of toxic species disposed to landfill, to provide pragmatic guidance for environmentally acceptable codisposal of industrial and municipal wastes. Seven-column experiments were set up to investigate the fate of eight metals (cadmium, chromium, copper, iron, lead, mercury, nickel, and zinc) released in solution onto pulverized municipal waste. Leachate and destructive analysis revealed that only nickel migrated through 0.5 m of waste, giving typical leachate concentrations of 20 mg/dm^3, although upwards of 80% of the nickel was removed from solution. The other metals were precipitated or absorbed very rapidly and were mainly found in the top 0.2 m. Fatty acid production was completely inhibited by the immobilized metals, leading to near-neutral pHs in these areas. Penetration of the dissolved metals was greater with a continuously applied dilute solution than with a brief dose of comparable loading.

KEY WORDS: municipal solid waste, landfill, leachate, attenuation, heavy metals, cadmium, chromium, copper, iron, lead, mercury, nickel, zinc, microbial inhibition, hazardous wastes, industrial wastes

When industrial wastes are disposed to landfill, there exists the potential for pollution of surface or groundwater resources. Controlled disposal is therefore required to ensure that the landfill leachates produced are either treated or attenuated to meet appropriate water quality standards. In order to control disposal effectively, and to assess the impact of a particular waste, information is required on the fate of the chemical species present in the waste. The Waste Research Unit of the Harwell Laboratory has developed several simple test methods to gather such information.

Many mechanisms serve to reduce contaminant concentrations in leachate

[1]Members of the Waste Research Unit, Harwell Laboratory, Oxon, OX11 ORA, England.

from those present in the original deposit of hazardous waste [*1*]. Much attention has been directed at extraction tests to quantify the attenuation factor obtained by the initial leaching from a waste (for example Refs *2*, *3*, and *4*). The Waste Research Unit has developed its own test using a high solid-to-liquid ratio, which was successfully validated against simulated field behavior [*5,6*]. The work described here is aimed at examining the subsequent interactions of leachate percolating through municipal waste. A total of eight metals were investigated, with the object of using the results in conjunction with the extraction test to provide guidance on the suitability of metal-bearing wastes for codisposal with municipal waste.

The column experiments discussed later, were aimed specifically at the fate of solubilized metals percolating through municipal waste, since this is an area in which information in the literature is limited.

Experimental Approach

The experiment was designed to examine the attenuation of solubilized heavy metals by a depth of 0.5 m of pulverized municipal waste, over a period of six months. Seven glass columns, each 50 cm high and 10 cm in diameter, were filled with 3.25 kg of fresh wet-pulverized municipal waste. The waste had a moisture content of about 20% and was packed to a density equivalent to 0.83 tonnes/m^3. The columns were initially flushed with nitrogen and then sealed to promote anaerobic conditions.

The seven columns were irrigated at a natural rainfall rate of 800 mm/year and were arranged as three pairs and a single experiment. Each pair consisted of one column dosed with heavy metals and a control column. For the first 18 days the dosed column of each pair received a solution of heavy metals while the control was irrigated with distilled water. Each pair of columns was irrigated with distilled water for the remainder of the experiment. The 18-day dose of metals was designed to mimic the effect of depositing a waste high in metal content on the municipal waste, for example, codisposal of neutralized pickling acid liquors. The first pair of columns were dismantled after two months of saturation, the second after four months, and the remaining pair after six months.

The heavy metal solution used contained eight metals: cadmium, chromium, copper, iron, lead, mercury, nickel, and zinc. All were present at concentrations of 1000 mg/dm^3 in the dosing solution, except for mercury, which was present at 100 mg/dm^3. The dosing solution was prepared using soluble nitrates or chlorides of the metals and had a pH of 2 to ensure that the metals remained in solution during dosing. The seventh column was irrigated for six months in the same way as the control, but using a solution containing 10 mg/dm^3 mercury and 100 mg/dm^3 of the other metals. The metal dose of the seventh column was therefore the same as for the other dosed columns but was spread over a period of six months instead of 18 days. The continuous release of metals more closely represents the case of codisposal with a solid metal-rich waste, which is then leached over an extensive period. The nominal dose of each metal was 310 mg, except for mercury (31 mg), although problems with the peristaltic pumps used for irrigation led to actual doses of between 238 and 340 mg.

Leachate samples were taken from the columns at twice-weekly intervals until dismantling. The columns were dismantled as quickly as possible to minimize drainage and precipitation of metals in solution and were separated into eight consecutive sections. Each section was split into three samples and stored at −20°C. One sample was used to establish the moisture content of the section. The second sample was centrifuged to remove the interstitial fluid for separate analysis. The third sample was homogenized by cryogenic grinding and stored at −20°C. A portion (5 g) of this ground sample was dissolved in aqua regia, and the resulting solution was analyzed for all eight metals by atomic absorption spectrophotometry. In addition, discrete metallic and other intractable items were removed during grinding and analyzed separately. The results of these analyses allowed a complete mass balance for each column for all eight metals to be carried out. Further portions of the ground waste are currently being used to investigate the speciation of the metals held within the columns.

The experimental approach adopted was aimed at producing information on three aspects of the attenuation of heavy metals in municipal waste:

(*a*) the effect of the metal dose on leachate quality after passage through 0.5 m of municipal waste,
(*b*) the distribution of each metal within the column of waste, and between the solid and liquid phases, at two monthly intervals following application of the dosing solution, and
(*c*) the effect of a dilute and prolonged loading of metal compared with an equivalent shorter, more concentrated release.

In addition, the mass balance work gave important information on the concentrations of metals in pulverized waste and on the variation between different samples of waste. The current work on speciation of the metals is aimed at establishing the permanence of the metal retention observed and investigating possible remobilization by pH changes.

Results and Discussion

Leachate samples were collected twice a week from the seven columns. The columns reached saturation after about 13 days of irrigation and were then irrigated for a further two months (Columns 1 and 2), four months (Columns 3 and 4), or six months (Columns 5, 6, and 7). The differences between the columns are summarized in Table 1. The leachates were analyzed for pH, fatty acids, chloride, and all the eight metals in the dosing solution.

TABLE 1—*Summary of column experiments.*

Column No.	Months of Saturation Before Dismantling	Dosing Period
1	2	18 days
2	2	none
3	4	18 days
4	4	none
5	6	18 days
6	6	none
7	6	continuous (dilute solution)

Leachate Quality

The leachate quality was remarkably similar for all the columns. In particular, the agreement between the three control columns indicated that comparison between different columns would be valid, and subject to only limited error because of the inhomogeneity of the pulverized waste.

The quantities of cadmium, chromium, copper, iron, mercury, and lead in the leachates from the dosed columns were not significantly different from the corresponding controls. The levels of cadmium, copper, and mercury remained below 1 mg/dm^3 and, except for mercury, did not peak appreciably. Mercury was only detected during the first two months, with a maximum concentration of about 10 $\mu g/dm^3$. The detection limit using a flameless atomic absorption technique was 2 $\mu g/dm^3$. Lead was initially present at a concentration of 1 mg/dm^3, declining to 0.3 mg/dm^3 or less after six months. Although the continuously dosed Column 7 gave a slightly higher release toward the end of the experiment (0.3 mg/dm^3 compared to <0.15 mg/dm^3 in Columns 5 and 6), this was not thought to be significant given that pulverized waste alone had leached greater concentrations initially. Iron concentrations varied considerably between adjacent samples, but the broad trend of 600 mg/dm^3, increasing to 1200 mg/dm^3 after two months, and then returning to 600 mg/dm^3 after six months, was common to all the experiments except Column 7, which released about 600 mg/dm^3 throughout.

Zinc concentrations were widely different, although there was no evidence for elevated releases from the dosed columns. Initial concentrations were generally highest, reaching 70 mg/dm^3 in Columns 4 and 5, although Column 7 gave no early release of zinc. Apart from a small peak of 10 to 20 mg/dm^3 after about two months, the remaining levels were below 5 mg/dm^3. However, control Columns 1 and 3 did not give these low levels and maintained concentrations of about 20 mg/dm^3. This behavior may be due to discrete sources of zinc in the waste in these columns; subsequent analysis of the waste demonstrated wide variations in the amounts of zinc present.

Nickel concentrations in the leachate were elevated in all the dosed columns except Number 5 (Fig. 1). The control columns all released fairly steady levels of nickel between 5 and 0.5 mg/dm^3. Nickel concentrations were noticeably elevated in Column 1 after 40 days, in Column 3 after 50 days, and in Column 7 after about 30 days. Only the initial breakthrough of nickel was observed before the dismantling of Column 1, but levels in Columns 3 and 7 settled down to a steady value of about 20 mg/dm^3. The additional quantity of nickel released in the leachate from these dosed columns was 5% of that added to Column 3 and 16% of that added to Column 7, which operated for two months longer. Of particular note was the similarity in the concentrations of nickel from Columns 3 and 7, which did not reflect the difference in concentrations of their respective dosing solutions. The attenuation of an initial dose of 1000 mg/dm^3 nickel and a continuous dose of 100 mg/dm^3 nickel had reduced the concentration to a similar level. One interpretation of this behavior would be that the nickel concentration in the leachate was being controlled by the availability of a mobilizing mechanism, which maintained a restricted quantity of nickel in the liquid phase.

The leachates were also analyzed for pH, chloride, and fatty acids. The pH of all the leachates was very stable, and no column gave leachate outside the range of pH 5.5 to 6. Chloride levels were elevated in the dosed columns, and the excess chloride was almost certainly derived from the heavy metal solutions. All the columns initially gave high chloride levels of 2000 to 3000 mg/dm^3, but the controls all reduced in chloride concentration rapidly, reaching 500 mg/dm^3 after four months and 50 mg/dm^3 after six months. From two months onwards, about the time of the nickel release, chloride levels were elevated by between 600 and 1200 mg/dm^3 over the corresponding control. This included Column 5, which had not released any nickel after six months. Column 7 gave a much slower decrease in chloride levels; starting at 2100 mg/dm^3, the concentration reached 900 mg/dm^3 after four months with very little further decrease. This more uniform release possibly reflects the continuous application of the dilute dosing solution.

The fatty acid results showed consistently higher concentrations in the controls than in the dosed columns. The results for Columns 3 and 4 are shown in Fig. 2 as an example. Following the first 20 days after leachate breakthrough, the four acids shown all fell in concentration, but with a consistent depletion in Column 3.

The indications of the leachate results in isolation are that the heavy metal applications have led to a small reduction in fatty acid production, although the pH has not been altered. The increase in chloride levels may be significant in mobilizing nickel, the only metal to penetrate 0.5 m of waste in any quantity, although the relationship between these two species is unclear.

Interstitial Fluid Analysis

In order to understand more about the movement of the metals in the liquid phase, the interstitial fluids from the columns were examined immediately af-

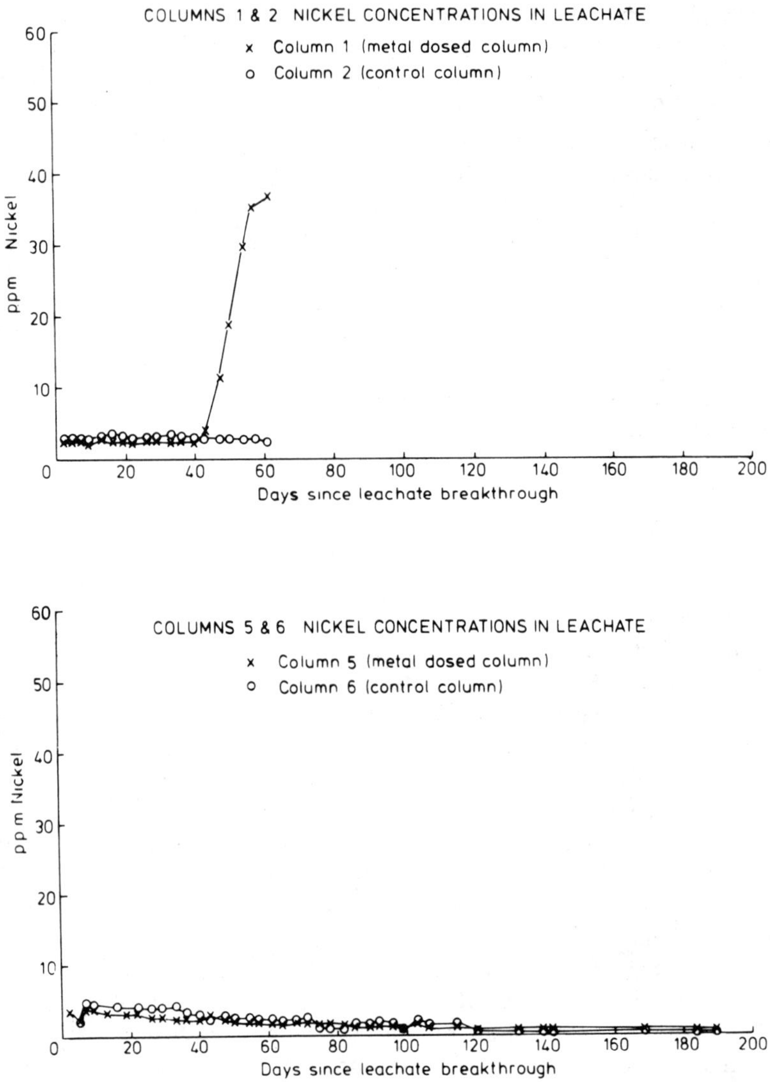

FIG. 1—*Nickel concentrations in the*

ter irrigation ceased. The columns were split into eight consecutive sections to allow investigation of any profiles within the pulverized waste. The results from the first six columns will be considered first, since these experiments indicate the fate of a discrete concentrated dose of metal into the liquid phase of the waste.

The distribution of nickel in the interstitial fluids was consistent with the be-

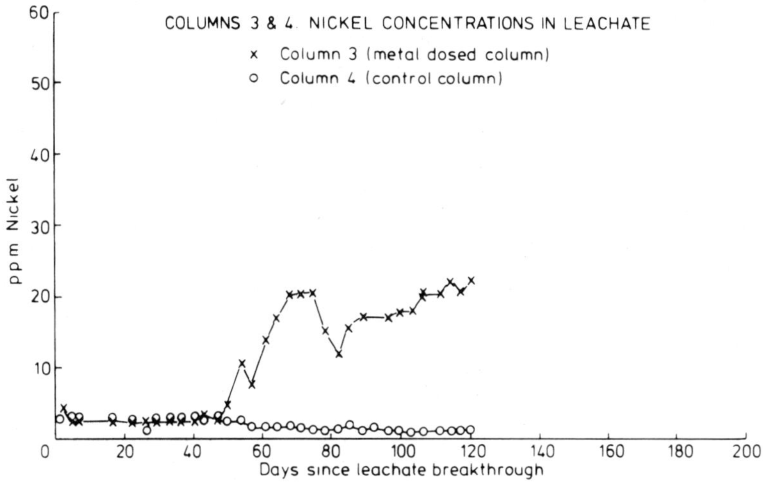

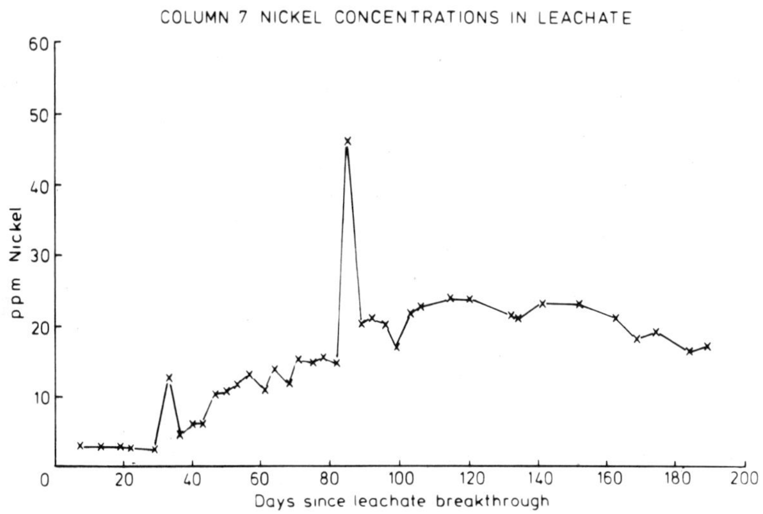

column leachates (1 ppm = 1 mg/dm^3).

havior observed in the leachates. The levels in the controls were uniformly low. Columns 1 and 3 showed a broad peak, rising to 54 mg/dm^3 0.2 m below the surface of Column 1 and 16 mg/dm^3 near the base of Column 3. Concentrations at the base were elevated, although only about half that observed in the final samples of leachate. The peak in Column 5 was sharper, rising to 30 mg/dm^3 at a depth of 0.2 m. The bottom two sections (0.1 m) of Column 5

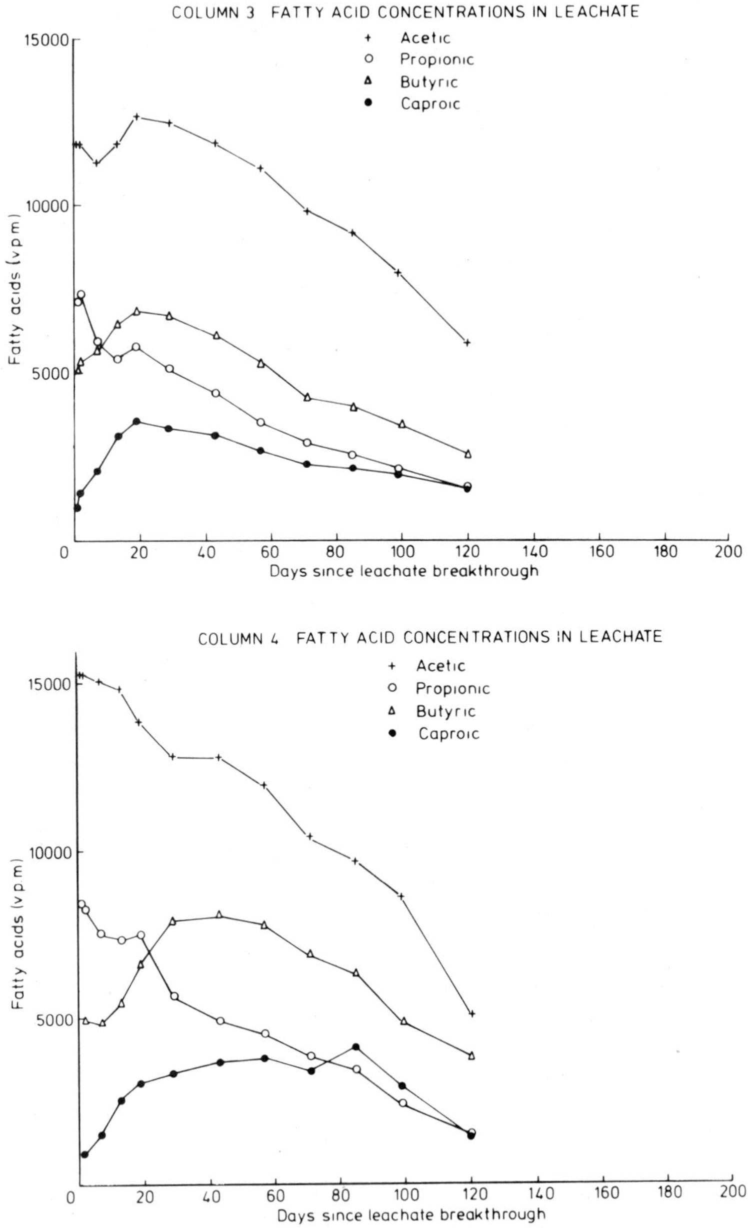

FIG. 2—*Fatty acid concentrations in the leachates from Columns 3 and 4 (1 volume per million ≃ 1 mg/dm^3).*

did not contain elevated levels of nickel, indicating that the metal had traveled more slowly in this column and confirming the observation of no leachate contamination. Expressed as a percentage of the nickel originally added to the columns, 8.6% remained in solution in Column 1, 2.8% in Column 3, and 1.8% in Column 5. This suggests that the nickel continued to be removed from solution throughout the first six months, and that after one year less than 1% might be expected to have remained in solution providing sufficient waste was available to prevent breakthrough in the leachate.

The chloride levels in the interstitial fluids were elevated in all of Column 1, in the bottom 0.3 m of Column 3, and in the bottom 0.2 m of Column 5, indicating that the chloride released within the dosing solution was moving at a slightly faster speed than the nickel through the pulverized waste. Thus, the largest excess of chloride in Column 5 was at the base, and very little was present at the depth of the peak nickel concentrations. Although the chloride may be involved in mobilizing the nickel, an insufficient amount was available in the interstitial fluids of Column 5 for the nickel to be present as the nickel tetrachloride soluble complex $[(NiCl_4)^{2-}]$.

The only other metal to show any persistence in a soluble form was cadmium (Fig. 3). Columns 1 to 6 show that cadmium was removed from the liquid phase steadily, largely in the top 0.25 m of the columns. There is some evidence of a second smaller peak moving toward the base of the columns, although Column 5 suggests that this would be attenuated to background levels before it reached the column base and that elevated levels of cadmium in leachate 0.5 m below the release are unlikely. After two, four, and six months, the total quantity of cadmium in the liquid phase represents only 1.5%, 0.4%, and 0.2% of that originally added. The results from Column 7 indicate that almost 5% of the cadmium added over the six-month period was still in solution but that levels were still not significantly elevated in interstitial liquid toward the base of the column.

The results from Columns 1 to 6 for the remaining six metals investigated showed that a negligible proportion was present in the liquid phase. Elevated levels were only evident toward the top of the dosed columns, and these decreased with time to insignificant proportions after six months. For example, the top section of Column 1 contained about 90 $\mu g/dm^3$ of mercury and 5 mg/dm^3 of both copper and chromium, but the levels of these metals were similar in the top sections of Columns 5 and 6.

The continuous release of 100 mg/dm^3 of metal onto Column 7 led to a larger proportion in the interstitial liquids. However, attenuation was still rapid, with particularly high levels restricted to the top two sections. The distribution of copper, chromium, nickel, and zinc is shown in Fig. 4. Copper and chromium were removed from the liquid phase very rapidly. Only 0.46% of the copper dose was still in solution and 0.04% of the chromium. Nickel and zinc exhibited greater penetration through the columns, although only nickel gave

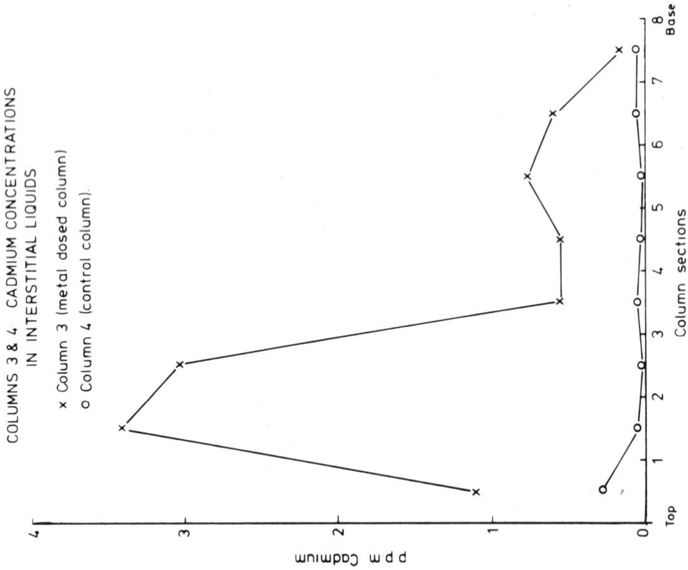
COLUMNS 3 & 4 CADMIUM CONCENTRATIONS
IN INTERSTITIAL LIQUIDS
x Column 3 (metal dosed column)
o Column 4 (control column)
p p m Cadmium
Column sections
Top
Base

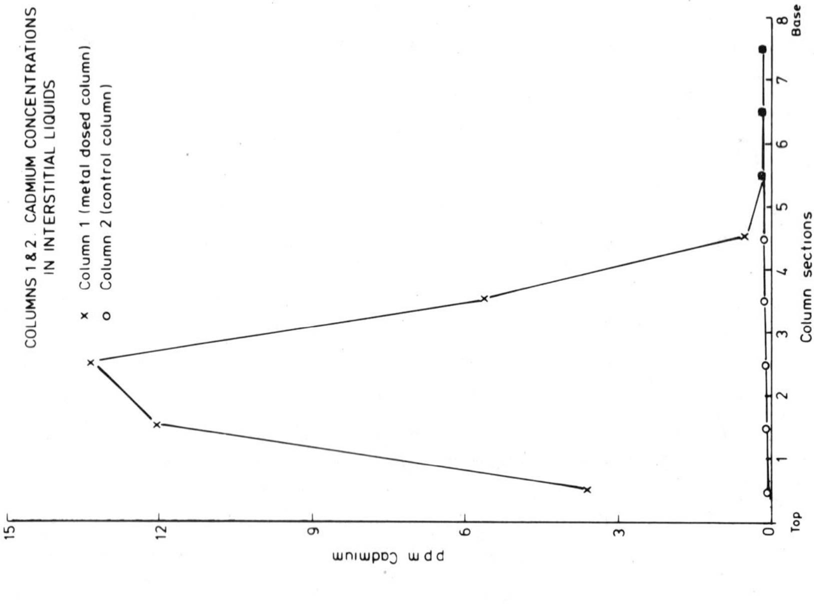
COLUMNS 1 & 2. CADMIUM CONCENTRATIONS
IN INTERSTITIAL LIQUIDS
x Column 1 (metal dosed column)
o Column 2 (control column)
p p m Cadmium
Column sections
Top
Base

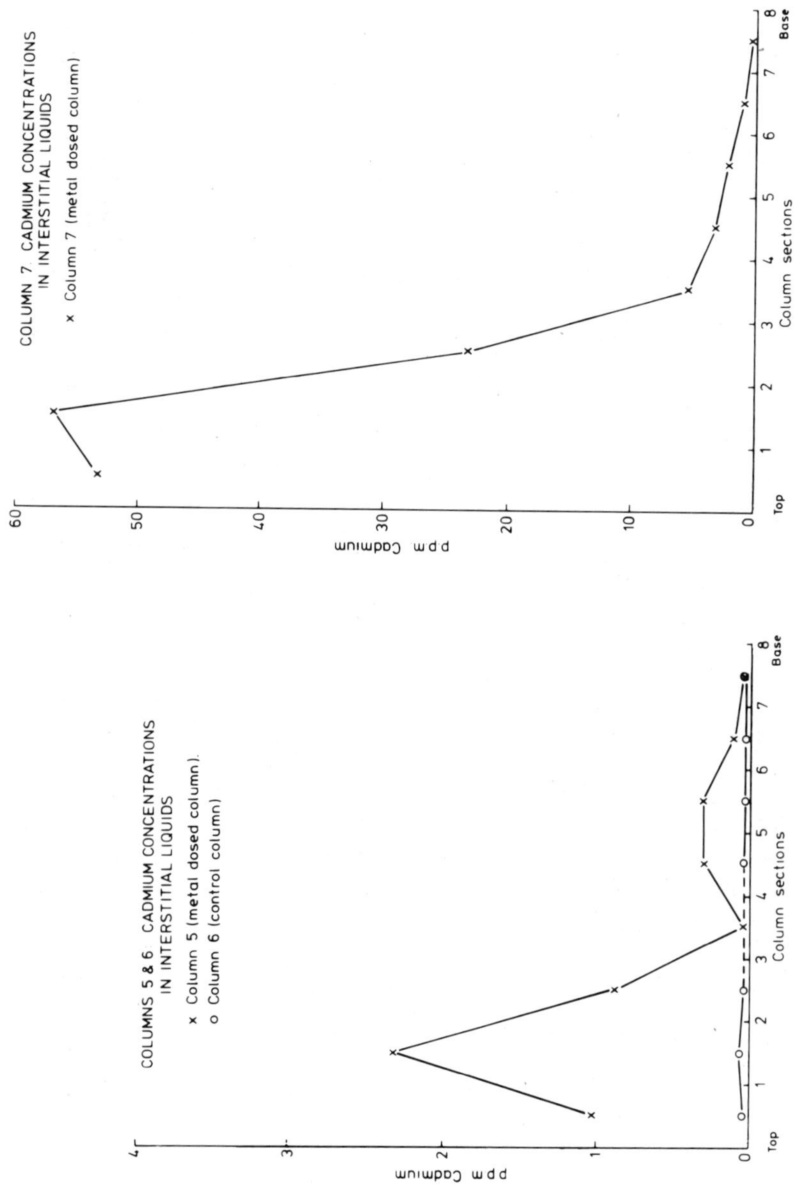

FIG. 3—*Cadmium concentrations in the interstitial liquids of columns (1 ppm = 1 mg/dm^3).*

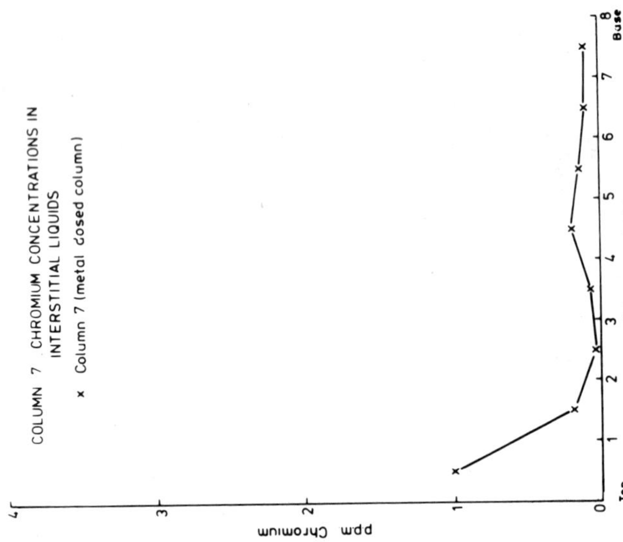
COLUMN 7 CHROMIUM CONCENTRATIONS IN INTERSTITIAL LIQUIDS
x Column 7 (metal dosed column)
ppm Chromium
0
1
2
3
4
Top
1
2
3
4
5
6
7
8
Base
Column sections

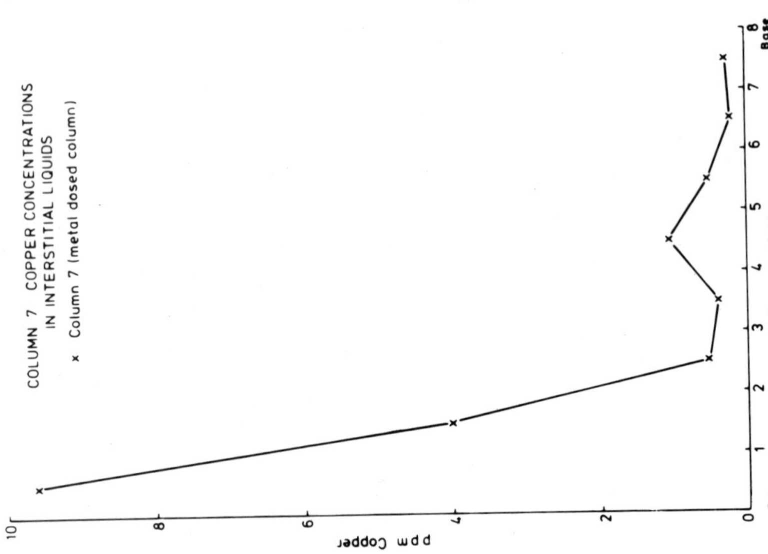
COLUMN 7 COPPER CONCENTRATIONS IN INTERSTITIAL LIQUIDS
x Column 7 (metal dosed column)
ppm Copper
0
2
4
6
8
10
Top
1
2
3
4
5
6
7
8
Base
Column sections

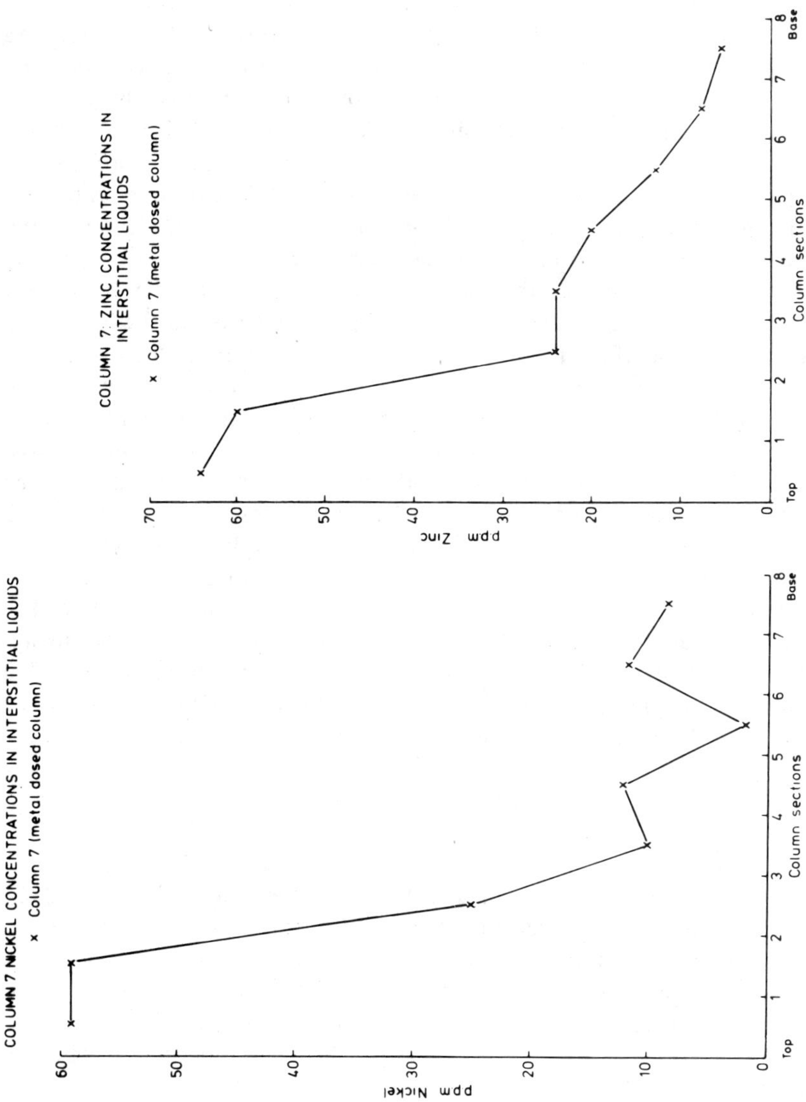

FIG. 4—*Selected metal concentrations in the interstitial liquid of Column 7 (1 ppm = 1 mg/dm^3).*

elevated leachate concentrations. Using Column 6 as a control, the excess nickel represents 5.8% of that added, and the zinc 6.4%.

The pH results were particularly interesting in view of the acidic nature of the dosing solution. All the dosed columns showed a tendency toward higher pHs toward the top of the waste. The control columns were reasonably consistent with similar pHs in the range of 5 to 6. In Columns 3 and 5 the pH reached neutral (pH 7) 0.1 to 0.2 m below the surface, whereas in Columns 1 and 7 the effect was less marked, with maximum pHs of 6.2 nd 6.3, respectively.

The analysis of the interstitial liquids for fatty acids provided an explanation for this behavior. All the dosed columns showed severe inhibition of fatty acid production in the top layers, although levels were only slightly lower at the column base, reflecting the final leachate samples. The results for Columns 3 and 4 are shown in Fig. 5. Analysis was also performed for isobutyric, isovaleric, valeric, heptanoic, and octanoic acid, and these all showed similar profiles. The complete absence of fatty acids in the top 0.2 m of Column 3 will lead to a relatively clean leachate and facilitated the near-neutral pHs observed. The presence of fatty acids normally buffers the liquid phase to a pH between 5 and 6, as was observed toward the base of Column 3 and in the control Column 4. The observed pHs at the top of Columns 1 and 7 were probably a result of residual ions from the dosing solution maintaining a pH slightly more acid than neutral. The fatty acids were absent from the top of these columns, with the effect penetrating to a similar depth in the continuously dosed Column 7.

The interstitial results, therefore, confirmed the leaching behavior and indicated that cadmium may also migrate a short distance through the refuse. The largest effect of the removal of the metals from the liquid phase appears to be a complete disruption of the decomposition of organics to fatty acids. The pH of the interstitial fluid tended toward neutral in the absence of the fatty acids, but the waste continued to neutralize the relatively acidic dosing solution.

Mass Balance Results

The analysis of a solid sample of waste from each section of each column after dismantling allowed a mass balance for all the metals to be carried out. The results of this mass balance work were most useful where the quantities of metals derived from the pulverized waste were both small in relation to the dose and consistent between different samples. Under these circumstances the excess of the dosed column section over the corresponding control column section gives a good indication of the quantity of the metal added which is present in that section.

Mercury was effectively absent from the control columns, but only a few milligrams was present in the top of the dosed columns. Some loss in the extraction procedure would seem likely since a separate extraction using perchloric acid was not carried out for mercury analysis. However, these results suggest that a substantial portion of the mercury added may have been lost to

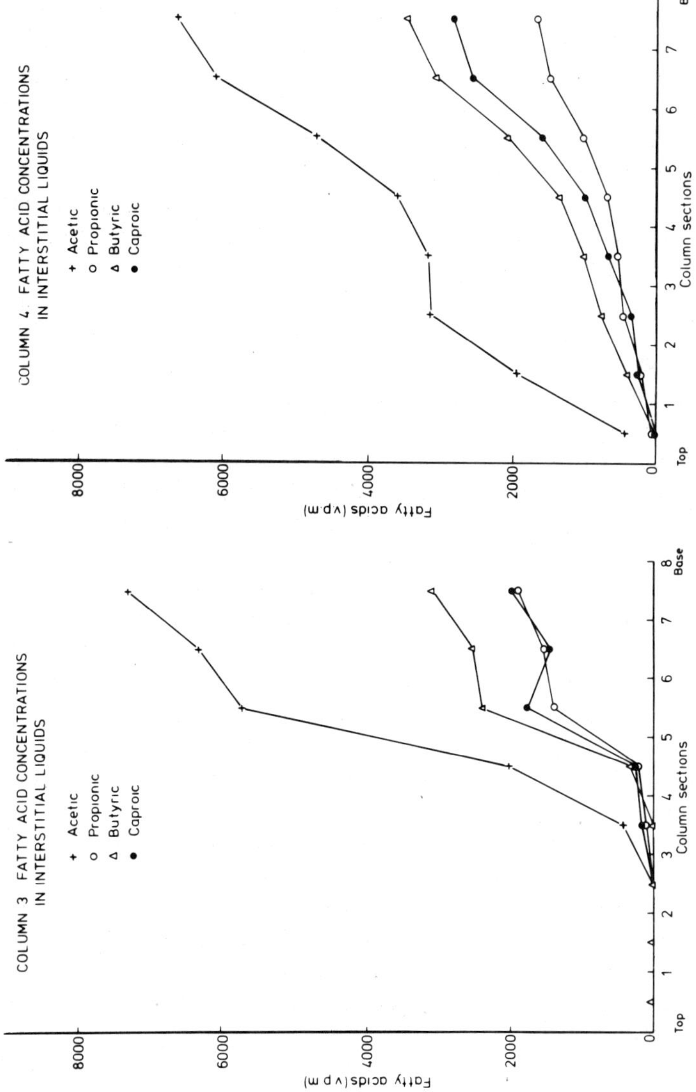

FIG. 5—*Fatty acid concentrations in the interstitial liquids of Columns 3 and 4 (1 volume per million ≃ 1 mg/dm³).*

the atmosphere from the top of the columns. Further work is intended on the fate of mercury released into municipal waste.

The cadmium results were subject to the least interference from background levels, and the distribution of cadmium in the dosed columns is shown in Table 2. The quantity released in the leachates and associated with the intractable items which could not be homogenized effectively are included but are small in comparison with the quantities derived from the dosing solution. The total excess of cadmium in each of the dose columns can be compared with that calculated from irrigation rates during the application of the dosing solution.

The results in Table 2 show two important points. First, the agreement between the quantities of cadmium recovered and that added are within 20%. The remaining disparity is largely due to a combination of two uncertainties. A series of double extractions in duplicate demonstrated that the single extractions used for these samples are liable to leave up to 10% of the metal in the solid phase. Generally the recovery is much better, but the single extraction procedure does lead to slight underestimates. A comparison between duplicates showed that the use of 5-g portions of cryogenically ground waste is normally representative. Agreement between most duplicates was good (within 10%), but an occasional sample may be unrepresentative because of a few small particles of relatively pure metal derived from a discrete source. The low recoveries of cadmium in Columns 1, 3, and 5 are probably largely due to the use of a single extraction, whereas an unrepresentative sample may have contributed to the apparent excess of cadmium in Column 7.

The second conclusion from Table 2 is that the cadmium was largely deposited in the same areas as those with a high concentration in the interstitial fluid. The majority of cadmium in Columns 1, 3, and 5 was found in the top 0.2 m of waste. Although the peak concentration in Column 1 was in the top section, comparison of Columns 3 and 5 indicates that no further migration of cadmium occurred after four months. Distribution of cadmium in Column 7 was wider, although the peak was still in the second section, and the bottom quarter contained insignificant amounts.

The distribution of the other metals reflected the interstitial fluid results, with only nickel showing significant excesses over the whole column. The remaining metals were apparently deposited in the top 0.25 m of waste, with the majority in either the first or second section. However, with some metals or columns, and in particular with iron, the quantities present in the refuse far exceeded those added in the dosing solution, and the distribution of the dosed metal was therefore unclear.

The analysis of the control columns also provided valuable information on the variability and range of metal concentrations in the pulverized municipal waste used. This waste represents about 60% by weight of the municipal waste collected in the area of the pulverizer. The remainder of the waste is made up either of large bulky items such as bicycle frames or mattresses or of pieces of ferrous metal larger than a few square centimetres, which are removed by a

TABLE 2—*The distribution of dosed cadmium (dosed column minus control column).*

Column Section	Excess in Column 1 over Column 2, mg	Excess in Column 3 over Column 4, mg	Excess in Column 5 over Column 6, mg	Excess in Column 7 over Column 6, mg
Section 1 (top)	88.9	66.7	61.4	82.5
Section 2	65.1	111.1	131.8	176.0
Section 3	43.2	53.4	72.6	68.7
Section 4	6.5	48.8	0.2	33.8
Section 5	2.0	6.7	0.4	23.8
Section 6	0.1	0.2	−0.6	8.4
Section 7	0.3	0.0	−0.2	0.2
Section 8 (bottom)	0.0	0.0	0.5	0.1
Leachate	0.02	0.00	0.02	0.08
Intractables	0.0	0.0	0.8	0.4
Total excess	206.1	286.9	266.9	394.0
Calculated dose	238	301	302	340

magnet. The waste is therefore perhaps a little low in metal content compared with that landfilled without pulverization treatment, but it is a convenient source of waste which can be handled in kilogram quantities.

The range of metal concentrations in the control column samples are given in Table 3. The range represented is that from two standard deviations below the mean to two standard deviations above the mean. This leads to apparent minimum quantities of zero which are unrealistic because the distribution of concentrations is not symmetrical at the extremes, with trace quantities of all metals always being present, although in the case of mercury these were not detectable. In general the levels of metal were similar between the different columns, although there were notable exceptions in the high chromium levels in Column 2 and the low nickel levels in Column 6. Comparison between the columns is thought to be valid, since the quantities released in the leachates were negligible, and there was no evidence that the metals were being leached toward the base of the columns in significant amounts. The concentrations observed in the control columns indicate that the range for different metals varies widely, but this range is reasonably consistent between different batches of about 3 kg. This suggests that applications of metal quantities comparable to those present in the waste are likely to be lost in the background. However, providing the variations within the waste are taken into account, comparison of pairs of columns seems valid, since the different control columns behaved in a similar way. The similarity in the leachates and interstitial liquid results for the control columns also indicates that differences in the behavior of the dosed columns may be attributable to the metals added.

Conclusions

Pulverized municipal waste has been shown to have a considerable ability to remove solubilized metals and immobilize them. Quantities of cadmium,

TABLE 3—*Range (for the 95% confidence interval) of metal concentrations in control column samples.*

	Range, mg/kg		
Metal	Column 2	Column 4	Column 6
Cadmium	0 to 3.5	0 to 2.3	0 to 2.9
Chromium	69 to 240	26 to 90	18 to 40
Copper	0 to 190	9 to 200	0 to 130
Iron, g/kg	0 to 44	0 to 41	0 to 34
Lead	52 to 160	11 to 220	62 to 150
Mercury	<0.002	<0.002	<0.002
Nickel	27 to 130	0 to 270	13 to 41
Zinc	0 to 2100	40 to 1000	0 to 950

chromium, copper, iron, lead, and zinc, equivalent to loading rates of about 40 g of each dissolved metal per square metre, were found to be deposited entirely within a depth of 0.5 m of waste. Penetration through the waste was greater when the metal loading was spread over six months, although leachate quality at a 0.5 m depth was still unaffected. Mercury levels equivalent to about 4 g/m^2 also had no effect on leachate quality, but subsequent analysis of the solid phase suggested that a substantial proportion had been lost from the column experiments, possibly by volatilization to atmosphere. Only nickel was found to migrate through the 0.5 m of refuse, with up to about 20% emerging in the leachate when added over a six-month period. Concentrations in leachate were typically 20 mg/dm^3 and were apparently independent of the dosing solution concentration.

Analysis of the municipal waste in sections has shown that most of the metals are deposited in the top 0.1 or 0.2 m of the waste and are no longer migrating downwards. The consequent increase in metal concentrations completely halted fatty acid production in these areas, allowing the pH to reach neutral in some cases. The waste is still capable of buffering the acidic dosing solution effectively even where fatty acids were depleted.

Further work is being done to establish the permanence of the attenuation observed and interpret the results to provide practical guidance for landfill practice. Some conclusions, however, can already be drawn about preferred landfill practices. Providing fissure flow is avoided, it appears that the relatively rapid release of metal from a liquid waste, for example, may be attenuated more effectively and in a shorter distance than an equivalent loading from the prolonged leaching of a metal-rich solid waste. This is providing that the pH of the refuse is not reduced to below 5 by excessive deposits of acid. When several metals are present, nickel is likely to be the most mobile and may be a useful indicator of potential metal release in the field. Cadmium is the next most mobile element of the eight studied. Finally, the assessment of the usefulness of the small-scale laboratory columns used suggests that reproducible behavior can be achieved. More important, the significance of the metal loadings in the field must be compared with the quantities of those metals already present in the wastes deposited and with the range of releases these wastes may cause in isolation.

Acknowledgments

This work has been sponsored by the U.K. Department of the Environment. The authors would like to express their thanks to W. Temple, D. J. Darke, and D. Jenkins for carrying out some of the sample analyses.

This paper is published by kind permission of both the Department of the Environment and the director of the Harwell Laboratory.

References

[1] Department of the Environment, "Co-operative Programme of Research on the Behaviour of Hazardous Wastes in Landfill Sites," Her Majesty's Stationery Office, London, 1978.

[2] Lowenbach, W., "Compilation and Evaluation of Leaching Test Methods," EPA 600/2-78-095, U.S. Environmental Protection Agency, McClean, Virginia, 1978.

[3] Anderson, M. A., Ham, R. K., Stegmann, R., and Stanforth, R., *Toxic and Hazardous Waste Disposal*, Vol. 2, R. B. Projasek, Ed., Ann Arbor Science Publishers, Ann Arbor, Mich., 1979, Chap. 8.

[4] Perket, C. L. and Webster, W. C. in *Hazardous Solid Waste Testing: First Conference*, ASTM STP 760, American Society for Testing and Materials, Philadelphia, 1981.

[5] Wilson, D. C. and Young, P. J., "Testing Methods for Hazardous Wastes Prior to Landfill Disposal," Paper presented at the Oak Ridge Fourth Life Sciences Symposium: The Environment and Solid Wastes: Characterization, Treatment and Disposal, Gatlinburg, Tennessee, 4–8 October 1981.

[6] Young, P. J. and Wilson, D. C., "Testing of Hazardous Wastes to Assess Their Suitability for Landfill Disposal," AERE R-10737, Her Majesty's Stationery Office, London, November 1982.

James S. Falcone, Jr.,[1] *Robert W. Spencer,*[1]
Richard H. Reifsnyder,[1] *and E. P. Lester Katsanis*[1]

Chemical Interactions of Soluble Silicates in the Management of Hazardous Wastes

---

REFERENCE: Falcone, J. S., Jr., Spencer, R. W., Reifsnyder, R. H., and Katsanis, E. P. L., **"Chemical Interactions of Soluble Silicates in the Management of Hazardous Wastes,"** *Hazardous and Industrial Waste Management and Testing: Third Symposium, ASTM STP 851,* Larry P. Jackson, Alan R. Rohlik, and Richard A. Conway, Eds., American Society for Testing and Materials, Philadelphia, 1984, pp. 213-229.

ABSTRACT: Batch and flow leaching studies were carried out using a lime-precipitated plating waste sludge to study the influence of soluble silicates on metal ion leachability. A reduction in metal ion leaching was observed when soluble silicates were used either alone or with a portland cement/fly ash combination. These results suggest that sodium silicate reduces the solubility of metal ions in the waste matrix in addition to protecting the matrix from mild acidic breakdown.

KEY WORDS: hazardous wastes, industrial wastes, soluble silicates, solidification, leaching, cement, fly ash

The use of various inorganic materials in the consolidation and control of wastes, particularly those containing leachable metal ions, has been a subject of great interest in the PQ Corp. Research and Development Center. In a recent paper [*1*], the authors summarized the literature of the Chemfix process[2] and other methods and focused developmental efforts on lime and silicate and cement-based systems. These systems, which are generally based on inexpensive alkaline inorganic materials, are effective in reducing the solubility of metal ions, the degree of which depends on the properties of the metal ions and their hydrolysis products. Besides solubility, there are other important factors

[1]Technical manager, Industrial Chemicals Research and Development, research and development chemist, technical service representative, and senior chemist, respectively, The PQ Corp., Research and Development Center, Lafayette Hill, Pa. 19444.

[2]Chemfix process is a registered trademark of Chemfix Technologies, Inc., Kenner, La.

which may vary depending upon the nature of the waste and the chemical used—for example, the handling characteristics of the treated waste, the total final volume of waste, and the solid waste matrix permeability.

This report summarizes a study of the effect of soluble silicates on the leaching characteristics of a lime-precipitated plating bath sludge containing toxic metal ions. This waste was chosen because of the suggestions in the literature that such solids are susceptible to degradation in acidic environments [*2*]. This effect can be so extreme that the waste matrix might actually deteriorate, leading to accelerated leaching of the components. The use of sodium silicates with acids, cement, lime, slags, and various other sources of multivalent metal ions which can act as initiators of silicate precipitation or gelation has been known for over a century. These reactions are the basis of many commercial technologies, such as acid-resistant cements and coatings, soil grouting, and catalysts. Research in the PQ Corp. Research and Development Center [*3*] has suggested that sodium silicates are useful in enhanced oil recovery systems, in part because of their ability to block fluid flow in the more permeable zones of an oil-bearing substrate.

Other results for selected multivalent metal ions [*4*][3] have shown that the precipitation domains are significantly different when soluble siliceous materials are present in the system. Furthermore, solubilities of these metal ions were reduced by orders of magnitude.

Based on this background evidence for the potential value of a soluble silicate as an acid-resistant matrix former and metal ion activity suppressor, the authors chose to study carefully the batch and continuous acidic leaching characteristics of a metal-plating waste sludge treated with acid-gelled sodium silicate and a cement/fly ash mixture with and without soluble silicate addition.

Experimental Procedures

A metal plating waste selected for treatment and extraction using standard batch and column methods [*5–8*] modified to produce severe acid leaching conditions. The waste solidification methods represent techniques widely used in industry to treat this type of waste material. The reagents used in this study were of analytical grade, and all of the solutions were prepared with deionized water. Throughout this study portland cement–Type I (PCI), special fly ash No. 258 (SFA), and Type N® 3.22-ratio sodium silicate[4] (silicate) were used in the treatments. All mixing and reagent additions were done in a 4-L Hobart mixer.

[3]E. P. Katsanis, PQ Corp., unpublished results, 1983.
[4]N® is a registered trademark of the PQ Corp.

Plating Waste

The plating waste was an alkaline-precipitated metal sludge generated by an industrial plater. An 18-L sample was collected on site from five randomly selected drums of stockpiled sludge. This waste was kept sealed in a plastic container, and all the sampling procedures were carried out as described in the ASTM Shake Extraction of Solid Wastes with Water (D 3987-81).

The chemical analysis of a representative sample is shown in Table 1. The pH value, solids content, and other physical characteristics were determined for a sample of fresh sludge. Metal concentrations were determined by atomic absorption (AA), plasma emission (PE), and energy dispersive spectroscopy (EDS). Zinc and chromium are the major toxic components, thus this study focused on these ions. The samples were treated according to the methods listed in Tables 2 and 3. Leachability was evaluated using batch and column methods.

Column Leaching

Six extraction columns were prepared from 2.54-cm diameter Pyrex glass tubing. Each column was cut to a length of 1 m and scored on the inside walls to prevent uneven flow down the column sides. All the columns contained 100 g of the original waste material. The columns were packed evenly to prevent nonuniform flow through the samples. The pasty samples were tamped in place in the tubes, using a glass rod to eliminate air pockets. A layer of sand over a glass wool plug supported the sample in the tube. The solidified wastes,

TABLE 1—*Plating waste sludge analysis (pH, 10.83; percent solids, 9.68).*

Element	Concentration, g/100 g Sludge	Analytical Method Used
Major element components		
Zinc	1.27	AA
Chromium	0.98	AA
Sodium	1.1	PES
Calcium	1.1	PES
Magnesium	1.1	PES
Manganese	0.7	PES
Molybdenum	0.7	PES
Iron	0.7	PES
Aluminum	0.5	PES
Acid-insoluble material (Si > > S, Cl)	1.6	EDAX
Minor element components (by EDS)		
Ni > P > Sr > Cd > Sn > Pb > Cu		
Trace element components (by EDS)		
Ba, Pd, Sb		

TABLE 2—*Experimental conditions for column leaching (basis 100 g sludge).*

Column No.	Treatment Materials	Leaching Solution[a]
1	none	acetic
2	none	citric
3	gelled with 14 g silicate and 0.5 M H_2SO_4 to a pH value of 7.0	acetic
4	same as Column 3	citric
5	mixed with 42 g PCI and 15 g SFA;[b] ground to $-100+200$ mesh, and tamped into column	citric
6	gelled as in Column 3, then mixed with 42 g PCI and 15 g SFA; ground and tamped as in Column 5	citric

[a]Citric acid (0.1 M) buffered to a pH value of 5.5 with 0.2 M disodium phosphate; 5% acetic acid adjusted to a pH value of 5.5 with 0.1 M NaOH.

[b]PCI = portland cement I; SFA = special fly ash.

TABLE 3—*Experimental conditions for batch leaching (basis 100 g sludge).*[a]

Batch No.	Treatment Materials	Leaching Solution
A	none	citric
B	mixed with 14 g silicate	citric
C	gelled with 14 g silicate and 0.5 M H_2SO_4 to a pH value of 7.0; fresh gel	citric
D	gelled as in Batch C; gel aged 1 month before extraction	citric
E	mixed with 42 g PCI and 15 g SFA; ground to $-100+200$ mesh	citric
F	gelled as in Batches C and D, then mixed with 42 g PCI and 15 g SFA; ground to $-100 + 200$ mesh	citric

[a]See footnotes for Table 2.

in Columns 5 and 6, were ground to -100 $+200$ mesh, and were tamped in layers so as to have similar initial flow rates. The columns with the pasty samples do not have a pore volume in the ordinary sense. However, the estimated hydrodynamic void volumes, based on an assumed average density of 2.5 g/cm^3 for the solid hydrous oxides, were calculated to be 89 mL for Columns 1 and 2 and 93 mL for Columns 3 and 4. The pore volumes for Columns 5 and 6, which contained packed solids, were approximately 71 mL each.

An aqueous 5% solution of acetic acid adjusted to a pH value of 5.5 with 0.1 M NaOH and a citric acid solution buffered to a pH value of 5.5 [*9*] were selected for the column extractions. A 100-mL volume of the leaching solution was added to each column at the beginning of the study, and this head volume was maintained for the duration of the experiment. For the six samples studied over a period of about 60 days, the total volume of acid added to any one column ranged between about 225 and 675 mL, depending on the flow rate of the liquid through the column (see Figs. 1 through 6).

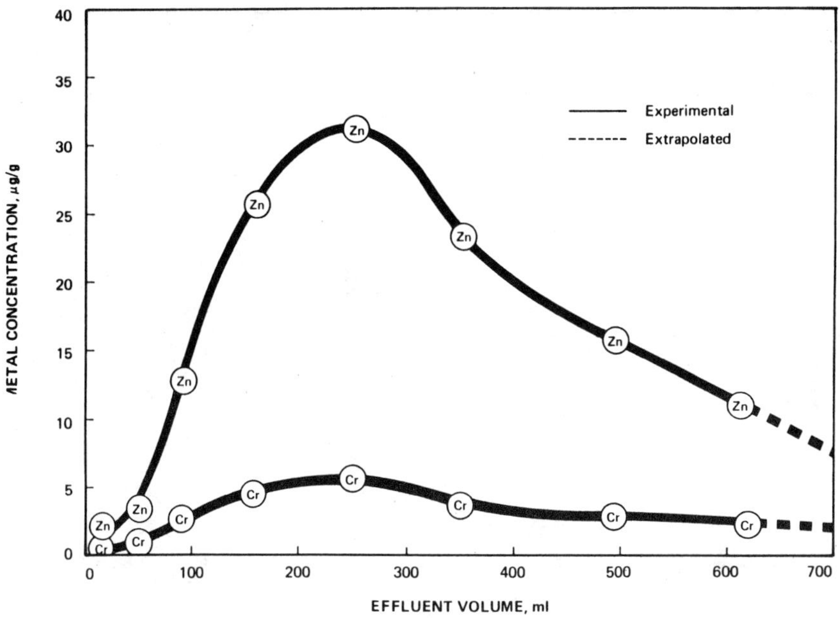

FIG. 1—*The elution profile for zinc and chromium from untreated sludge leached with acetic acid; Column 1.*

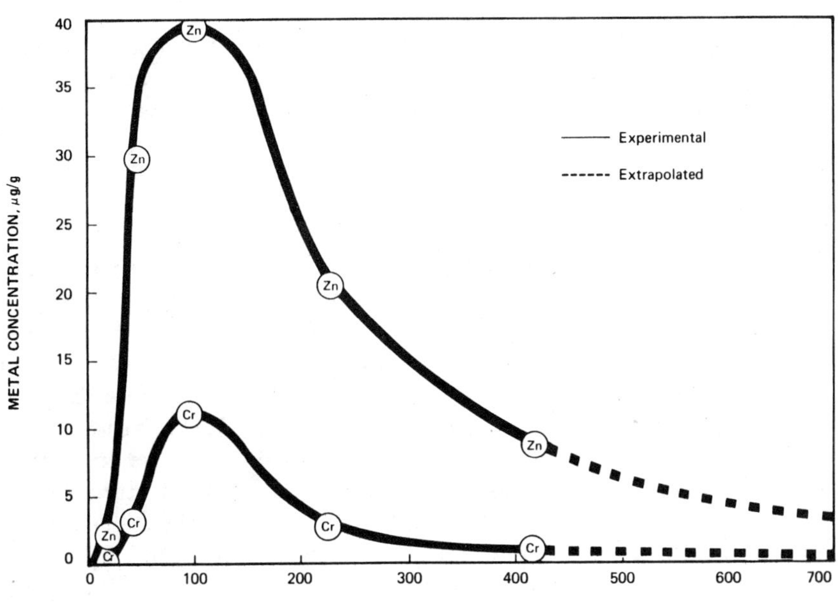

FIG. 2—*The elution profile for zinc and chromium from untreated sludge leached with citric acid; Column 2.*

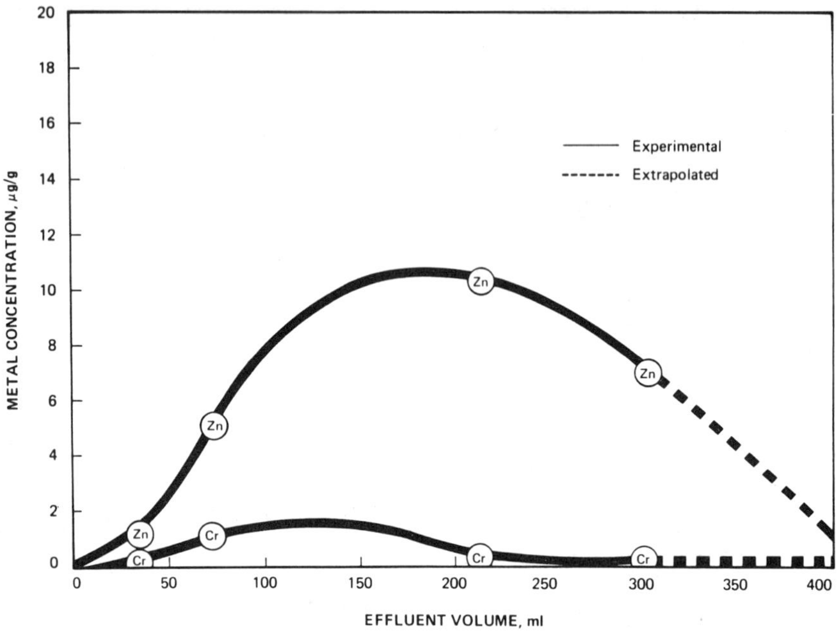

FIG. 3—*The elution profile for zinc and chromium from gelled sludge ($H_2SO_4/Na_2O:3.22\ SiO_2$) leached with acetic acid; Column 3.*

Samples of leachate were collected in sealed graduated cylinders to prevent evaporation loss. The collected samples were acidified to a pH value of 2.0 with concentrated nitric acid (HNO_3) and analyzed for metal content. The low pH was necessary to keep the metals in solution.

Batch Leaching

Batch extractions were performed on the untreated and five treated waste samples using the buffered citric acid solution as the leaching medium. Extractions were carried out in Nalgene bottles using a variable-speed reciprocating shaker set at 80 cycles per minute. A sludge-to-acid ratio of 1:6 (weight/weight) was used in all cases, based on a total weight of sludge of 100 g. Batch sampling was done at 24, 48, 96, 384, 472, and 720 h. At each of these times 50 g of slurry was removed with a syringe and filtered under pressure using 0.4-μm filter paper. The solids removed by filtration were returned to the extraction vessel, and enough citric acid buffer was then added to make up the lost weight. The pH value of the batch was adjusted to 5.0 with 5% HNO_3, if needed, after each sampling step to maintain the pH value. The filtrates were acidified to pH 2.0 with concentrated HNO_3 and stored in plastic bottles, for analysis of the heavy metals by AA.

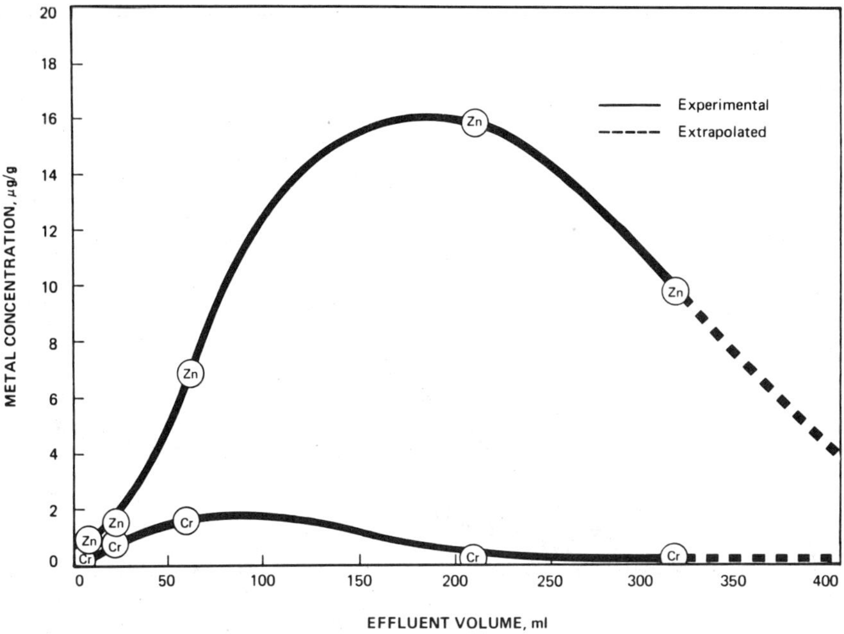

FIG. 4—*The elution profile for zinc and chromium from gelled sludge* ($H_2SO_4/Na_2O:3.22$ SiO_2) *leached with citric acid; Column 4.*

Experimental Results

Column Leaching Experiments

The leaching profiles for the different columns are shown in Figs. 1 to 5. Figures 1 and 2 show the leaching of zinc and chromium from the untreated sludge, with acetic and citric acids, respectively, as the leaching solutions. Since the elution profile in these figures approaches the abscissa, we were able to estimate the area under the curves which gave the total weight of the metals extracted. These results are given in Table 4. Estimates of zinc and chromium leached during column extraction on a constant volume basis are given in Table 5.

Figures 3 and 4 show the corresponding results for portions of sludge solidified with gelled sodium silicate. Estimates of total metal ion eluted are also given in Table 4 for Columns 3 and 4. Estimates of zinc and chromium leached during column extraction on a constant volume basis are given in Table 5. Finally, extraction of the metals from the cement and cement/silicate columns are shown in Fig. 5. Because of the low rate of metal ion extraction and low flow rate, full elution profiles were not obtained.

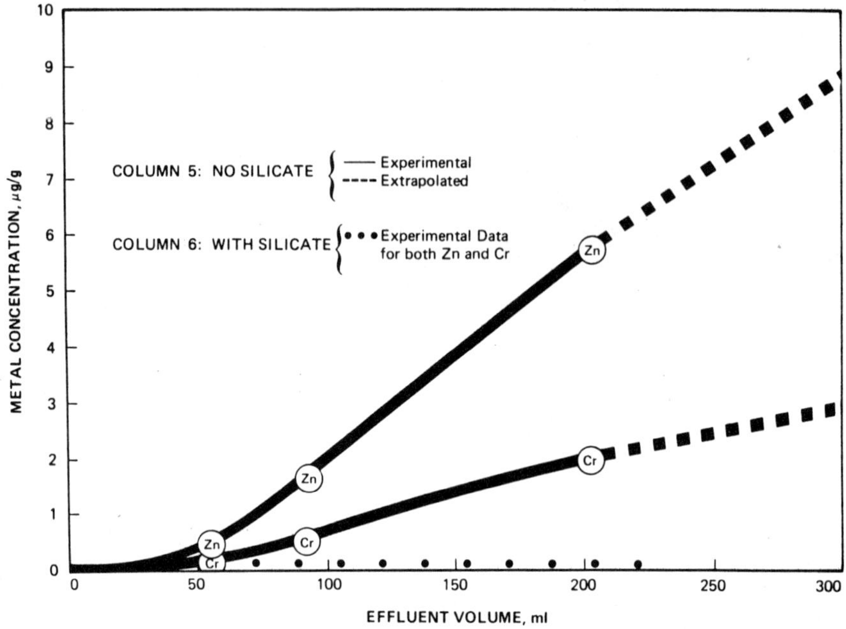

FIG. 5—*The elution profile for zinc and chromium from sludge consolidated with portland cement-Type I and fly ash, with and without* $Na_2O:3.22\ SiO_2$. *Leached with citric acid.*

The flow rates through the columns are shown in Fig. 6. The initial rates were in the range of 1 to 5 mL/day, but within 20 days the structure of the untreated sludge was visibly broken down, and correspondingly the flow rates went up sharply. The same effect occurred in the gelled columns in about 35 days, but the increase in flow rate was less pronounced. In the cement and cement/silicate columns there was no evidence of breakdown after 60 days based on flow rates or visual evidence.

Figure 7 shows the pH levels of the leachate solutions throughout the leaching process.

Batch Experiments

Figure 8 shows leaching profiles for zinc from the various solidification systems. The points on the curve represent zinc concentrations in the 50-g sample removed from the 700 g of sludge/buffered citric acid combination. The sludge/buffered citric acid replacement method gradually depleted the divalent metal ion, Me^{2+}, in the sludge and increased its H^+/Me^{2+} ratio. The influences of the various treatment methods are summarized in Table 6. The total concentration was calculated from the concentrations of the dissolved species at 30 days plus the sum of those removed during periodic sampling.

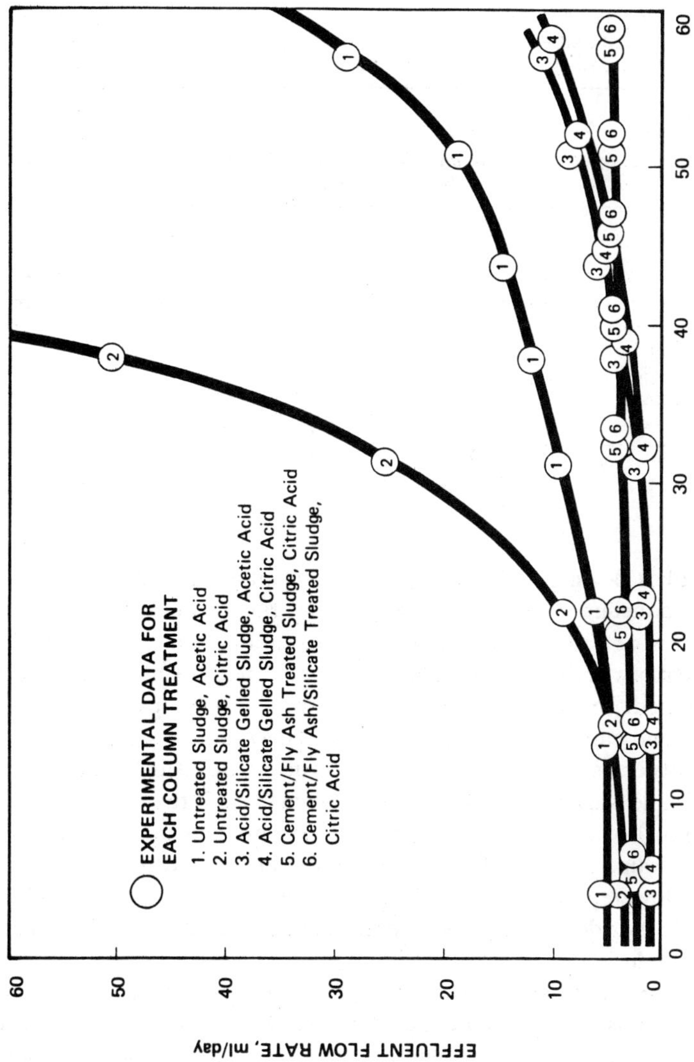

FIG. 6—The leachate flow rate profiles for Columns 1 through 6.

TABLE 4—*Estimates of the total amounts of zinc and chromium leached during column extractions.*

Column No.	Condition	Extraction Medium	Zinc, μg	Chromium, μg
1	untreated sludge	acetic acid	14 600	3 000
2	untreated sludge	citric acid	10 700	1 900
3	gelled sludge	acetic acid	4 100	180
4	gelled sludge	citric acid	5 600	250

TABLE 5—*Estimates of zinc and chromium leached during column extractions on a constant leachate volume basis.*

Column No.	Leachate	Leachate Volume	Time Required for Leachate Volume, days	Micrograms Metal per Gram Leachate	
				Zinc	Chromium
1	acetic	360	41	20.4	3.7
2	acetic	360	68	6.2	0.26
3	citric	375	35	23.0	4.2
4	citric	375	68	11.3	0.55

Discussion

Many studies dealing with homogeneous precipitation of metal (hydrous) oxides from aqueous solutions have shown that the chemical and structural composition as well as the morphology of the resulting solids are affected by the type of cations and anions present and the pH value at which such precipitates are formed. In some cases the ions are present in the particles as "contamination" and can be readily removed [*10*]. However, for some systems the ions are incorporated stoichiometrically into the particles and cannot be removed by simple washing [*11*]. It has been shown [*12–17*] that the solubility of metal (hydrous) oxides is strongly dependent on pH and that for each solid there is a pH value at which the solubility of the metal (hydrous) oxide is at a minimum. However, adjusting that pH for the minimum solubility of one metal (hydrous) oxide may result in appreciable concentrations of other metals remaining in solution.

Our data show that the use of alkaline silicates strongly influences the solubility of metal ions to make consolidated wastes less leachable. These conclusions can be derived from the column elution data shown in Figs. 1 through 5 and Tables 4 through 6. The heights of the peaks in Figs. 1 and 3 show that the maximum amount of zinc leached during elution with acetic acid is about 33 μg/g for the untreated sludge, but only about 11 μg/g for sludge gelled with silicate. This represents a threefold decrease. Under these conditions the concentration of chromium also decreased threefold, but the actual values

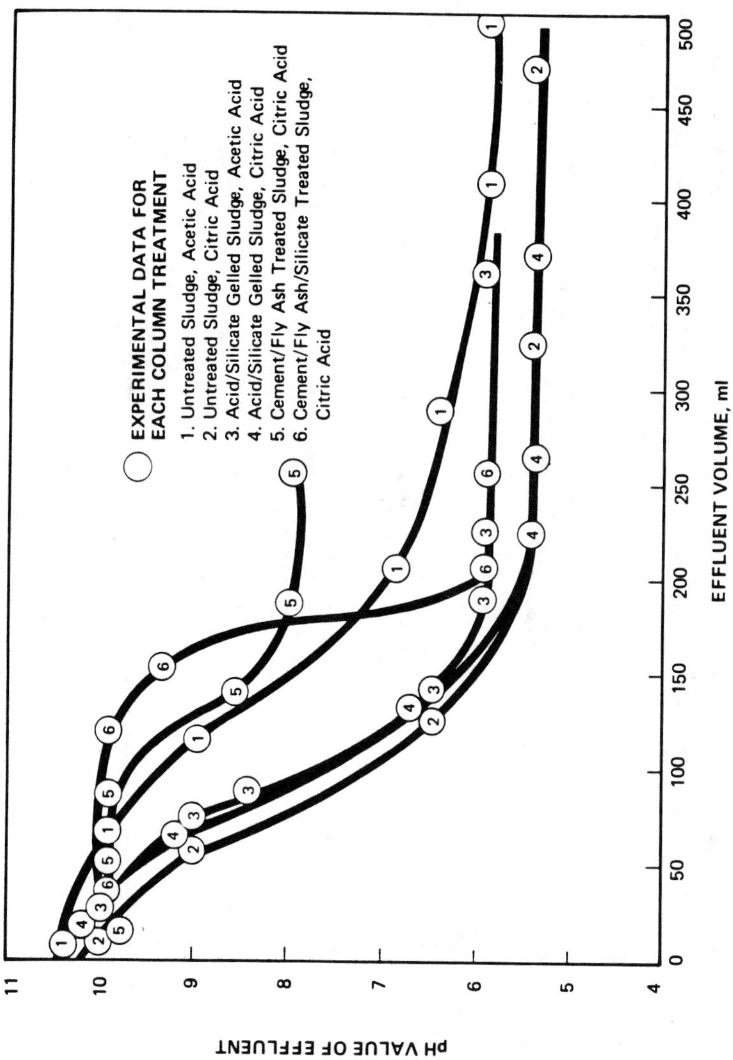

FIG. 7—*The pH value profiles for the leachate solutions for Columns 1 through 6.*

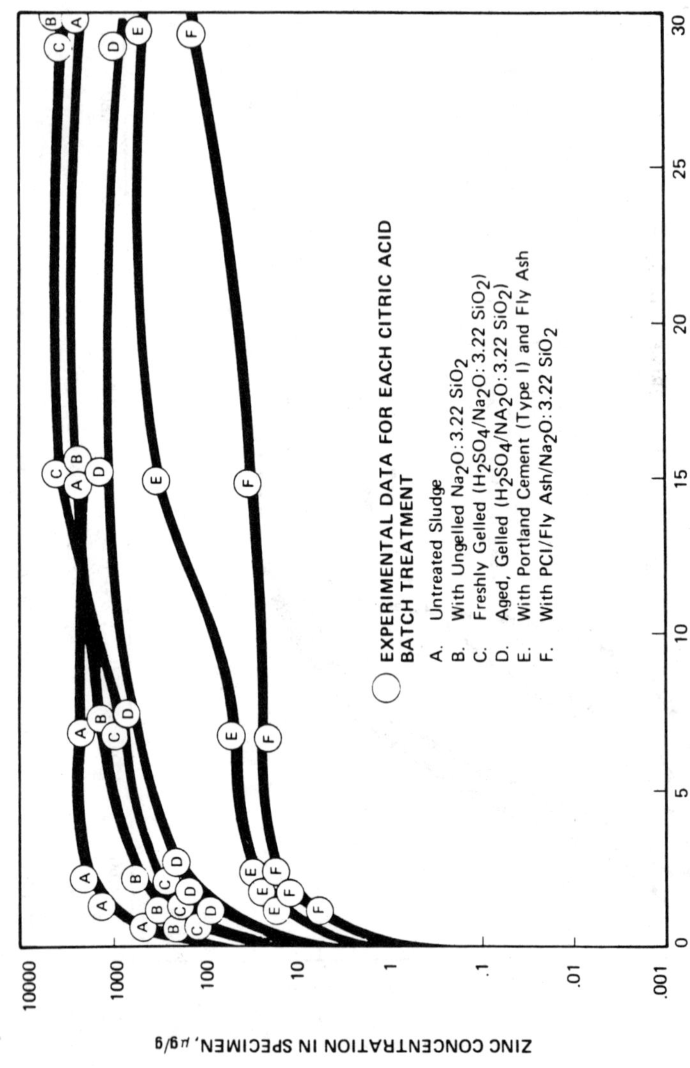

FIG. 8—*The batch extraction of zinc from treated sludges with citric acid.*

TABLE 6—*Total amount of zinc and chromium leached with buffered citric acid from treated plating sludges under batch conditions for 720 h.*[a]

		Zinc		Chromium	
Batch No.[b]	Sludge Treatment	Amount Leached, g	Percentage of Total	Amount Leached, g	Percentage of Total
...	total metal ion present per 100 g sludge[c]	1.27	...	0.98	...
A	none	1.2	95	0.16	16
B	ungelled sludge	1.2	94	0.13	13
C	fresh gel	1.1	84	0.10	10
D	aged gel	0.84	66	0.08	8
E	cement	0.44	35	0.05	5
F	cement/silicate	0.07	5	0.02	2

[a]All values are estimated to be within ±5%.
[b]For treatment details see Table 3.
[c]Determined by chemical analysis.

were about seven times smaller than the zinc values. In the presence of citric acid, similar correlations were observed for the leaching of zinc and chromium from untreated sludge and sludge gelled with silicate. When silicate was used to gel the sludge, the maximum amounts leached during elution were reduced by a factor of 2.5 for zinc and 6.1 for chromium, compared with leaching of the untreated sludge (Figs. 2 and 4). The estimated values of total zinc and chromium leached during column extraction, as determined from the areas under the curve for Figs. 1 through 4, also exhibit similar correlations on the effect of silicate in reducing metal ion solubility (Table 4). For sludge gelled with silicate, the total amount leached by acetic acid was reduced by a factor of 3.6 for zinc and 16.7 for chromium compared with untreated sludge. Similarly, the total amount of zinc and chromium leached by citric acid from sludge gelled with silicate was reduced by factors of 1.9 and 7.6 for zinc and chromium, respectively.

These correlations are also true when comparisons are done on a constant leachate volume basis, as shown in Table 5. For example, it took 68 days to pass 375 mL of citric acid buffer through Column 4 (gelled sludge) but only 35 days through Column 2 (untreated sludge). Under these conditions, the total amount of zinc leached from Column 2 was 2.2 times the amount leached from Column 4. The amount of chromium leached from Column 2 was 8 times that leached from Column 4. Similar observations were made for acetic acid leaching. It took 68 and 41 days to pass 360 mL of leachate through Column 3 (gelled sludge) and Column 1 (untreated sludge), respectively. In this volume of leachate, 3.3 times as much zinc and 14.2 times as much chromium were removed from the untreated sludge as from the gelled sludge.

The effect of silicate in reducing the amount of zinc and chromium leached from sludge treated with PCI and No. 258 SFA is demonstrated in Fig. 5. The citric acid elution curves for the two systems clearly show that the concentrations of zinc and chromium are reduced by many orders of magnitude when the sludge is gelled first with the silicate. For example, the total amount of zinc and chromium leached after 225 mL in Column 5 was 600 and 213 μg, respectively, as determined by the area under this portion of the leaching curves. When treated with gelled silicate/cement/fly ash (Column 6), no zinc or chromium was detected in the leachate. The limit of detection of our method was 0.01 μg/g.

Under batch conditions, the same effects of lowered metal ion solubilities were noted for sludges gelled with silicate. Although Fig. 8 illustrates these effects for the various sludge treatments studied, the results of Table 6 are more descriptive in that they represent the total amount of zinc and chromium leached rather than the rate of dissolution and approach to steady-state conditions. It is shown in Table 6 that the total amount of zinc and chromium leached from sludge gelled with silicate is less than that leached from untreated sludge under the test conditions. For instance, after 720 h exposure of a sludge treated with aged gel, the percentage of total zinc leached is 66%, versus the 95% leached from untreated sludge. The corresponding values for chromium are 8% and 16%. These effects are even more pronounced for sludge gelled with silicate and then solidified with PCI and SFA. Under these conditions, only 5% of the total zinc and 2% of the chromium were leached after 720 h, compared to 35% zinc and 5% chromium leached under the test conditions for a sludge treated only with PCI and fly ash.

A comparison of Tables 4 and 6 shows that the total amount of metal ions extracted from a 100-g sample of the waste by the batch method was greater by two to three orders of magnitude than that extracted by the column tests. This is in sharp contrast with what one would expect from the bell-shaped profiles of the leaching curves for Columns 1 through 4, that is, complete depletion of the metal ions. At least three factors contribute to this apparent discrepancy:

1. There was more intimate contact between leachate solution and the solids in the batch method because of the constant agitation and breakdown of particles to small sizes.

2. Nitric acid was added to the batch samples to maintain the pH value at 5.5. This amounted to 12 to 26 mL of 5% HNO_3 in the various samples (10 to 21 meq per 100 g of waste).

3. The possible formation of microchannels in the columns would allow the leachate flow to increase while diminishing the intimate contact between the leachate solution and the solid phase. This is consistent with Figs. 1 through 4 (Columns 1 through 4), where the curves are bell shaped. In each case a substantial increase in flow was observed (Fig. 6), along with a decrease in amount of metals leached. For Columns 5 and 6, not enough leachate passed through to delineate similarly shaped curves.

The flow rates for column leaching as a function of time, shown in Fig. 6, stay relatively low and constant even after about 55 days for the sludge consolidated with cement and fly ash and for the sludge first gelled with silicate before consolidation, as described earlier. This is in sharp contrast with the untreated sludges in which, after about 30 to 50 days, the flow rates increase sharply, indicating that the sludge matrix is breaking down. The flow rates for sludge gelled with silicate are initially low and do not increase as much even after 55 days. These results indicate that the silicate alone, the cement alone, or a silicate/cement combination are instrumental in consolidating the sludge and maintaining its integrity even after acid treatment.

The silicate is also believed to react with the cement surface to encapsulate it, at least in part, based on evidence that the amount of calcium leached was reduced by more than half (410 mg/g for Column 5 versus 160 μg/g for Column 6) when the silicate was present. Further support for this mechanism comes from Fig. 7, which clearly shows that the final pH value of the leachate for the cement/fly ash system is much higher than that for a similar system pretreated with silicate. This is probably due to the demineralization of cement by the acid. In all other cases the untreated sludge and the silicate-treated sludge reach a steady-state pH value of about 5.5, that is, the initial value of the acid solution used.

The proposed silicate-cement interactions are also consistent with precipitation domain data which show that calcium and magnesium ions at concentration levels of only 100 ppm each change the shape and position of the silicate domains by several pH units and several orders of magnitude in the concentration of silica (SiO_2) needed for precipitation to occur [*17*]. Obviously, the reactions between the hardness ions and the soluble silicates are very strong under these conditions.

In summary, we have found that silicate-based formulations reduced the amount of leachable metal ions compared with similar systems that do not contain silicate. This is readily illustrated by comparing the weights of extracted metals shown in Tables 4 and 6. The sludge solidified with gelled silicate, followed by portland cement, and fly ash retained the greatest proportion of the heavy metals in the stringent batch leaching test; this consolidation treatment was much better than cement and fly ash alone. There is also a rough correlation between the batch and column leaching results when equivalent sludge treatments are compared.

Finally, our data suggest that at least three mechanisms could be involved in reducing the solubility of chromium and zinc ions in the waste. These are as follows:

(*a*) the formation of insoluble chromium and zinc (hydrous) oxide silicates,

(*b*) encapsulation of chromium and zinc that involves the formation of insoluble calcium and magnesium (hydrous) oxide silicates on the sludge particle surface, and (*c*) encapsulation of chromium and zinc that involves the formation of a gelled silicate sheath over the sludge particles, which is then rendered insoluble by the addition of calcium ions, for example, from cement.

It is impossible to distinguish which of these three mechanisms applies to a given waste treatment, since calcium and magnesium are present in the sludge (Table 1) and also in the cement. One cannot determine whether the reduced leachability is the result of encapsulation by a calcium-magnesium (hydrous) silicate species or by formation of chromium-zinc (hydrous) silicate species that are less soluble than the corresponding metal (hydrous) oxides. The interactions of silicates with zinc metal ions have been reported in the literature [*18,19*], and our own unpublished results based on X-ray identification suggest that similar reactions could occur with chromium.[3]

It is likely, however, that all three mechanisms are involved to some extent. But, when cement is added to the pregelled sludge, the third mechanism appears to be dominant. This conclusion is consistent with the results of Table 6, where it is shown that for a cement/silicate treatment, only 5% of the total zinc and only 2% of the total chromium present are leached. In contrast, when an equivalent amount of silicate was added (relative to the waste) but without cement, the total amount of metal ion leached ranged between 66 and 84% for zinc and 8 and 10% for chromium, depending on how long the silicate-gelled waste was aged prior to extraction with citric acid. These results suggest that either insufficient amounts of calcium and magnesium were present in the sludge or that the calcium and magnesium in the sludge were less active, and therefore encapsulation by Mechanisms 2 and 3 was less significant. Possibly, under such conditions, the first mechanism could predominate.

Reference

[*1*] Spencer, R. W., Reifsnyder, R. H., and Falcone, J. S., Jr., *Proceedings*, Third National Conference and Exhibition on Management of Uncontrolled Hazardous Waste Sites. Washington, D.C., 29 Nov.-1 Dec. 1982, pp. 237-243.

[*2*] Pojasek, R. B., *Chemical Engineering*, 13 August 1979, p. 141.

[*3*] Krumrine, P. H., Falcone, J. S., Jr., and Campbell, T. C., *Society of Petroleum Engineers Journal*, August 1982, p. 503; December 1982, p. 983

[*4*] Katsanis, E. P., Krumrine P. H., and Falcone J. S., Jr., "Chemical Reactions in an Alkaline Flood," paper presented at the National Meeting of the American Chemical Society, Kansas City, 12-17 September 1982.

[*5*] Perket, C. L. and Webster, V. C., "Literature Review of Batch Laboratory Leaching and Extraction Procedures," in *Hazardous Solid Waste Testing: First Conference, ASTM STP 760*, American Society for Testing and Materials, Philadelphia, 1981, pp. 7-27.

[*6*] Welsh, S. K., Gagnon, J. E., and Elnabarawy, M. T., "Comparison of Three Acid Extraction-Leaching Test Protocols," in *Hazardous Solid Waste Testing: First Conference, ASTM STP 760*, American Society for Testing and Materials, Philadelphia, 1981, pp. 28-39.

[*7*] Sack, W. A., Boomer, B. A., Tarantino, J. T., Keefer, G. B., Seasl, R. K., and Miller, M., "Evaluation of Fly Ash Leachability Using Batch Leaching Procedures," *Hazardous Solid Waste Testing: First Conference, ASTM STP 760*, American Society for Testing and Materials, Philadelphia, 1981, pp. 61-82.

[*8*] Jackson, K., Benedik, J., and Jackson, L., "Comparison of Three Solid Waste Batch Leach

Testing Methods and a Column Leach Test Method," *Hazardous Solid Waste Testing: First Conference, ASTM STP 760*, American Society of Testing and Materials, Philadelphia, 1981, pp. 83–98.
[9] "McIlvane's Buffer Table," in *Handbook of Chemistry and Physics*, 25th ed., C. D. Hodgman, Ed., Chemical Rubber Publishing Co., Cleveland, Ohio, 1941, p. 1324.
[10] Brace, R. and Matijevic, J., *Journal of Inorganic and Nuclear Chemistry*, Vol. 35, 1973, p. 3961.
[11] Katsanis, E. P. and Matijevic, E., *Colloids and Surfaces*, Vol. 5, 1982, p. 43.
[12] Stumm, W. and Morgan, J. J., *Aquatic Chemistry*, Wiley–Interscience, New York, 1970.
[13] Hayden, P. L. and Rubin, A. J. in *Aqueous Environmental Chemistry of Metals*, A. J. Rubin, Ed., Ann Arbor, Mich., 1974, p. 317.
[14] Goldshmid, T. and Rubin, A. J. in *Chemistry of Wastewater Technology*, A. J. Rubin, Ed., Ann Arbor, Mich., 1978, pp. 59–80.
[15] La Iglesia, A. and Van Oosterwyck-Gastuche, M. C., *Clays and Clay Minerals*, Vol. 26, 1978, p. 397.
[16] Robinson, A. K., "Sulfide-vs.-Hydroxide Precipitation on Heavy Metals from Industrial Wastewater," *First Annual Conference on Advanced Pollution Control for the Metal Finishing Industry*, EPA-600/8-78-010, American Electroplaters' Society and U.S. Environmental Protection Agency. 1978, pp. 59–65.
[17] Katsanis, E. P., Krumrine, P. H., and Falcone, J. S., Jr., "Chemistry of Precipitation and Scale Formation in Geological Systems," SPE Paper No. 11802, International Symposium on Oilfield and Geothermal Chemistry, Denver, Colo., 1–3 June 1983.
[18] Boberski, W. G., Seiner, J. A., and Blasko, J. E., *Industrial and Engineering Chemistry Product Research Development*, Vol. 21, 1982, p. 528.
[19] Sprycha, R., *Colloids and Surfaces*, Vol. 5, 1982, p. 147.

Waste as Resource

Ming-Hung Wong,[1] *Wing-Cheung Chan*,[1] *and Lee-Man Chu*[1]

Food Processing Wastes as Nutrient Sources in Algal Growth

REFERENCE: Wong, M.-H., Chan, W.-C., and Chu, L.-M., **"Food Processing Wastes as Nutrient Sources in Algal Growth,"** *Hazardous and Industrial Waste Management and Testing: Third Symposium, ASTM STP 851*, Larry P. Jackson, Alan R. Rohlik, and Richard A. Conway, Eds., American Society for Testing and Materials, Philadelphia, 1984, pp. 230–247.

ABSTRACT: Utilization of food processing wastes for biological production will ease part of the disposal problem, especially the potential hazards of eutrophication, and at the same time recycle the inherently rich plant nutrients in the waste materials.

The present investigation is an attempt to study the feasibility of using five food processing wastes, including carrot, coconut, eggshell, soybean, and sugarcane, for culturing *Chlorella pyrenoidosa* (a unicellular green alga).

When using test tubes as culture tubes, all the concentrations (0.5, 1, 3, and 5% weight/volume aqueous extracts) of eggshell waste had similar growth rates, whereas all the concentrations of sugarcane and carrot wastes had higher growth rates of algal cells compared with the control group using Bristol solution as the culture medium. However, only 0.5 and 1% soybean waste and 3 and 5% coconut waste had higher algal growth than the control.

When larger containers were used, similar results were obtained except for 5% soybean waste, which produced the highest growth rate. This was possibly due to the modified growing conditions. *Chlorella pyrenoidosa* harvested from soybean and sugarcane wastes had higher cell numbers, dry weight, and protein contents than those cultivated in the Bristol solution.

These waste materials contained high levels of essential plant nutrients including nitrogen, phosphorus, and potassium. However, the rather high levels of organic carbon and soluble carbohydrate in some of the wastes, such as coconut and carrot wastes, seemed to affect the contents of protein and chlorophyll of the waste-grown algae.

The fact that these waste materials contained no hazardous substances would render the algal products safe for subsequent usage (as animal feeds or even for human consumption), unlike the sewage-grown algae which accumulated a substantial level of trace metals.

It is suggested that the process of algal culture could be incorporated in large-scale food processing industries for the dual purpose of treating the wastes and mass producing edible protein.

KEY WORDS: food processing wastes, coconut, carrot, eggshell, soybean, sugarcane, *Chlorella pyrenoidosa*, plant nutrients, trace elements, protein, hazardous wastes, industrial wastes

[1] Senior lecturer, technician, and teaching assistant, respectively, Department of Biology, The Chinese University of Hong Kong, Shatin, Hong Kong.

Disposal of industrial wastes is a universal problem. This is further intensified in a densely populated area such as Hong Kong where the industrial establishments are interdispersed with residential areas. Food processing wastes often contain a high portion of organic materials which are unstable and readily oxidized, and will consequently deplete the dissolved oxygen in water by slow bacterial breakdown. Eutrophication will then be induced because of the large amount of available plant nutrients released upon decomposition of the wastes.

Recycling of these waste materials not only solves part of the pollution problem but also reutilizes the inherently rich nutrients for producing edible protein. The latter context is important because of the ever-increasing population, the decline in arable land, and the decreases in per capita food production throughout the world.

Attempts have been made to recycle food processing wastes as animal feed, in the irrigation of agricultural land, and for the production of single-cell proteins. In the production of protein with microorganisms, most of the studies have been concerned with the cultivation of yeast in various industrial wastes, such as whey and carrobean sugars [*1*], natural gas and mineral oil, methane, methanol, ethanol, paraffin, and gas oil [*1,2*]. Some studies have also looked into the feasibility of cultivating fungi in cellulose [*3*], for example, fibers from sprint mills [*4*] and corncob waste [*5*].

In mass cultivation of microorganisms for protein production, bacterial cells are not selected because of their high nucleic acid content and small size, which makes them difficult to harvest although photosynthetic bacteria have been cultivated in sewage, livestock manure, and other agricultural wastes [*6*].

Unicellular green algae such as *Chlorella* sp. have been suggested as a new source of protein because of their superior efficiency in converting nitrogen to protein [*7*] and the high contents of vitamins and fats [*8*] and all the essential amino acids similar to those in common use [*9*].

Mass cultivation of microalgae using inorganic media was regarded as a highly economical type of agriculture. A considerable number of experiments have been conducted on the production of microalgae in sewage effluent, both in the laboratory and outdoors, [*10–12*]. Recently, it has been demonstrated that aqueous extracts of sewage sludge also support excellent algal growth [*13–16*]. The growth rates and protein contents of *Chlorella pyrenoidosa* cultivated in 2% aqueous extracts of sewage sludge were similar to those cultivated in sewage effluent of the same treatment plant [*16*].

A significant correlation was found between the trace metal concentration in the algal cells and the amount of metal in their surrounding waters [*17–20*]. Algae accumulated higher contents of heavy metals when reared in sewage and animal manure than in artificial media [*21,22*]. Concentrated metal content in algae might restrict their usage as animal feeds. Therefore, propagation of algae in wastes containing no hazardous substances is desirable. Some pilot experiments have indicated that *Chlorella* grew well in blood waste [*23*], brewery and soybean wastes [*24,25*], and palm oil mill sludge [*26*].

The present investigation is aimed at studying the feasibility of culturing *C. pyrenoidosa* in different food processing by-products, namely, coconut waste (CN), carrot waste (CR), eggshell waste (ES), soybean waste (SB), and sugarcane waste (SC).

Materials and Methods

The waste materials were collected from food processing factories dealing with carrots (for producing carrot juice), coconuts (for coconut juice), eggs (for incubation for chicks), soybeans (for making beancurd), and sugarcane (for sugarcane juice). In order to compare the properties of different wastes and their effects on algal growth, all the wastes were air dried for two weeks and ground into powder form. All the analyses were based on dry weight.

The powder of different wastes was subjected to extraction with distilled water for 2 h by shaking using a rotary shaker at 100 rpm [*27*] at 25°C. The suspensions were then filtered with a Whatman filter paper No. 42, and the supernatants were autoclaved. The sterilized extracts were used as test media for the cultivation of *Chlorella pyrenoidosa* (*26*). The Bristol solution, which was recommended for growing freshwater Chlorophyceae [*28*], was prepared and subsequently used as a control medium for comparison.

A preliminary investigation was conducted in order to search for the appropriate range of concentrations of wastes suitable for growing *C. pyrenoidosa.* Pure culture (approximately 1×10^6 cells/mL *Chlorella* suspension) was inoculated into 20 mL of test medium using 24 by 150-mm test tubes as culture tubes after they had been autoclaved. The tubes were then placed into the growth chamber at 22 ± 4°C for 14 days with a 16-h light (3000 lx)/8-h dark cycle. The number of cells per millilitre was counted every other day by a Neubauer counter after the suspension had been thoroughly stirred. Each value was a mean of four replicates.

The properties of the waste extracts were analyzed for the following items: pH (pH meter), ammonia-nitrogen (ammonia electrode connected to a 701A pH Orion ionalyzer), nitrate-nitrogen (2,4-xylenol method) [*29*], total phosphorus [digested by concentrated sulfuric acid (H_2SO_4) and tested by molybdenum blue method] [*30*], water-soluble phosphorus (molybdenum blue method), and water-soluble contents of metals including potassium, magnesium, calcium, zinc, manganese, and iron (atomic absorption spectrophotometry) [*29*].

In order to test the algal products produced by various waste materials, a larger-scale cultivation was conducted (using 0.5 and 5% weight/volume of the five wastes). Extraction of the wastes was carried out using the same method described previously. After filtering the waste extracts with Whatman filter paper No. 42, some of the extracts (SB and CN) were subjected to another filtration process using a millipore filter paper with a 0.45-μm diameter pore size because of their cloudy appearance. The extracts were then autoclaved before adding (approximately 1×10^6 cells/mL) pure culture of *Chlorella* suspension to each flask. Thereafter, the flasks (500-mL conical flasks) were placed under the same conditions as the previous culture except they were shaken daily for an hour using a rotary shaker at 150 rpm. Cell counts were conducted at the end of 14 days.

The algal suspensions were then centrifuged (5000 rpm for 10 min). The residues were washed three times with distilled water before they were analyzed for the following items: oven-dried weight (obtained by drying at 105°C until a constant weight), contents of protein (Folin phenol reagent) [*31*], chlorophyll (acetone extraction method) [*32*], soluble carbohydrate (anthrone reagent) [*33*], and lipid (chloroform-methanol extraction) [*34*].

Results and Discussion

Growth Rates of C. Pyrenoidosa *Cultivated in Different Waste Extracts*

The growth rates of *C. pyrenoidosa* cultivated in different types of waste extracts are shown in Figs. 1 through 5. The results were compared with those growing in Bristol solution, which is commonly used for vigorously growing organisms of the *Chlorella* type [*28*].

Carrot Waste—In general, all the concentrations of CR (Fig. 1) yielded higher algal growth than Bristol solution. However, a higher concentration of CR resulted in poorer yield in terms of the number of cells. Decreased growth rate was observed in 3 and 5% extracts at the end of the culture period.

Coconut Waste—Higher concentrations of CN (Fig. 2) resulted in a higher number of cells. Bristol solution was slightly inferior to 5% extract but produced a better growth than the rest of the waste extracts. The cell number also declined at the end of the experiment in all the concentrations used.

Eggshell—Three and 5% extracts of ES (Fig. 3) produced better algal growth for the first half of the growing period, but higher growth rates were observed from the lower concentrations at the end of the experiment. On the 14th day, 1% extract gave the highest cell number, followed by 0.5%, Bristol solution, and 3 and 5% extracts.

Soybean Waste—All the concentrations of SB (Fig. 4) supported a higher growth rate than Bristol solution during the whole culture period. The number of cells was higher in the higher concentrations for the first four days, but the reverse was true for the rest of the culture period. In general, growth rates declined as the concentrations of the waste extracts increased.

Sugarcane Waste—All the concentrations of SC (Fig. 5) provided better algal growth than Bristol solution, except on Day 11. The 5% extract gave the best growth rate, followed by the 3, 0.5, and 1% extracts.

Higher concentrations of SC and CN supported better growth of *C.*

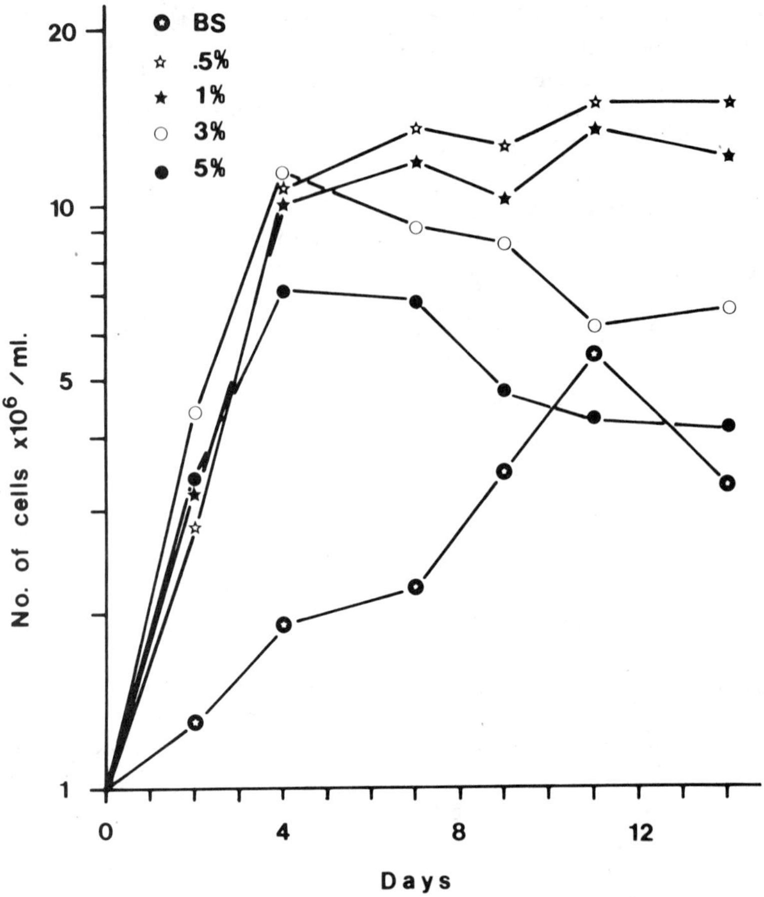

FIG. 1—*The growth rates of* C. pyrenoidosa *in different concentrations of carrot waste compared with those cultivated in Bristol solution (BS).*

pyrenoidosa. On the other hand, the 3 and 5% extracts of ES, SB, and CR gave lower yields in terms of cell count, although these concentrations initially had higher growth rates.

A comparison of the growth rates of *C. pyrenoidosa* in various waste extracts on the 14th day is shown in Fig. 6.

Properties of Different Waste Extracts

Table 1 shows the results of pH values, inorganic nitrogen [as ammonia (NH_3) and as nitrate (NO_3)], and phosphorus (water-soluble phosphorus and total phosphorus) in different extracts of waste materials.

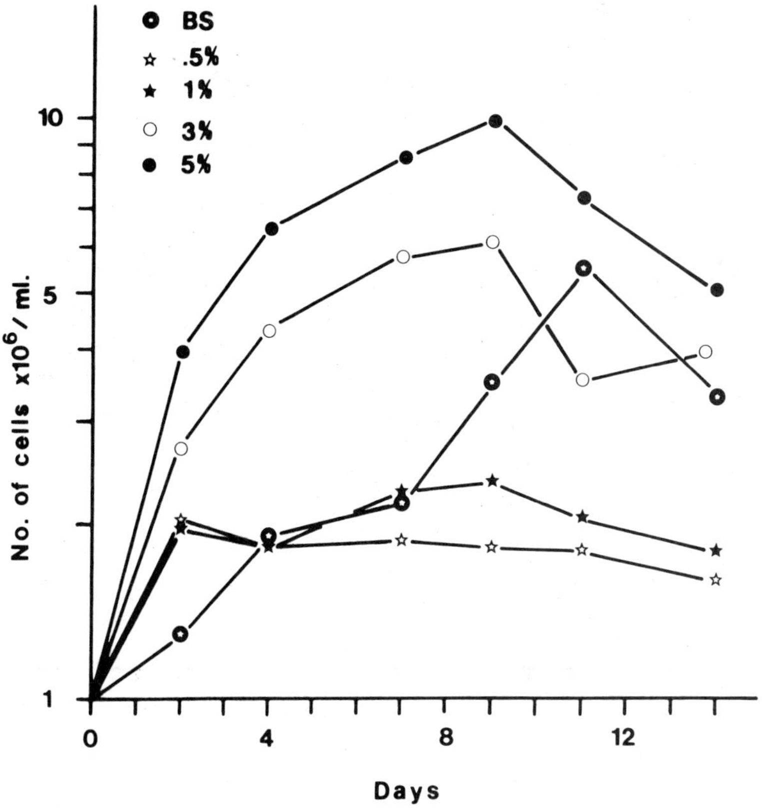

FIG. 2—*The growth rates of* C. pyrenoidosa *in different concentrations of coconut waste compared with those cultivated in Bristol solution.*

Most of the waste extracts were acidic (pH below 6) except the ES extracts, which were alkaline with pH ranging between 7.1 and 8.1. Bristol solution possessed a pH of 6.4. The vast majority of freshwater algae can grow over the pH range of 5.5 to 9, which is mostly found under natural conditions. Although there are species of *Chlorella* that grow at pHs below 4, even down to just below pH 2 [*35*], the pH of the medium for *Chlorella* should be between 5 and 7 [*36*]. When comparing the effect of three different pH values (3, 6.4, and 7) on the growth of *C. pyrenoidosa*, it was noted that Bristol solution adjusted to pH 7 had the highest growth rate, followed by pH 6.4, and the least growth was found in pH 3 [*25*]. Therefore, the lowest and highest pH values obtained in 5% CR and ES (4.2 and 8.1, respectively) might exert harmful effects on algal growth.

The SB waste extract had the highest levels of nitrogen (as NH_3) (6.9 to 44.6 mg/L), followed by CR (1.8 to 12.9 mg/L), ES (1.1 to 6.3 mg/L), SC (0.8 to 5.7 mg/L), and CN (0.35 to 0.64 mg/L). The levels of nitrogen as NH_3 was un-

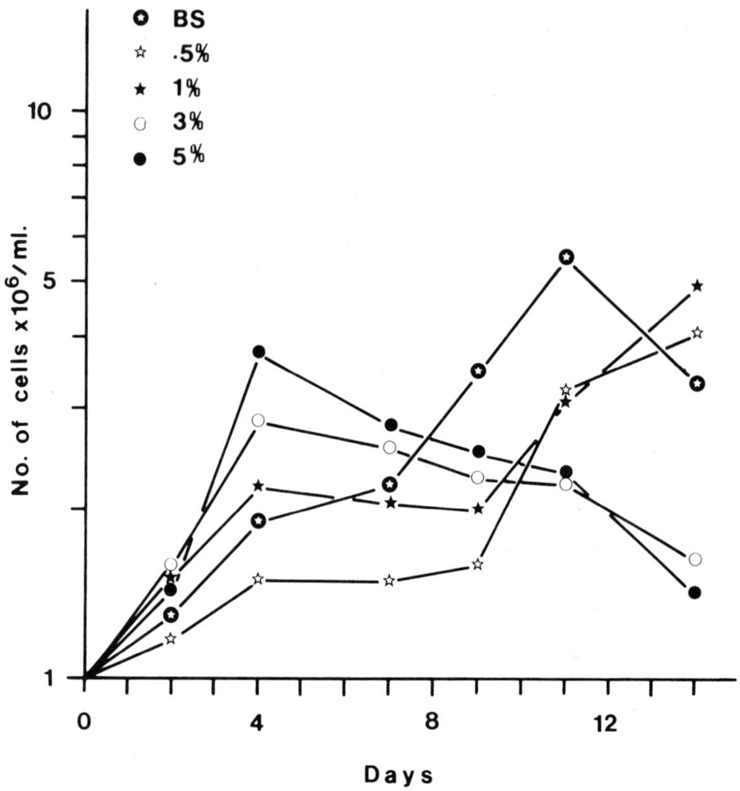

FIG. 3—*The growth rates of* C. pyrenoidosa *in different concentrations of eggshell waste compared with those cultivated in Bristol solution.*

detectable in Bristol solution. Only 1% of various extracts were chosen for the measurement of nitrogen as NO_3. All the waste extracts contained rather low levels (1.1 to 4.5 mg/L), whereas Bristol solution had a high value of 50.8 mg/L as it contained NO_3 as the source of inorganic nitrogen.

Chlorella is able to use many forms of nitrogen. In fact, most species of *Chlorella* are supported in growth by nitrogen as both NH_3 and NO_3 [*37*]. It has been discovered that nitrogen as NH_3 at a concentration greater than 31.5 mg/L inhibited the growth of *C. pyrenoidosa* [*38*], although it has also been noted that *C. vulgaris* grew quite well in wastes derived from a nitrogen fertilizer plant containing 600 mg nitrogen as NH_3 per litre [*39*]. If *C. pyrenoidosa* was more sensitive to nitrogen as NH_3, the rather high level of 44.6 mg/L found in the 5% SB might exert harmful effects.

Bristol solution contained the highest level of water-soluble phosphorus (53 mg/L), whereas CN and CR had moderate levels (2 to 33 mg/L and 2.8 to 32.6 mg/L, respectively). The rest contained rather low levels (trace amount to 4

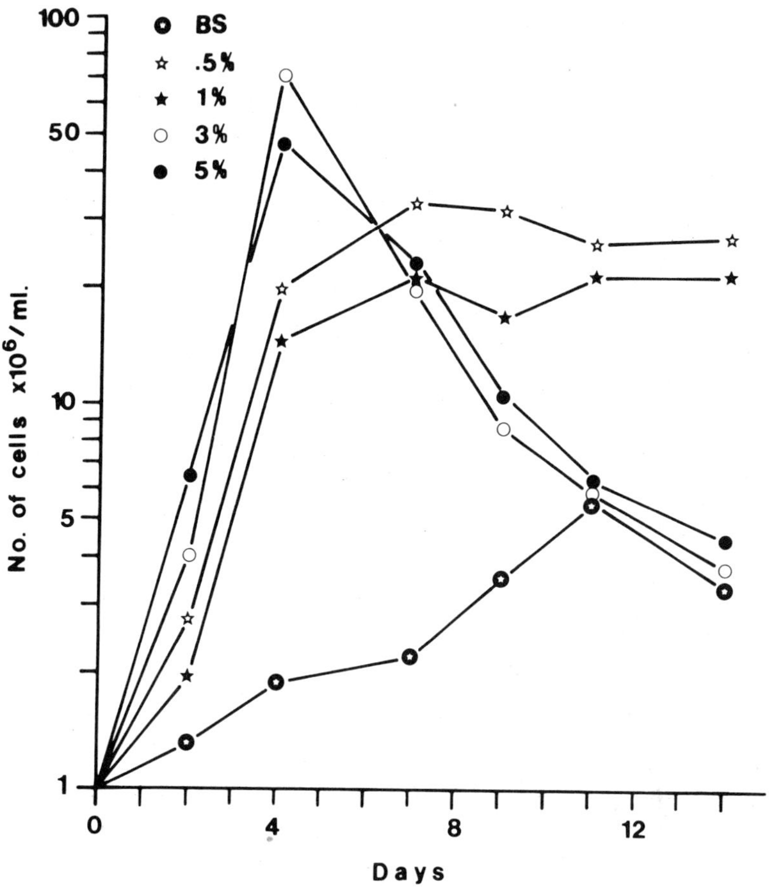

FIG. 4—*The growth rates of* C. pyrenoidosa *in different concentrations of soybean waste compared with those cultivated in Bristol solution.*

mg/L), especially SC. However, all the waste materials contained rather high levels of total phosphorus, ranging from 1350 to 1950 mg/kg, with the sole exception of ES which contained a low level of 470 mg/kg. It was assumed that all the phosphorus in Bristol solution was in inorganic form, and the level of total phosphorus was therefore not tested.

Algae are capable of absorbing organic as well as inorganic phosphorus compounds [*36*]. In fact, because it has been claimed that inorganic phosphorus is used so quickly and can be stored in excess of what is immediately used, the amount of total phosphorus may be a better index of water fertility [*35*]. Therefore, the waste materials used in the present study would provide an adequate phosphorus supply.

Table 2 lists the contents of organic carbon and soluble carbohydrate in the

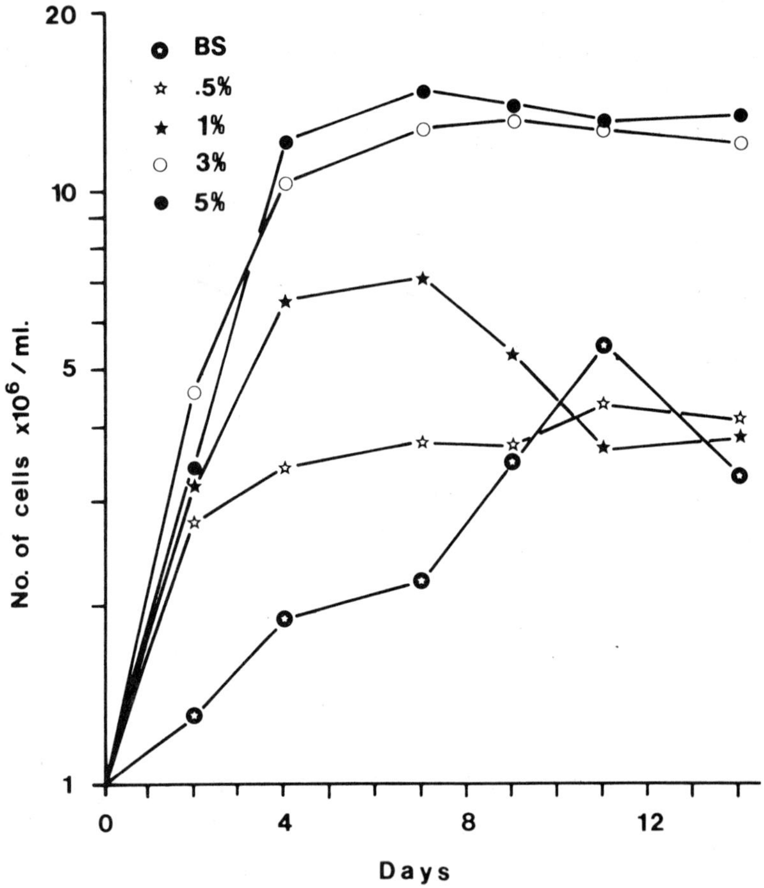

FIG. 5—*The growth rates of* C. pyrenoidosa *in different concentrations of sugarcane waste compared with those cultivated in Bristol solution.*

waste extracts. In general, CN had the highest levels of these two items (for example, the 5% extract had 12 495 mg/L organic carbon and 28 790 mg/L soluble carbohydrate), followed by SC, CR, SB, and ES. These two parameters had very close relationships according to the analysis of correlation coefficient ($r = 0.93$, with $P < 0.001$ for organic carbon and soluble carbohydrate). Very low contents of these items were noted in the Bristol solution, as expected.

Various organic compounds, including carbohydrate, can be used as carbon or energy sources for algae [*36*]. However, media containing sugars and other organic compounds may have become inactivated or even reacted to form toxic substances during autoclaving [*40*].

C. pyrenoidosa (strains 241, 252, and 395) will grow in the dark in a mineral

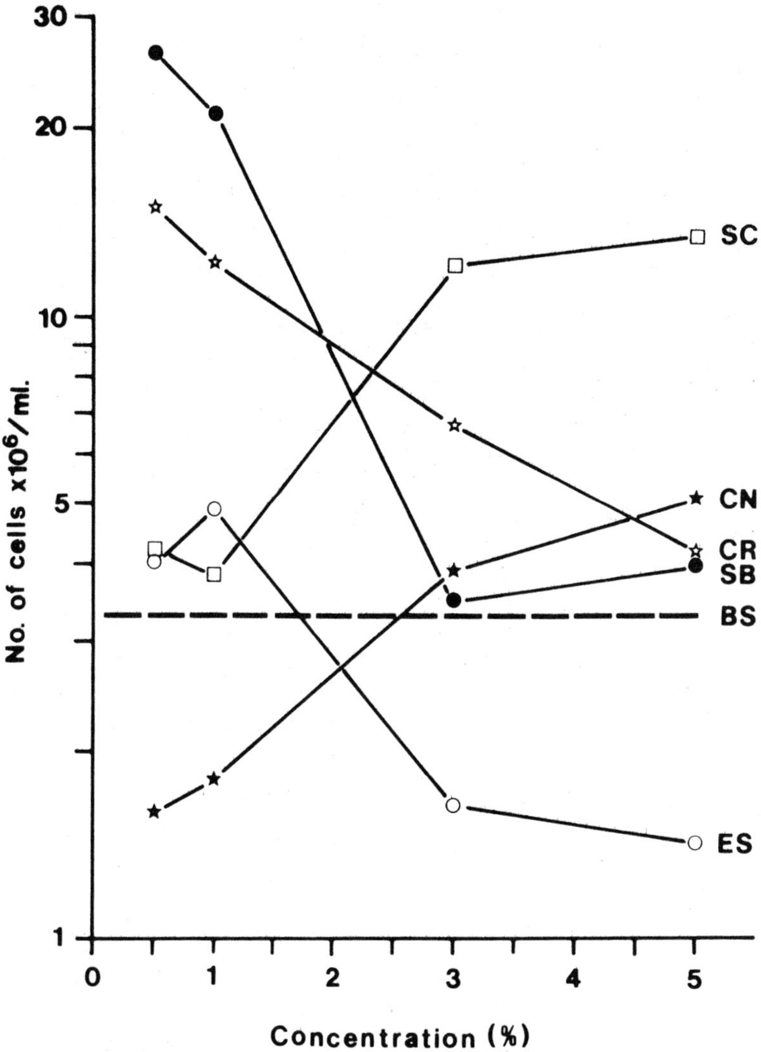

FIG. 6—*Comparison of the number of cells produced by different waste extracts on the 14th day (CN = coconut, CR = carrot, EG = eggshell, SC = sugarcane, SB = soybean, and BS = Bristol solution).*

medium supplemented with glucose or galactose (1% weight/volume) [*41*], whereas in the light, glucose inhibits carbon dioxide (CO_2) assimilation [*42*]. However, Shihira and Krauss [*37*] noted that *C. pyrenoidosa* (458) is not stimulated by glucose. In this study, the chlorophyll in algal cells cultivated in waste extracts with high soluble carbohydrate contents, for example, SC, CN, and CR, disappeared; the algae appeared yellowish even whitish, indicating

TABLE 1—*The contents of pH value, nitrogen (as NH_3 and as NO_3), and phosphorus (water-soluble and total phosphorus) in different wastes (mean of triplicates).*

Concentration, %(weight/volume)	pH				Nitrogen as NH_3, mg/L				Nitrogen as NO_3, mg/L				Water-Soluble Phosphorus, mg/L				Total Phosphorus, mg/kg
	0.5	1	3	5	0.5	1	3	5	0.5	1	3	5	0.5	1	3	5	
Carrot	4.5	4.4	4.4	4.2	1.8	2.8	7.5	12.9	...	1.1	...	...	0.5	0.8	2.5	4.0	1810
Coconut	5.7	5.2	4.8	4.7	0.4	0.4	0.4	0.6	...	2.7	...	...	2.8	5.4	17.5	32.6	470
Eggshell	7.1	7.4	7.8	8.1	1.1	1.7	4.4	6.3	...	2.3	...	...	0.3	0.5	1.6	2.7	1350
Soybean	5.6	5.3	5.2	5.1	6.9	9.8	27.5	44.6	...	1.4	...	...	2.0	4.5	13.0	20.5	1950
Sugarcane	5.0	4.8	4.6	4.5	0.8	1.6	4.1	5.7	...	4.5	...	...	t[a]	0.2	0.6	0.9	1690
Bristol solution	6.4				t				50.8				53.3				nt[b]

[a] t = trace amount.
[b] nt = not tested.

high levels of carbohydrate, which posed a serious problem for *C. pyrenoidosa* (*26*).

Table 3 shows the metal contents of the waste extracts. There was no general pattern related to metal contents in different waste materials. CR had the highest levels of potassium, iron, zinc, and lead, while SB had the highest levels of magnesium and copper. SC possessed moderate levels of all the metals, whereas CN and ES had lower levels, except for calcium in ES. Zinc appeared in trace amounts whereas lead was undetected in most extracts except CR.

It is commonly known that the chemical compositions of the algae are largely related to the composition of the culture media, especially heavy metal contents [*19*]. Magnesium at and below levels of 2.8 mg/L has been noted to be insufficient for prolonged growth of *Chlorella* cultivated on glucose in the light [*36*]. It has been found that the growth of *C. pyrenoidosa* was depressed with less than 7.8 mg/L potassium in the medium, resulting in the deficiency of potassium and phosphorus contents of the algal cells [*43*]. Calcium is not essential for the growth of *Chlorella*, and many cultures omit calcium [*44*]. Bollard and Bulter [*45*] and Walker [*46*] pointed out that calcium should be regarded as a micronutrient as the production of 1 g dry weight only required 10 μg of calcium. Myers [*47*] found that concentrations of iron ranging from 0.01 to 7.4 mg/L gave adequate growth of *Chlorella.* Most of the waste extracts had sufficient contents of magnesium, potassium, and iron but seemed to be deficient in the mineral requirement of 2.8 mg/L magnesium at the lower concentrations (0.5%).

Trace elements such as copper, manganese, and zinc are also essential for normal algal growth. However, only a small amount is needed. It has been noted that trace amounts up to 0.6 mg/L copper, 0.1 to 1.2 mg/L zinc, and 3 to 50 mg/L manganese inhibited the cell division rate, photosynthesis, carbon uptake, and respiration as well as causing chlorosis and even cell lysis of *Chlorella* cells [*25,48–52*].

In the present experiment, manganese, lead, and zinc were either undetected or existed in trace amounts in most of the waste extracts except the lead content in 5% CR (0.5 mg/L) and the zinc contents in 5% SB, CN, and CR (0.02, 0.13, and 0.49 mg/L, respectively). These seemed to be within the upper limits, which produced adverse effects on the growth of *Chlorella*. Perhaps ethylenediaminetetraacetic acid (EDTA) could be added in the food processing wastes in order to verify the effects of trace metals as the addition of 10^{-4} and 10^{-3} M of EDTA greatly increased the growth of *C. pyrenoidosa* in activated sludge (6.3% weight/volume), which contained 1.8 mg/L zinc, 0.13 mg/L copper, 0.9 mg/L manganese, and 1.59 mg/L iron [*53*].

Properties of Waste-Grown Algae

The results of the number of cells at 14 days obtained in the second experiment using larger containers were largely the same as those of the previous test

TABLE 2—*The contents of organic carbon and soluble carbohydrate in waste extracts (mean of triplicates).*

Concentration, %(weight/volume)	Organic Carbon, mg/L				Soluble Carbohydrate, mg/L			
	0.5	1	3	5	0.5	1	3	5
Carrot	490	1415	4050	7470	550	1360	3 775	6 690
Coconut	1610	2490	7465	12495	3380	8718	18 538	28 790
Eggshell	98	196	274	293	6	8	14	15
Soybean	150	440	1025	2010	36	65	165	390
Sugarcane	345	1220	3230	5425	1430	3105	7 888	11 648
Bristol solution	5				13			

TABLE 3—*The metal contents in different concentrations of waste extracts (mean of triplicates).*

Concentration, %(weight/volume)	K, mg/L		Ca, mg/L		Mg, mg/L		Fe, mg/L		Cu, mg/L		Zn, mg/L		Pb, mg/L	
	0.5	5.0	0.5	5.0	0.5	5.0	0.5	5.0	0.5	5.0	0.5	5.0	0.5	5.0
Carrot	17.49	20.35	3.59	37.81	0.16	5.38	1.16	33.50	nd[b]	0.14	t	0.49	nd	0.50
Coconut	6.24	18.19	0.70	5.04	t[a]	8.54	nd	2.01	nd	0.08	t	0.13	nd	nd
Eggshell	2.80	15.81	10.10	44.20	t	5.97	nd	2.00	nd	0.05	t	t	nd	nd
Soybean	17.49	19.55	1.30	13.59	1.89	16.14	0.30	2.00	0.07	0.23	t	0.02	nd	nd
Sugarcane	7.62	18.66	1.84	18.95	t	6.21	nd	2.05	nd	0.12	t	t	nd	nd
Bristol solution	276.80		3.60		5.57		0.11		0.03		0.04		nd	

[a] t = trace amount.
[b] nd = nondetectable.

using test tubes as culture tubes. However, 5% SB had the highest cell number, possibly because of the modified growing conditions—that is, millipore filtration of the extracts before autoclaving and daily mechanical shaking of the culture. The former might have excluded some growth inhibitors, and the latter provide aeration and dispersion, resulting in higher algal yield.

Table 4 illustrates the properties of algal products, including yield (in terms of dry weight) and the contents of protein, chlorophyll, carbohydrate, and lipid. The results of Student's *t*-test of each parameter between the control group (Bristol solution) and other treatment groups (waste extracts) were also listed. Most of the algal products harvested from the wastes had a significantly higher cell number, oven-dried weight, and protein contents; comparable carbohydrate and lipid contents; and lower chlorophyll content compared with the control group. SB waste extract, especially the 5% extract, seemed to be the most promising among the five waste materials; it produced algal products with a cell number 7.5, dry weight 8, protein 1.5, and chlorophyll 3 times greater than those cultivated in Bristol solution, although it only contained half the carbohydrate and similar lipid contents.

The analysis of correlation coefficient further pinpointed the effect of the various parameters of the waste materials on the algal quantity and quality. Nitrogen as NH_3 and magnesium in the waste extracts had significant negative relationships with protein and lipid contents of the harvested algal cells.

Pollution caused by food processing wastes, such as sugar factory waste, has been revealed frequently [*54*], and measures have been taken to safeguard against water pollution in the food processing industries. Direct protein recovery from these wastes would be an uneconomical and energy-consuming process, for example, heat precipitation or coagulation, which involves heating the wastes to 70 to 120°C although lower temperatures at 18 to 22°C have been attempted [*55*]. Food processing wastewater has been used for irrigation [*56*]. Higher hay yield and improvement of quality were noted when sugar beet wastes were used for irrigation [*57*]. Our study also indicated the alternative of using these wastes for algal culture.

However, the present experiment only demonstrated the possibility of using these wastes for propagating algae in a laboratory scale. Further research is desirable to pinpoint the exact concentrations of the waste materials and the most suitable growing conditions for the optimal algal yield and quality in an outdoor scale.

The choice of algal species is also important for producing algal products with high protein contents and other qualities. The protein content of different algal species have been compared and *C. pyrenoidosa* is one of the algae with a high protein content of 57% per dried cell [*58,59*]. When comparing this figure with the results obtained in the present study, it is apparent that the food processing wastes produced algae with a high protein content. Spoehr and Milner [*38*] and Gordon [*60*] also showed that the chemical composition of *C. pyrenoidosa* could be influenced by the culture conditions employed, and it

TABLE 4—*Yield and composition of algal products cultivated in different waste extracts (mean of four replicates).*

	Concentration, %(weight/volume)	Cell No. x 10^6 Cells/mL	Dry Weight, μg/mL	Protein %	Chlorophyll, %	Carbohydrate, %	Lipid, %
Carrot	0.5	8.42***[a]	68.40*	55.32*	0.44***	10.90	10.28*
	5.0	4.90	15.85***	41.32	t[b]	16.02	12.68
Coconut	0.5	2.22***	25.50*	31.33	t	11.54	5.23***
	5.0	4.90	28.70	24.55*	t	7.65*	4.74***
Eggshell	0.5	6.11***	59.95***	50.23	1.68	11.29	10.31
	5.0	2.03***	41.95*	65.82***	0.90	6.35***	19.52
Soybean	0.5	14.81***	115.45***	58.78*	3.24***	11.51	15.88
	5.0	32.66***	282.95***	57.12*	3.92***	7.52**	15.89
Sugarcane	0.5	6.84***	56.80**	53.97	0.33***	17.06	11.94
	5.0	15.55***	89.05***	66.87*	0.66*	18.10	9.25*
Bristol solution		4.34	34.25	40.79	1.21	14.69	16.14

[a] Student's *t*-test comparing the control group (Bristol solution) with different treatment groups: $*P < 0.05$, $**P < 0.01$, $***P < 0.005$.
[b] t = trace amount.

is possible to control these conditions so that a protein content of about 50% can be maintained.

Chlorella is also very tolerant of variation in nutrient levels, the major salts, for example, nitrogen and phosphorus, could be varied by dilution of the entire medium 20-fold without affecting growth or photosynthetic efficiency [*47*]. Its different modes of nutrition also enables *Chlorella* to thrive in different materials that are organic in nature including sewage effluent, aqueous extracts of sewage sludge, animal manure, and other food processing wastes. Finding existing algal species which already flourish in the wastes might be an answer for choosing the right species to adapt such waste materials as well as local growth conditions.

However, the rather high levels of organic carbon and soluble carbohydrate found in some of the wastes, which adversely affected algal growth, should be amended. Perhaps pretreatment of the wastes using percolating filtration or aeration treatment would lower the contents of organic carbon and associated substances. Recycling of nutrients with synthesis of single-cell protein and the elimination of noxious by-products could be achieved simultaneously by constructing algal ponds next to food processing factories. Further studies should be conducted on the untreated wastes to simulate natural conditions. It is also possible that the drying, autoclaving, and sterilizing of the wastes would denature and destroy much of the available nutrients needed for good algal growth.

It is commonly known that wastes such as sewage effluent contain a large number of complex and potentially useful organic molecules and can be used as the substrate for the growth of single-celled algae [*10,12,61,62*]. Sewage oxidation ponds are also known to be feasible for the mass production of single-cell protein because they have a plentiful supply of nitrogen, and considerable carbon dioxide is produced by bacterial decomposition of the sewage [*63*]. Oxidation ponds can also be constructed for the dual purpose of single-cell protein production and purification of the wastes. Furthermore, the public has reacted swiftly against waste-grown foods with potential objectionable inputs, especially those cultivated in domestic sewage and animal manure [*64–67*]. It is envisioned that products derived from food processing wastes would face less resistance.

References

[*1*] Senez, J. C. in *Single-Cell Proteins*, Symposium d'Aix-en-Provence, 15 February 1972.

[*2*] Shacklady, C. A. and Gatumel, E. in *Livre jubilaire en l'honneur du Professeur Dr. H. C. Albert de Vuyst*, Louvain, Belgium, 1972, pp. 203–212.

[*3*] Dunlap, C. E., *Food Technology*, Vol. 29, 1975, pp. 62–67.

[*4*] Eriksson, K. E. and Larsson, K., *Biotechnology and Bioengineering*, Vol. 17, 1975, pp. 327–348.

[*5*] Rajagopal, M. V., *Journal of Food Technology*, Vol. 12, 1977, pp. 633–637.

[*6*] Smith, J. E., *Biotechnology*, Edward Arnold, London, 1981.

[*7*] Wassink, E. C., Kok, B., and van Dorschot, J. C. P. in *Algal Culture, from Laboratory to Pilot Plant*, J. S. Burlew, Ed., Publication No. 600, Carnegie Institute of Washington, Wash., 1964, pp. 55–62.

[8] Provasoli, L. in *Algal Physiology and Biochemistery*, W. D. P. Stewart, Ed., Blackwell, London, 1974, pp. 741-787.
[9] Milner, H. W., *Scientific American*, Vol. 189, 1953, pp. 31-35.
[10] Dunstan, W. M. and Menzel, D. W., *Limnology and Oceanography*, Vol. 16, 1971, pp. 623-632.
[11] Goldman, J. C., Ryther, R. H., and Williams, L. D., *Nature* (London), Vol. 254, 1975, pp. 594-595.
[12] Ludwig, H. F. and Oswald, W. J., *Scientific Monthly*, Vol. 74, 1952, pp. 3-6.
[13] Wong, M. H., *Environmental Pollution*, Vol. 14, 1977, pp. 207-211.
[14] Wong, M. H., Yip, S. W., and Fan, K. Y., *Environmental Pollution*, Vol. 12, 1977, pp. 205-207.
[15] Tam, F. Y. and Wong, M. H., *Archiv. für Hydrobiologie*, Vol. 96, 1983, pp. 475-485.
[16] Yip, S. W. and Wong, M. H., *Archiv. für Hydrobiologie*, Vol. 84, 1978, pp. 368-380.
[17] Bradfield, R. E. N., Kingsburg, R. W. S. M., and Rees, C. P., *Marine Pollution Bulletin*, Vol. 7, 1976, pp. 187-193.
[18] Butterworth, J., Lester, P., and Nicklers, G., *Marine Pollution Bulletin*, Vol. 3, 1972, pp. 72-74.
[19] Seelinger, U. and Edwards, P., *Marine Pollution Bulletin*, Vol. 8, 1977, pp. 16-18.
[20] Trollope, D. R. and Evans, B., *Environmental Pollution*, Vol. 11, 1976, pp. 109-116.
[21] Cheung, Y. H. and Wong, M. H., *Agricultural Wastes*, Vol. 3, 1981, pp. 109-122.
[22] Wong, M. H., *Environmental Pollution*, Vol. A21, 1980, pp. 121-130.
[23] Wong, M. H., *Conservation and Recycling*, Vol. 4, 1981, pp. 9-14.
[24] Wong, M. H., *Journal of Environmental Science and Health*, Vol. A12, 1977, pp. 567-574.
[25] Wong, M. H. and Lay, C. C., *Environmental Pollution*, Vol. A23, 1980, pp. 247-259.
[26] Sivalingham, P. M., *Progress in Water Technology*, Vol. 11, 1979, pp. 69-87.
[27] Wong, M. H., *Archiv. für Hydrobiologie*, Vol. 88, 1980, pp. 426-438.
[28] Starr, R. C., *American Journal of Botany*, Vol. 47, 1960, pp. 67-86.
[29] Allen, S. E., Grimshaw, H. M., Parkinson, J. A., and Quarmby, C., *Chemical Analysis of Ecological Materials*, Blackwell, London, 1974.
[30] Watanabe, F. S. and Olsen, S. R., *Soil Science*, Vol. 93, 1962, pp. 183-188.
[31] Lowry, D. H., Resebrough, N. J., Farr, A. C., and Randell, R. J., *Journal of Biological Chemistry*, Vol. 193, 1951, pp. 265-275.
[32] Arnon, D. I., *Plant Physiology*, Vol. 24, 1949, pp. 1-15.
[33] Dreywood, R., *Industrial Engineering Chemistry Analytical Edition*, Vol. 18, 1946, p. 499.
[34] Bligh, E. G. and Dyer, W. J., *Canadian Journal of Biochemistry Physiology*, Vol. 37, 1959, p. 911.
[35] Lund, J. W. G., *Biological Reviews*, Vol. 40, 1965, pp. 231-293.
[36] Krauss, R. W., *Annual Review of Plant Physiology*, Vol. 9, 1958, pp. 207-245.
[37] Shihira, I. and Krauss, R. W., *Chlorella, Physiology and Taxonomy of Forty Isolates*, Port City Press, Baltimore, 1965.
[38] Spoehr, H. A. and Milner, H. W., *Plant Physiology*, Vol. 24, 1949, pp. 120-150.
[39] Matusiak, K., *Acta Microbiologica Polonica*, Vol. 8, 1976, pp. 233-242.
[40] Patton, A. R. and Hill, E. G., *Science*, Vol. 107, 1948, pp. 68-69.
[41] Samejima, H. and Myers, J., *Journal of General Microbiology*, Vol. 18, 1958, pp. 107-117.
[42] Myers, J., *Plant Physiology*, Vol. 32, Supplement xvi, 1957.
[43] Scott, G. T., *Journal of Cellular and Comparative Physiology*, Vol. 21, 1943, pp. 327-338.
[44] Ketchum, B. H., *Annual Review of Plant Physiology*, Vol. 5, 1954, pp. 55-75.
[45] Bollard, E. G. and Bulter, G. W., *Annual Review of Plant Physiology*, Vol. 17, 1966, pp. 77-112.
[46] Walker, J. B., *Archives of Biochemistry and Biophysics*, Vol. 46, 1953, pp. 1-11.
[47] Myers, J., *Annual Review of Microbiology*, Vol. 5, 1951, pp. 157-180.
[48] DeFillips, L. F. and Pallaghy, C. K., *Zeitschrift für Pflanzenphysiologie*, Vol. 78, 1976, pp. 197-207.
[49] DeFillips, L. F. and Pallaghy, C. K., *Zeitschrift für Pflanzenphysiologie*, Vol. 78, 1976, pp. 314-322.
[50] Hassall, K. A., *Physiologia Plantarum*, Vol. 16, 1963, pp. 323-333.
[51] Kanazawa, T. and Kanazawa, K., *Plant Cell Physiology*, Vol. 10, 1969, pp. 495-502.
[52] McBrien, D. C. H. and Hassall, K. A., *Physiologia Plantarum*, Vol. 20, 1976, pp. 113-117.

[53] Wong, M. H., Chu, L. M., and Chan, W. C., *Environmental Pollution, Series A*, Vol. 34, pp. 55-71.
[54] Mariani, E., *Sugar Industry Abstracts*, Vol. 34, 1972, p. 186.
[55] Knorr, D., *Compost Science/Land Utilization*, Vol. 20, 1979, pp. 22-23.
[56] Pearson, G. A., *Journal of Environmental Quality*, Vol. 1, 1972, pp. 394-397.
[57] Brandyk, T., *Sugar Industry Abstracts*, Vol. 34, 1970, p. 140.
[58] Fisher, A. W., Jr., and Burlew, J. S. in *Algal Culture, from Laboratory to Pilot Plant*, J. S. Burlew, Ed., Publication No. 600, Carnegie Institute of Washington, Wash., 1964, pp. 303-310.
[59] Parsons, T. R., Stephens, K., and Strickland, J. D. H., *Journal of the Fisheries Research Board of Canada*, Vol. 18, 1961, pp. 1001-1016.
[60] Gordon, J. F. in *Algal Protein as Human Food*, R. A. Lawrie, Ed., Butterworth, London, 1970.
[61] Coldrick, J., *New Scientist*, Vol. 68, 1975, pp. 276-278.
[62] Ludwig, W. J., Gotass, H. B., and Lynch, V., *Sewage and Industrial Wastes*, Vol. 23, 1951, pp. 1337-1355.
[63] Burlew, J. S. in *Algal Culture, from Laboratory to Pilot Plant*, J. S. Burlew, Ed., Publication No. 600, Carnegie Institute of Washington, Wash., 1964, pp. 3-23.
[64] Barlow, E. W. R., Boersma, L., Phinney, H. K., and Miner, J. R., *Agriculture and the Environment*, Vol. 2, 1975, pp. 339-355.
[65] Bremer, H. J., *Archiv. für Hydrobiologie*, Beiheft, Ergebnisse der Limnolgie, Vol. 111, 1978. pp. 218-222.
[66] Dodd, J. C., *Agricultural Wastes*, Vol. 1, 1979, pp. 23-37.
[67] Wilson, M. and Houghton, J. A., Jr., *Journal of Agricultural Research*, Vol. 13, 1974, pp. 49-60.

Paolo Sanna[1] *and Marcello Camilli*[1]

Energy Recovery from Agroindustrial Wastes with Prevalently Solute Pollutants Using Fixed-Bed Anaerobic Reactors

REFERENCE: Sanna, P. and Camilli, M., **"Energy Recovery from Agroindustrial Wastes with Prevalently Solute Pollutants Using Fixed-Bed Anaerobic Reactors,"** *Hazardous and Industrial Waste Management and Testing: Third Symposium, ASTM STP 851*, Larry P. Jackson, Alan R. Rohlik, and Richard A. Conway, Eds., American Society for Testing and Materials, Philadelphia, 1984, pp. 248–259.

ABSTRACT: A process for anaerobic digestion of agroindustrial wastes containing mainly soluble pollutants has been developed utilizing fixed-bed reactors.

A pilot plant at a sugar refinery with two reactors in parallel, respectively of 10 and 50 m^3 operating volume, has functioned successfully in treating wastes produced in regenerating ion-exchange columns used to purify the sugar juices. With an hydraulic retention time (HRT) of less than one day, a maximum gas production of 11.5 volumes per operating volume of reactor per day was obtained together with a reduction of up to 80% of the biological oxygen demand for five days (BOD_5), and volatile solids and up to 70% of the chemical oxygen demand (COD).

The gas contained from 70 to 75% methane.

On the basis of these results, a full-scale industrial plant was built which has produced up to 7500 m^3 a day of biogas. The process has been found suitable for a large number of industrial wastes, especially when plants are seasonally operated.

KEY WORDS: digestion, anaerobic digestion, anaerobic fermentation, biogas, methane, anaerobic filter, fixed-bed anaerobic reactors, digester, wastewater treatment, agroindustrial wastes, hazardous wastes, industrial wastes

Effluents coming from agricultural food processing and canning industries often contain rather large amounts of pollutants, generally of the soluble type [*1,2*]. The elimination of such pollutants using conventional aerobic systems calls for construction of large, costly plants whose operation weighs heavily

[1]Biomass project leader and researcher, respectively. ASSORENI, ENI Group Research Association, Laboratori Microbiologici, 00015 Monterotondo, Rome, Italy.

upon the company's economics. The choice of anaerobic digestion as a pretreatment system for these kinds of wastes allows recovery of a large part of the energy potentially remaining in them. Moreover, this leaves to the aerobic section only the function of effluent finishing to obtain clarified water at a lesser cost and with the least possible consumption of energy. Last, in certain cases anaerobic digestion may represent the best system for wastewater treatment where the effluent is particularly refractory to normal aerobic-system treatment alone.

Conventional Anaerobic Reactors

Heretofore anaerobic digestion has been applied almost exclusively for treating primary and secondary sludges coming from aerobic clarification processes and, recently, for animal sewage (mainly pig and cattle). Because of the high content of suspended solids (SS) in such wastes and the significantly long time periods required for the digestion to produce methane, with retention times of generally 10 to 30 days, "batch" reactors were employed initially. Then it became popular to use single and two-stage continuous stirred reactors (CSTR) to reduce retention time [*3*].

The choice between single-stage and two-stage systems is generally based upon the need to guarantee a solids retention time (SRT), inside the reactor, greater than the hydraulic retention time (HRT), determined by the ratio of useful reactor volume to the rate of flow into the reactor. As a matter of fact, a high SRT value, causing an increment inside the system of the retention time of both suspended solids and anaerobic bacteria, contributes to the degree of organic substance conversion into biogas.

Moreover, a low HRT value means that the size of the reactor can be reduced, with the relative advantages this produces. Unfortunately, the difficulty in the settling of the anaerobic bacteria does not allow any decisive reduction in the HRT values of the two-stage system compared with the one-stage and, also, because of lesser system complexity, it is often preferrable, in terms of overall energy and economic balances, to use the single-stage system.

Fixed-Bed Anaerobic Reactors

Agroindustrial wastes, generally marked by high flow rates and soluble pollutants, call for the use of low-HRT and highly efficient reactors to ensure economy in their treatment. Having rejected using a two-stage CSTR reactor because they result in high HRT values, investigations turned to fixed-bed reactors. These are reactor vessels filled with inert material which has the purpose of entrapping the bacteria inside the system, through both adhesion of such microorganisms to the surfaces of the filler and simple mechanical effects, allowing the waste under treatment to flow freely through the reactor [*4*].

In this way any washout of the microbial flora is obstructed, and it becomes possible to feed the reactor at flow rates corresponding to HRT values which are lower than times of anaerobic microorganism duplications.

The HRT values in such systems depend upon the degradability characteristics of the waste, the type of filler material used, and, finally, the intended aim, usually purification or energy production.

Selection of Waste

Among sugar refinery wastes, the eluates from regenerating ion-exchange columns, utilized for purification of the sugar juices, appear especially interesting. In fact, they are difficult to clarify as they constitute about half of the effluent pollutant load from sugar refineries that use this processing technique.

The present experiment was carried out on wastes from a sugar beet processing factory at Argelato, near Bologna, Italy. The main characteristics of these wastes and their variables are shown in Table 1.

Based upon preliminary laboratory results, a pilot plant was built at the sugar factory mentioned.

The Pilot Plant

This pilot plant was built early in 1981, in accordance with the flow-sheet design shown in Fig. 1.

Its main features are two separate fixed-bed anaerobic reactors installed on parallel and independent process lines. The reactors differ from each other in form and size as well as in the materials of which they are built. The first, indicated as R-1, is cylindrical with a conical roof, is made of carbon steel, and has an operating volume of 10 m^3.

TABLE 1—*Sugar beet refinery effluents composition (ion exchange resin eluates).*[a]

Constituent[b]	Composition, kg/m^3
COD	9 to 40
BOD_5	5 to 25
N_{total}	0.4 to 1.8
NH_4^+	0.01 to 0.04
TS	12 to 60
VS	6.5 to 32

[a]Key to abbreviations:
COD = chemical oxygen demand.
BOD_5 = biological oxygen demand, five days.
N_{total} = total nitrogen.
NH_4^+ = ammonia nitrogen.
TS = total solids.
VS = volatile solids.

[b]At a pH of 6 to 12.

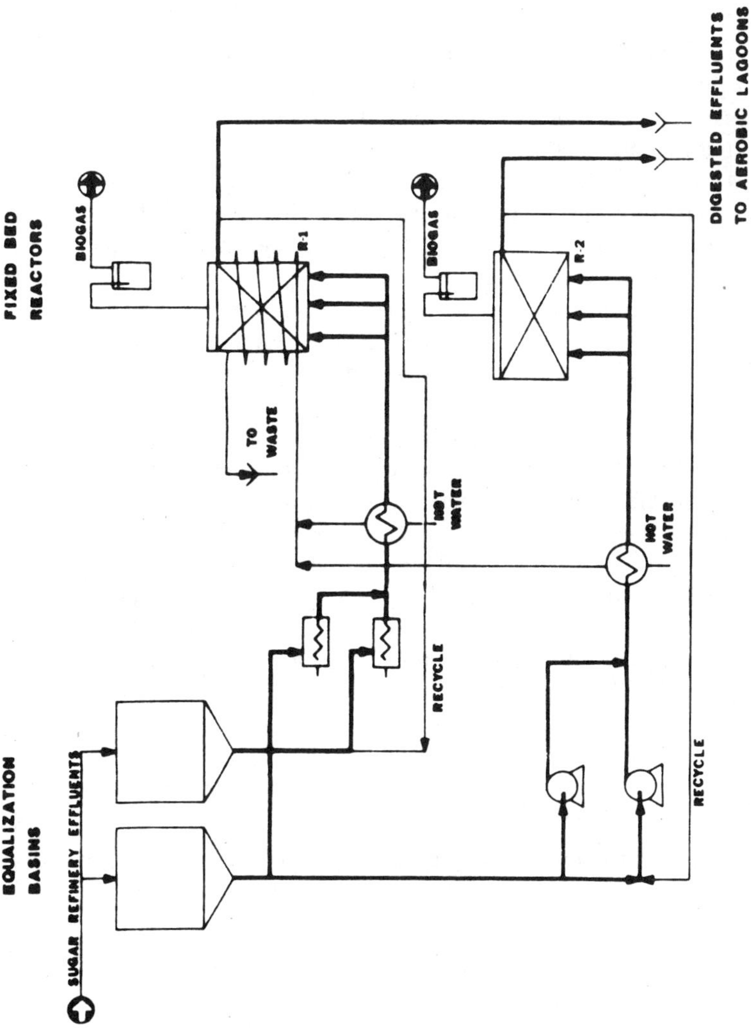

FIG. 1—*Flow sheet of the pilot plant.*

The second reactor, indicated as R-2, is rectangular with a flat roof. The sides of the reactor are made of prefabricated self-supporting cement and the roof is of carbon steel. The operating volume is 50 m^3.

The term *operating volume* is intended to mean the volume effectively occupied by the fermenting liquid in the reactor and is calculated by subtracting from the total geometric volume the space at the top occupied by the gas as well as the effective space taken up by the filler material.

Both reactors are insulated with sprayed-on polyurethane, the surface of which is protected by a reflective paint to prevent aging caused by ultraviolet rays. Because of its small size, Reactor R-1 was also provided with an outside thermal guard consisting of a serpentine coil of tubing, firmly attached to the wall, through which hot water flowed.

No protection was provided on the inside of the walls so that the corrosive effects of the anaerobic ambient on the construction materials employed for both reactors could be observed.

The waste coming from the sugar plant was first introduced into two equalization tanks for the purpose of attenuating the periodic variations in feed quality due to the cyclic nature of ion-exchange column regeneration at the plant.

The equalized waste was then fed to the two reactors through independent lines. Preheating of the feed to a temperature around 35°C was obtained by means of tube-sheet heat exchangers, through which hot water taken from the wastewater discharge line of the factory was circulated.

The digested waste issuing from the reactors was sent to the aerobic ponds of the factory for further purification until it was suitable for discharge into public waters.

The biogas produced by the anaerobic digestion process was measured by two separate gas meters and then discharged into the air through one of the stacks of the factory.

Data Acquisition Procedure

The pilot plant's main operating parameters, considered necessary both for controlling system functions and for data acquisition useful in defining the process are given in Table 2, together with the frequency of making surveys.

Waste analyses were made in accordance with American Public Health Association standard methods [5]. The percentages of carbon dioxide (CO_2) in the gas were measured in the field using a gas Fyrite instrument.

For measuring hydrogen sulfide (H_2S) a Draeger ampule was used.

The lack of suitable instrumentation at the sugar refinery prevented taking measurements of methane and other gases present in trace amounts, using the gas chromatography method normally employed at ENI laboratories. Therefore, the methane content in the gas was simply approximated as complemen-

TABLE 2—*Main controlled parameters.*

Frequency		
Daily	Weekly	Other
COD in	BOD_5 in	temperature in[a]
COD reactor	BOD_5 out	temperature reactors (3 levels)[a]
COD out	total solids in	feed flow rate[a]
pH in	total solids out	H_2S concentration[b]
pH reactor	volatile solids in	volatile acids[c]
pH out	volatile solids out	
rH in	phosphorus in	
rH out	phosphorus out	
Alkalinity in	N_{total} in	
Alkalinity out	N_{total} out	
Gas production	NH_4 in	
CO_2 concentration	NH_4 out	

[a]Continuously recorded.
[b]Two times a week.
[c]During start-up period and occasionally.

tary to the CO_2 content obtained using the gas Fyrite, except during the reactors' start-up period.

The gas was measured with normal household gas meters (dry type), which were found sufficiently accurate. Occasionally, it became necessary, however, to empty the meters of water accumulated because of the amount of moisture in the gas.

Results

The pilot plant was operated from September to December 1981, during the period when sugar manufacturing was in progress. Figure 2 shows the chemical oxygen demand (COD) values at the feed input and the waste discharge output of the R-1 reactor during the entire experimental period.

Figure 3 shows the same values for Reactor R-2.

Figures 4 and 5 show the daily production of biogas and its methane content, respectively, for Reactors R-1 and R-2. Biogas production, in both cases takes about one month to reach maximum levels, because this much time is needed for the anaerobic colonization of the fixed bed.

Generally the maximum efficiency values were shown by Reactor R-1, but probably this was because Reactor R-2 was operated for a shorter period of time because of some delays in its construction.

With regard to parameters not reported in the figures, it should be pointed out that the abatement of biological oxygen demand for five days (BOD_5) and volatile solids reached values of as much as 80%, when COD abatement was of about 70%.

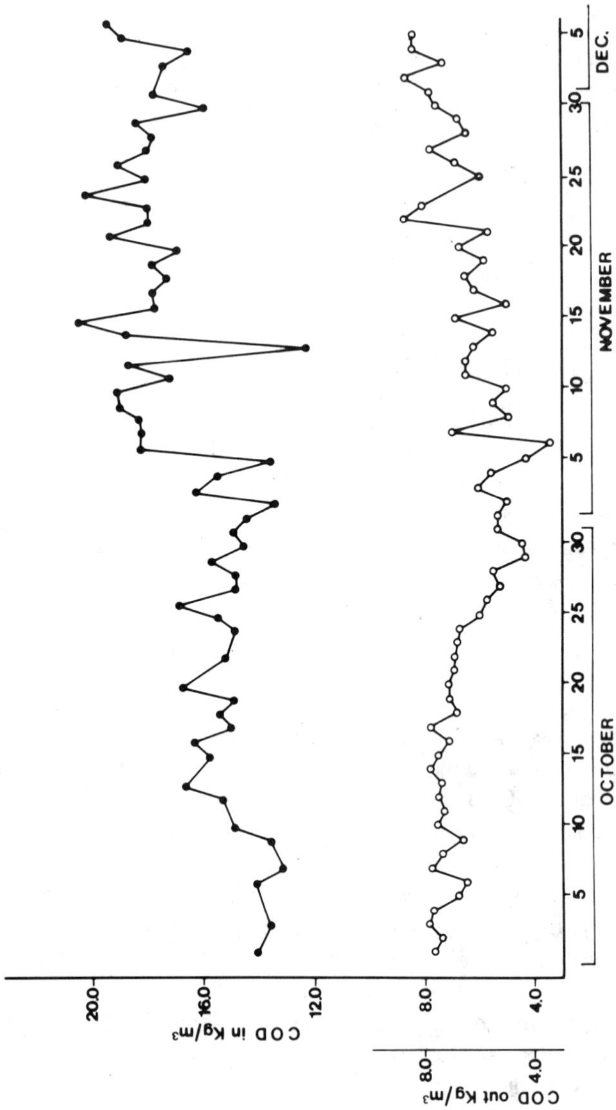

FIG. 2—*COD values at the feed input and waste-discharge output of the R-1 reactor.*

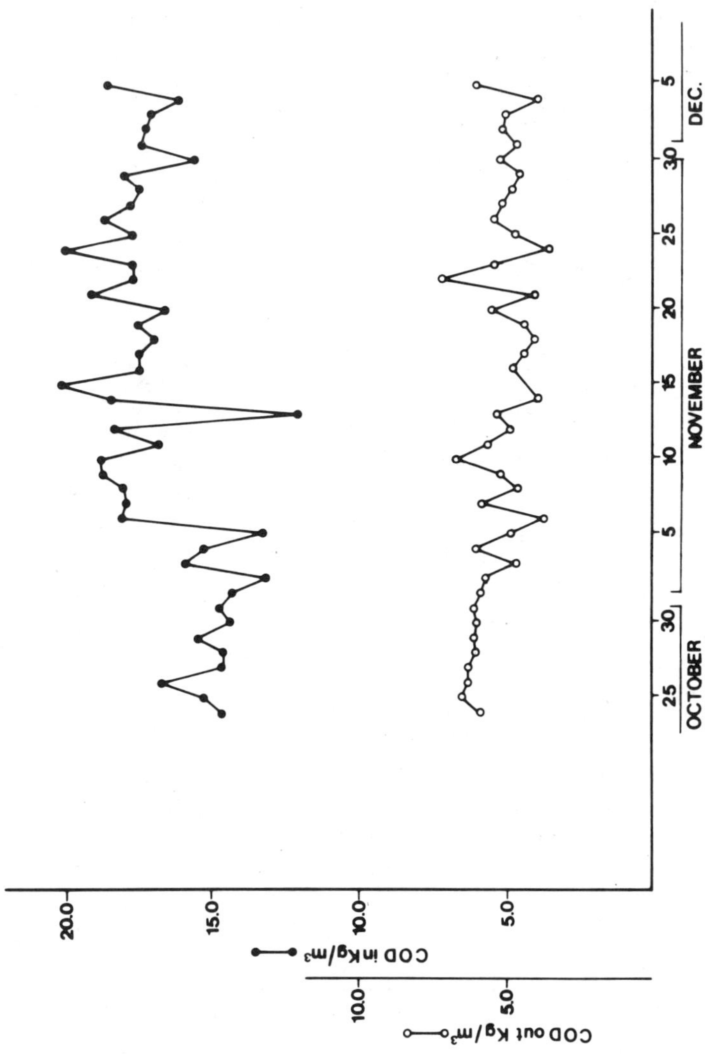

FIG. 3—*COD values at the feed input and waste-discharge output of the R-2 reactor.*

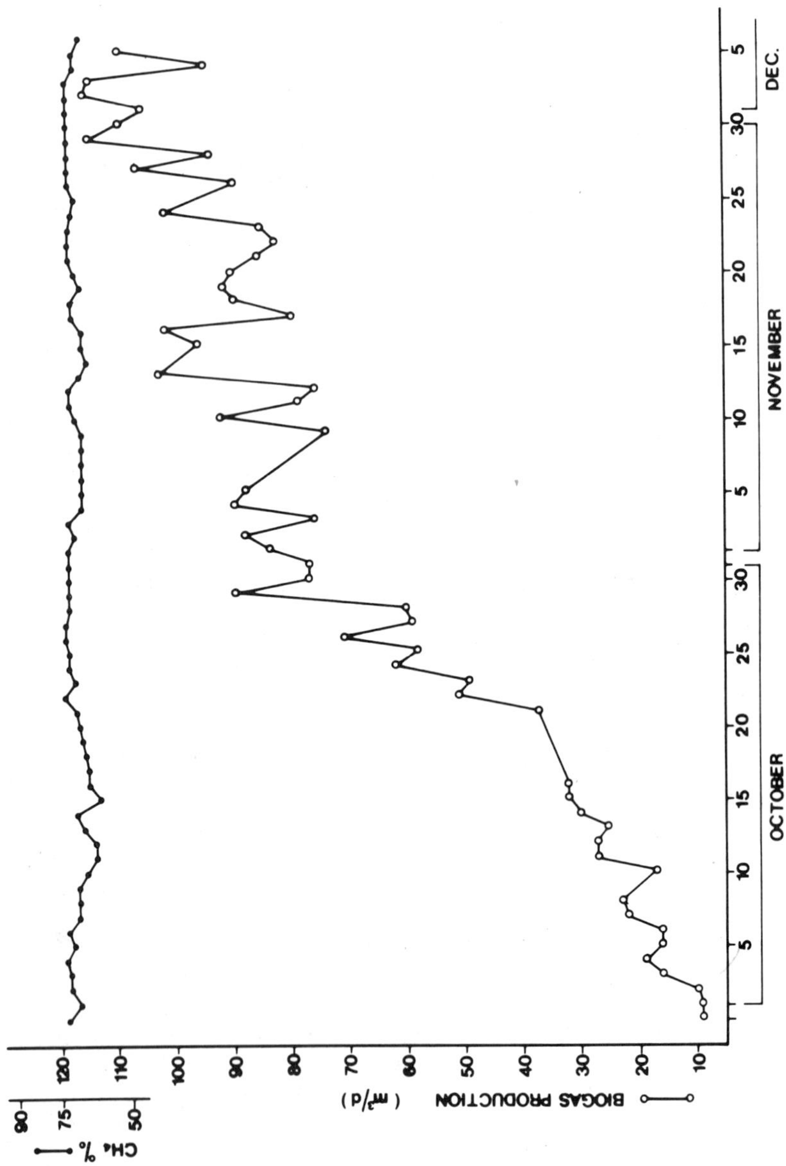

FIG. 4—*Production of biogas and its methane content for the R-1 reactor.*

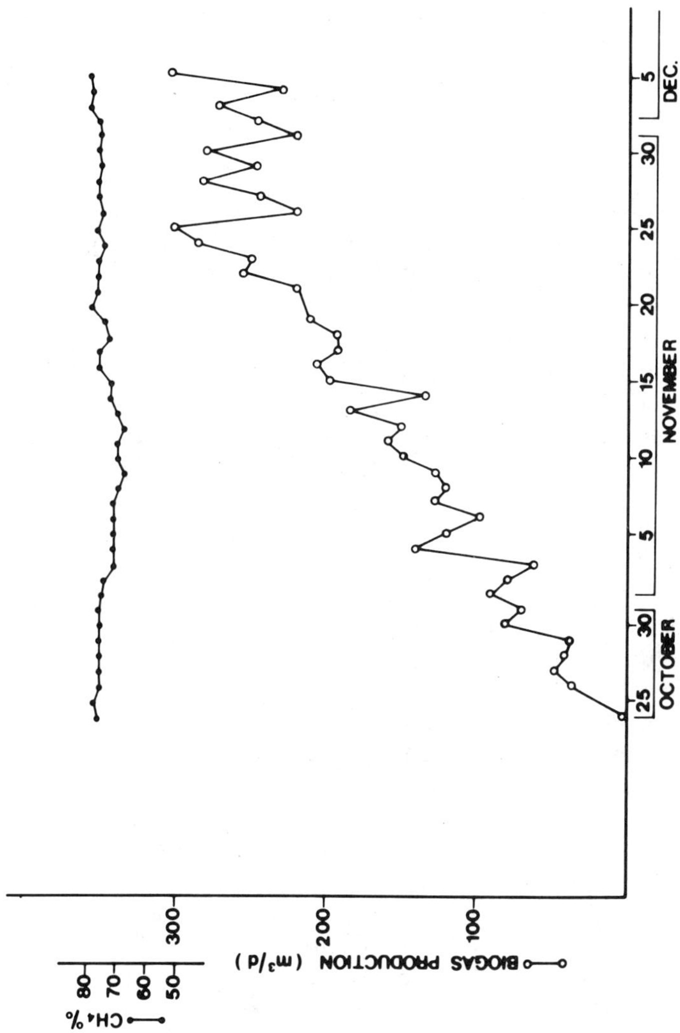

FIG. 5—*Production of biogas and its methane content for the R-2 reactor.*

In addition, there was a notable nitrogen shift to the ammoniacal form, so much so that the percentage of nitrogen as ammonia on the total Kjeldahl nitrogen in the digested waste issuing from the reactors was 90% on the average. During the experiment, pH values in both reactors ranged from 7.6 to 8.0.

Regarding the composition of the gas, the percentage of H_2S became stabilized at acceptable values of between 0.2 and 0.3%.

Reactor R-1 reached a peak production of biogas of fully 11.5 volumes per reactor operating volume per day. This was attained with an HRT of less than one day.

Discussion

The results obtained during operation of the pilot plant, despite its brevity, fully confirmed the preliminary data produced in the laboratory. In fact, the efficiency of the process increased significantly, perhaps because of the greater freshness of the waste available at the sugar refinery.

This waste had been indicated by the factory people as being particularly difficult to treat using the aerobic method. Consequently, the results obtained in treating the waste using the anaerobic system are to be considered satisfactory, keeping in mind that this system must always be seen as a pretreatment method.

The high specific production of gas obtained and its optimum quality also suggest that energy recovery was important. An experiment, originally aimed chiefly at solving an ambient pollution problem, turned out to be a valuable energy source. The full-scale anaerobic digestion plant built and started up during 1982, which reached a daily biogas production of 7500 m^3, has confirmed that this anaerobic pretreatment system for industrial wastes can also be a good investment.

Laboratory tests made on wastes of various origins have shown that fixed-bed reactors may be used advantageously in the treatment of a wide range of industrial effluents.

One very important characteristic of anaerobic processes compared with the aerobic systems lies in the possibility of interrupting flows of feed to reactors for quite long periods without compromising the process' efficiency.

Indeed, in our laboratories, we repeatedly checked that it is possible to return anaerobic reactors to full operating condition within a few days after they have been idle for many months. Restart periods are really remarkably shorter than the first start-up period, because the packing material has been already colonized.

In the case of the sugar refinery cited herein, for example, the factory is in operation for five months and is idle for seven months each year. A condition of this nature can quite decisively favor the choice of an anaerobic system for those industries in which certain effluents are characteristically seasonal.

Conclusion

The experiments carried out on a pilot-plant scale with a sugar refinery effluent have shown that with suitably designed reactors it is feasible to obtain anaerobic fermentation of large quantities of wastes containing soluble pollutants. At the present state of the art, anaerobic digestion is clearly a valid alternative to aerobic treatment of industrial wastes, particularly when the availability of such wastes is seasonal or when they are difficult to treat with an aerobic system.

The extension of knowledge in the field of treating wastes containing toxic substances, as well as in that of reducing process retention times and temperatures, opens to systems of anaerobic treatment increasingly broader ranges of application in the replacement of conventional and costly aerobic processes. This will contribute to better ways of solving environmental and ecological problems connected with the discharge of industrial wastes and reducing pollution abatement costs through recovery of the energy potential available in the effluents.

Acknowledgments

The authors wish to acknowledge the Commission of the European Communities and AGIP Nucleare who have sponsored the research project.

The authors also appreciate the cooperation, in the course of their work, of AGIP GIZA, which has engineered the full-scale plant.

References

[1] Institut National de la Recherche Agronomique, *Pollution by the Food Processing Industries in the EEC*, Graham & Trotman Ltd., London, 1977.
[2] Litchfield, J. H. in "Food and Bioengineering: Fundamental and Industrial Aspects," American Institute of Chemical Engineers Symposium Series, Vol. 67, 1971, pp. 164–172.
[3] Hobson, P. N., Bousfield, S., and Summers, R., *Methane Production from Agricultural and Domestic Wastes*, Applied Science Publishers Ltd., London, 1981.
[4] Young, J. C. and McCarty, P. L., "The Anaerobic Filter for Waste Treatment," Technical Report No. 87, Stanford University, Stanford, Calif., 1968.
[5] American Public Health Association, *Standard Methods for the Examination of Water and Wastewater*, 15th edition, Washington, D.C., 1980.

Giuseppe Ferraiolo,[1] *Marco del Borghi,*[1] *Carlo Solisio,*[1] *and Rosanna Gardi*[1]

Optimization Criteria for the Stabilization of Sewage Sludge and Biogas Production Through Anaerobic Digestion: An Example of an Environmental Biotechnology Application

REFERENCE: Ferraiolo, G., del Borghi, M., Solisio, C., and Gardi, R., **"Optimization Criteria for the Stabilization of Sewage Sludge and Biogas Production Through Anaerobic Digestion: An Example of an Environmental Biotechnology Application,"** *Hazardous and Industrial Waste Management and Testing: Third Symposium, ASTM STP 851,* Larry P. Jackson, Alan R. Rohlik, and Richard A. Conway, Eds., American Society for Testing and Materials, Philadelphia, 1984, pp. 260–277.

ABSTRACT: According to environmental protection and energy conservation principles, anaerobic digestion of activated sludges should have two main purposes: to stabilize the sludge within allowable limits and to increase biogas production as much as possible, in order to meet ecological requirement without neglecting a particular energy source. This implies optimization procedures for the design of the process, based upon its actual kinetics.

This optimization has been developed in the present paper on the basis of experimental results on anaerobic digestion kinetics achieved in previous research.

In particular, optimal biogas production has been determined under ecological constraint due to sludge stabilization.

This procedure has proved that the volatile solids (VS) concentration in the digester feed must be kept above an allowable limit, which has been determined in the authors' research.

If the process design is also required to keep the pollutant concentrations of the outlet supernatant liquor within allowable limits, optimization of biogas "production factors" is no longer feasible because, in this instance, a single digester would work as a purifier rather than a biogas productive unit.

KEY WORDS: anaerobic digestion, sludge stabilization, gas production, optimization, hazardous wastes, industrial wastes

[1]Professor, associate professor, and researchers, respectively, Instituto di Scienze e Tecnologie dell'Ingegneria Chimica, Facolta di Ingegneria-Universita di Genova, Genoa, Italy.

Nomenclature

a, K, m, n	Kinetic constant and exponents
L	Loading rate, kg volatile solids/m^3/day
P	Global biogas production, m^3/day
P_R	Reactor productivity, m^3/m^3 day
P_S	Specific biogas production, m^3/kg volatile solids
Q	Flow rate, m^3/day
X	Concentration, mg/L
α	Constant
γ	Biogas specific gravity at ambient conditions, kg/m^3
η	Efficiency
μ	Specific reaction rate, 1/day
τ	Time, day
V_r	Reactor volume, m^3
σ	Standard deviation
M	Mean deviation

Subscripts

i	1, 2, 3, 4, 5 = Volatile solids, supernatant five-day biochemical oxygen demand (BOD_5), supernatant chemical oxygen demand (COD), total BOD_5, total COD
0	Inlet stream
f	Outlet stream

As is well known, the disposal of activated sludges is one of the main problems occurring in the biological treatment processes, since it requires proper stabilization of the sludges themselves in order to minimize environmental implications.

The problem is particularly critical in Italy because of the increasing number of biological plants erected or planned for treating sewage from urban areas. About 7.2×10^6 metric tons/year of activated sludges are produced in Italy, but a general agreement on acceptable solutions has not yet been achieved [*1*].

Strictly speaking, an activated sludge is to be considered stabilized when its disposal does not affect the environment. Practically, however, it is commonly agreed that a satisfactory stabilization is reached when at least 40% of the organic matter has been removed from the sludge [*2*].

Actually, the residual amount of organic matter within the sludges does not represent the only ecological constraint, but even the absence of certain germs and viruses may be enforced by local laws and regulations, at least in Italy.

The removal of those microorganisms, which again represents an actual concern, is rather unlikely through the usual low-temperature stabilization processes, such as aerobic and anaerobic fermentation; sterilization may be requested or even required.

On the other hand, anaerobic digestion is one of the most convenient ways of stabilizing sludges. It produces a certain amount of biological methane, so that sterilization, if required, may be achieved through the produced biogas. The whole process can be rendered independent of other energy supplies.

Therefore, anaerobic digestion can aim at a double objective, namely, environmental protection and energy conservation. This implies a proper design of the process, which, in turn, requires the availability of suitable analytical expressions for the involved kinetics.

The latter have been derived in this paper on the basis of experimental data previously obtained [*3*]. They have been used to analyze and predict the role of the main operating variables on process performance, so as to rationalize the project design procedure. They may also achieve an optimal balance between main requirements, that is, an allowable degree of sludge stabilization and the highest obtainable production of biogas.

Kinetics and optimization are typical well-known chemical engineering elements involved in biotechnology when they are applied to the biological area. They become environmental biotechnology (EBT) if mainly the environment is involved.

The present work aims at providing an example of EBT applications, as will be shown in the following sections.

Experimental Procedure

The experiments were carried out using the sludge produced by a biological purification plant for municipal wastewaters.

Well-mixed samples were collected from the settler of the sludge and diluted or concentrated according to the loading rate programed for the experimental digester.

The digester was a 5000-mL capacity and 3000-mL working volume glass vessel maintained at constant 35°C temperature and placed on a magnetic stirrer [*3–5*].

Gas production was measured through a collection Winkler tube.

Because of difficulties in pumping small volumes of effluent at very low flow rates, a once-a-day loading was adopted. All together the whole system behaved as a continuous stirred tank reactor (CSTR) since no significant variations in the composition of the reacting liquor inside the digester occurred within each experimental run.

Analytical Methods

Before the whole waste was analyzed, the sludge was stirred and samples were taken.

The supernatants were analyzed by pipetting the clarified liquor after 1 h in 1-L measuring cylinders; volatile solids (VS), chemical oxygen demand (COD), and five-day biochemical oxygen demand (BOD_5) were estimated according to the standard methods of the American Public Health Association [*6*].

Biogas composition was determined by means of a Perkin Elmer gas chromatograph with a flame ionization detector.

Procedures and Results

All the experimental results were checked by measuring VS, BOD_5, and COD in the digester inlet and outlet for both sludges and supernatant liquor.

The measures were expressed as concentration X_{i0}, in milligrams per litre, and X_{if}, in milligrams per litre, where $i = (1, 2, 3, 4, 5)$ stands for VS, supernatant BOD_5 and COD, and total BOD_5 and COD. In turn, 0 and f mean inlet and outlet [*7–10*].

The system was tested for a year using four residence times (20, 15, 10, and 8 days). For each residence time, experiments were carried out at different VS concentrations in the load.

In this way, a wide range of loading rates, L (kilograms of volatile solids per cubic metre per day) were tested (from 1.35 to 7.5 kg VS/m^3 per day).

A period of two turnover times at each loading rate was allowed to ensure equilibrium in the digester.

Each loading situation was checked three times at equilibrium by analyzing the inlet and outlet sludge and gas production and composition. All the experimental data employed in the following calculations are the means of three analytical results. The mean standard deviation of the experimental data is less than 10%.

Kinetics of Sludge Stabilization Through Anaerobic Digestion

According to the experimental results of previous research [*3,5,10*], obtained by the procedure briefly recalled in the preceding section, the specific reaction rates for the removal of the various pollutants versus the actual concentration of the pollutant in the reactor are calculated and reported in Figs. 1 and 2.

The experimental results shown in the Figs. 1 and 2 can be extrapolated by a curve for each parameter.

Owing to the data plotted in Figs. 1 and 2, the specific reduction rates can be analytically expressed according to the following general empirical equation

$$\mu_i(X_{if}) = \frac{1}{X_{1f}} \frac{dX_{if}}{dt} = a_i X_{if}^{mi} \exp(-K_i X_{if}^{ni}) \qquad (1)$$

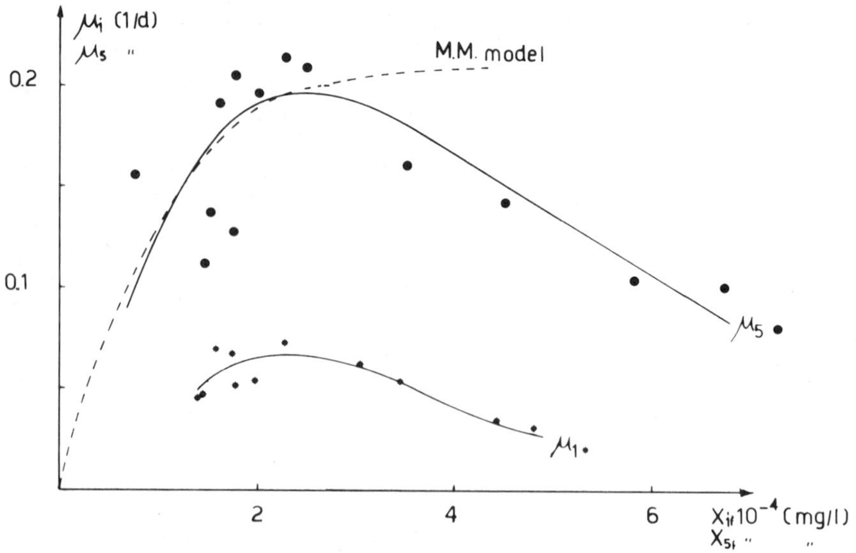

FIG. 1—*Effect of substrate concentration on specific growth rate: volatile solids specific removal rate* = μ_1 *(one per day) versus volatile solids concentration, outlet,* X_{1f} *(in milligrams per litre) (*); total COD specific removal rate* = μ_5 *(one per day) versus total COD concentration, outlet,* X_{5f} *(in milligrams per litre) (●); Michaelis and Menten model (----).*

where

$$\begin{aligned} \mu_i &= \text{specific reaction rate, 1/day} \\ X_{if} &= \text{outlet concentration of parameter, mg/L,} \\ t &= \text{time,} \\ a_i, m_i, K_i, n_i &= \text{kinetic constants, and} \\ i = 1, 2, 3, 4, 5 &= \text{VS, supernatant BOD}_5\text{, supernatant COD, total BOD}_5\text{, and total COD.} \end{aligned}$$

The reduction rates follow a different pattern from the well-known Michaelis and Menten model

$$\mu_i(X_{if}) = \frac{1}{X_{1f}} \frac{dX_{if}}{dt} = \frac{KX_{if}}{K_M + X_{if}}$$

where K_M is the Michaelis and Menten constant depicted in Fig. 1 by a dotted line for total COD.

The difference, analyzed in detail in the previous paper [*3*], is caused by inhibition due to high pollutant concentration in the reactor feed.

The specific reaction rate given by Eq 1 reaches its maximum value at

$$X_{if\max} = \left(\frac{m_i}{K_i n_i}\right)^{1/n_i} \tag{2}$$

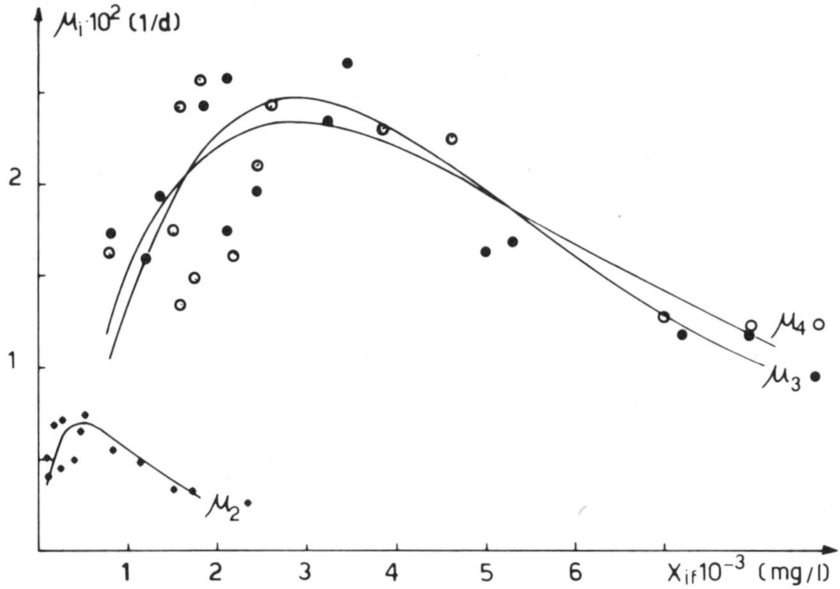

FIG. 2—*Effect of substrate concentration on specific growth rate: supernatant BOD$_5$ specific removal rate* = μ_2 *(one per day) versus supernatant BOD$_5$ concentration outlet,* X_{2f} *(in milligrams per litre) (*); supernatant COD specific removal rate* = μ_3 *(one per day) versus supernatant COD concentration, outlet,* X_{3f} *(in milligrams per litre) (●); total BOD$_5$ specific removal rate* = μ_4 *(one per day) versus total BOD$_5$ concentration, outlet* X_{4f} *(in milligrams per litre) (○).*

so that

$$\mu_{i\max} = a_i \left(\frac{m_i}{n_i K_i}\right)^{m_i/n_i} \exp\left(-\frac{m_i}{n_i}\right) \qquad (3)$$

Through the experimental data, the numerical values of a_i, m_i, n_i, and K_i have been calculated, according to the results listed in Table 1, by means of the least-square method [*10*].

According to the experimental diagrams of Fig. 3, the following relationship can also be derived

$$X_{i0} = \alpha_i X_{10} \qquad (4)$$

where X_{i0} is the inlet concentration of parameter i (in milligrams per litre).

The values of the constants α_i are listed in Fig. 3. It is worth noting that the experimental results, plotted in Fig. 3, were obtained from real sewage.

Once the kinetics of the digestion process are known, the analytical expression for the pollutant removal efficiency can be determined. To this purpose, the following procedure can be adopted, provided that the digester can be

TABLE 1—*Numerical values of the constants* a_i, m_i, K_i, n_i *in the equation* $a_i X_{if}^{mi}$ *exp*$(-K_i X_{if}^{ni})$, *and the mean and standard deviations.*

	a_i	m_i	K_i	n_i	σ, Standard Deviation 1/day	M, Mean Deviation 1/day
X_1 = volatile solids	0.135 10^{-8}	1.9	0.0127 10^{-4}	1.38	0.006 7	0.010
X_2 = supernatant BOD_5	0.0439 10^{-3}	1.2	0.083	0.54	0.009 9	0.000 74
X_3 = supernatant COD	1.183 10^{-7}	1.86	0.0830 10^{-2}	0.72	0.003 1	0.002 7
X_4 = total BOD_5	1.962 10^{-6}	1.44	8.22 10^{-3}	0.695	0.003 7	0.002 9
X_5 = total COD	0.105 10^{-6}	1.63	0.692 10^{-3}	0.79	0.033	0.033

visualized as a CSTR with negligible variation of flow rate Q (in cubic metres per day) between the inlet and outlet streams.

In this instance, the removal efficiency η_i can be written as follows

$$\eta_i = \frac{X_{i0} - X_{if}}{X_{i0}} \tag{5}$$

In turn, the loading rate L_i is given by

$$L_i = 10^{-3}\,\frac{X_{i0}}{\tau_{0i}} = 10^{-3}\,\frac{X_{i0}Q}{V_{ri}} \tag{6}$$

V_{ri} being the reactor working volume, measured in cubic metres, and τ_{0i} the residence time measured in days.

At steady conditions, the hydraulic residence time is expressed through the following relationship

$$\tau_{0i} = \frac{V_{ri}}{Q} = \frac{X_{i0} - X_{if}}{X_{1f}\,\mu(X_{if})} \tag{7}$$

By combining Eqs 1, 5, and 6, one obtains

$$\eta_i = \frac{X_{1f}\mu_i(X_{if}) \times 10^{-3}}{L_i}$$

$$= \frac{a_i(X_{i0}(1-\eta_i))^{m_i}\exp(-K_i(X_{i0}(1-\eta_i))^{n_i})}{L_i} \times 10^{-3} \tag{8}$$

From Eq 8, the efficiency η_i cannot be obtained explicitly, but only the implicit function

$$f(X_{i0}, L_i, \eta_i) = \eta_i - L_i$$

$$- a_i[X_{i0}(1-\eta_i)]^{m_i}\exp\{K_i[X_{i0}(1-\eta_i)]^{n_i}\} = 0 \tag{9}$$

may be written. This function is represented by the diagrams at constant X_{i0} in Figs. 4, 5, and 6, where a comparison is also established with the experimental results obtained in previous research [3]. The latter prove to be in good agreement with the values predicted by the function in Eq 9.

As already mentioned, each data point represents the mean of three experimental results obtained after the steady state was reached in the reactor. The deviation from the mean was less than 10%.

The variation of L was obtained by changing the residence time in the reactor at every value of the parameter concentration in the reactor feed X_{i0} (in milligrams per litre).

Maximum Values of η_i

According to Eq 8, the various η_i values reach a maximum for any given L_i, when

$$(1 - \eta_{i\max})X_{i0} = \left(\frac{m_i + 1}{n_i K_i}\right)^{1/n_i} = X_{if}^* \tag{10a}$$

$$i = 1$$

$$(1 - \eta_{i\max})X_{i0} = \left(\frac{m_i}{n_i K_i}\right)^{1/n_i} = X_{if}^* \tag{10b}$$

$$i = 2, 3, 4, 5$$

so that Eq 8 becomes

$$\eta_{i\max} = \frac{X_{1f}^* \mu_i(X_{if}^*)}{L_i} \times 10^{-3} \tag{11}$$

Referring to VS, that is, to biogas production and sludge stabilization, Eqs 10 and 11 allow the calculation of maximum removal efficiency and the loading rate, L_1, in correspondence with any given value for X_{10}. Then, the required volume, V_{ri}, is also determined by Eq 6.

In turn, the values for X_{10} and L_1, required to attain the maximum efficiency, η_i, are given by the equations

$$X_{i0} = \frac{X_{if}^*}{1 - \eta_{i\max}} = \frac{X_{if}^*}{1 - \dfrac{X_{1f}^* \mu_i(X_{if}^*)}{L_i}} \tag{12}$$

where $i = 1, 2, \ldots 5$.

As for the other pollutants, the preceding procedure is still applicable. As a matter of fact, the maximum values for removal efficiency are unlikely for all

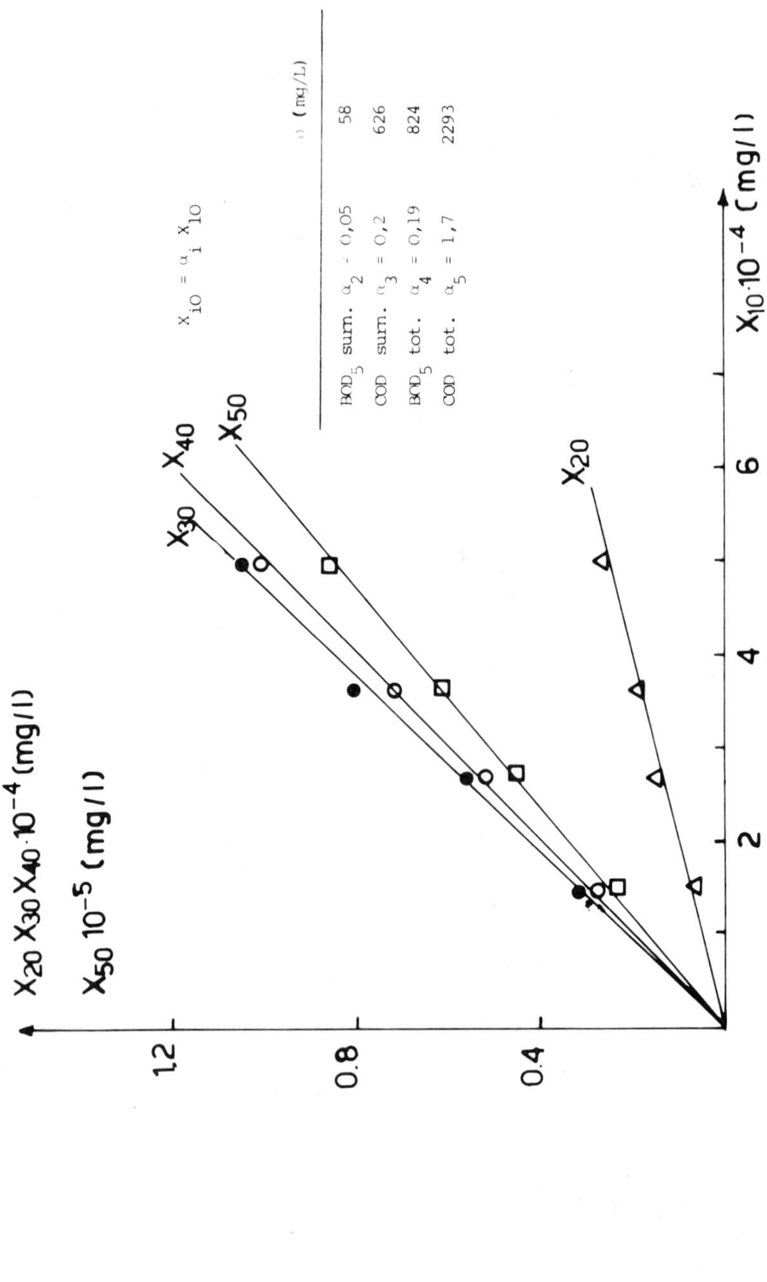

FIG. 3—*Effect of substrate concentration on specific growth rate: relationship between supernatant* BOD_5 *inlet concentration* = X_{20}, *supernatant COD inlet concentration* = X_{30}, *total* BOD_5 *inlet concentration* = X_{40}, *total COD inlet concentration* = X_{50}, *and volatile solids inlet concentration* = X_{10} *(all in milligrams per litre).*

pollutants, but particularly for a single one. The values of the variables maximizing the case $i = 1$ are inconsistent with the maximum conditions required by the other cases, and vice versa, (for instance, $L_1 \neq L_2$; $V_{r1} \neq V_{r2}$).

Optimization of Biogas Production and Ecological Constraints

Single Ecological Constraint

As has been mentioned, sludges must be stabilized so as to minimize their environmental implication.

Practically, it is commonly agreed that a satisfactory stabilization is reached when at least 40% of the organic matter has been removed from the sludge.

In mathematical terms, this means that the volatile solids removal efficiency value η_1 must be kept above the value 40%.

In considering production of biogas, with the only ecological requirement involving sludge stabilization, the following "production factors" are to be introduced

$$P_S = \eta_1/\gamma = \frac{X_{1f}\mu_1(X_{1f})}{\gamma L_1} \times 10^{-3}$$

$$= \text{specific biogas production, m}^3/\text{kg VS} \qquad (13)$$

$$P_R = \frac{\eta_1 L_1}{\gamma} = \frac{X_{1f}\mu_1(X_{1f}) \times 10^{-3}}{\gamma}$$

$$= \text{reactor productivity, (m}^3/\text{m}^3\text{ day)} \qquad (14)$$

$$P = \frac{QX_{10}\eta_1}{\gamma} = P_S QX_{10} = P_r V_{r1} = Q\frac{X_{10} - X_{1f}}{\gamma}$$

$$= \text{global biogas production, m}^3/\text{day} \qquad (15)$$

where γ (in kilograms per cubic metre) stands for biogas specific gravity ambient conditions.

Since η_1 reaches maximum values that can be expressed in the functions X_{10} and L_1, according to Eqs 10 and 11, these production factors can be optimized for practical purposes, provided that the ecological constraint

$$(\eta_{1,\max})_A > 0.4 \qquad (16)$$

stating the allowable removal efficiency, is taken into account.

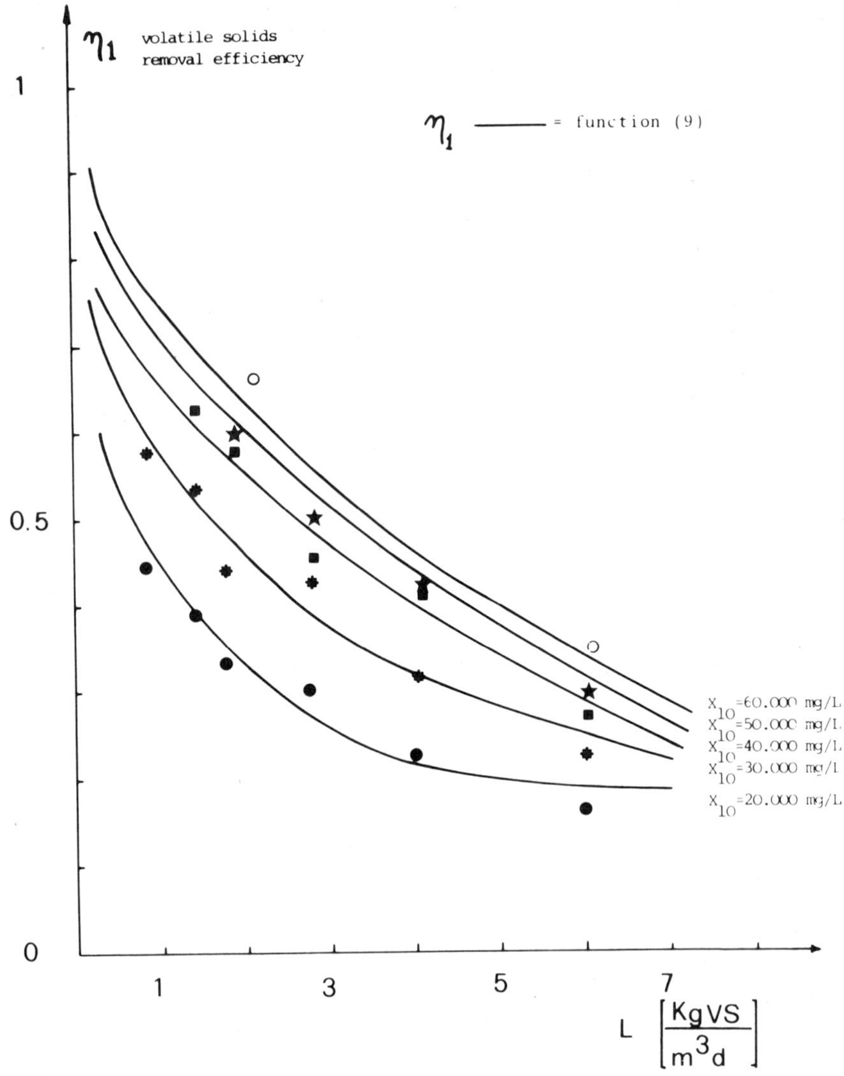

FIG. 4—*Volatile solids removal efficiency (η_i) versus loading rate,* L, *at constant* X_{10} *(in milligrams per litre).*

When this is done, the following limiting conditions can be derived

$$L_{1A} = \frac{10^{-3} X_{1f} \mu_1 (X_{1f}^*)}{(\eta_{1,\max})_A} < 4.8, \text{ kg VS/m}^3 \text{ day} \tag{17}$$

$$X_{10,A} = \frac{X_{1f}^*}{1 - (\eta_{1,\max})_A} > 53\ 000, \text{ mg/L} \tag{18}$$

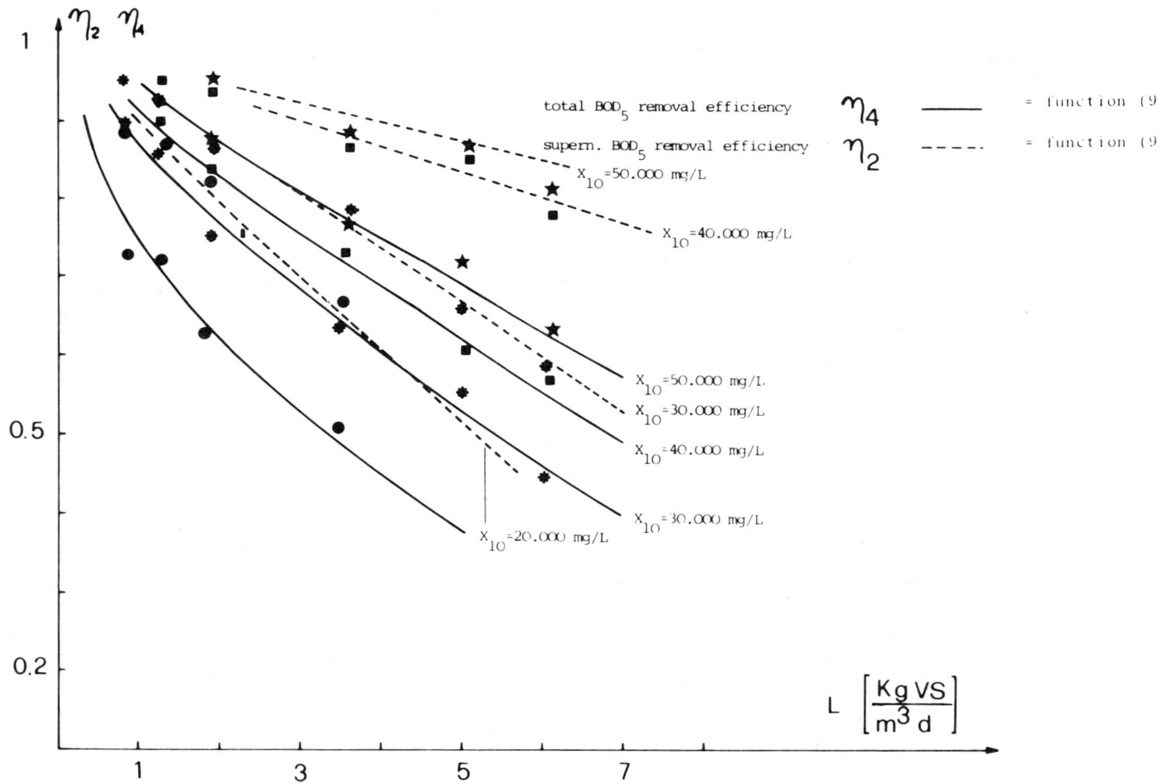

FIG. 5—*Total and supernatant BOD_5 removal efficiency (η_2, η_4) versus loading rate,* L, *at constant* X_{10}.

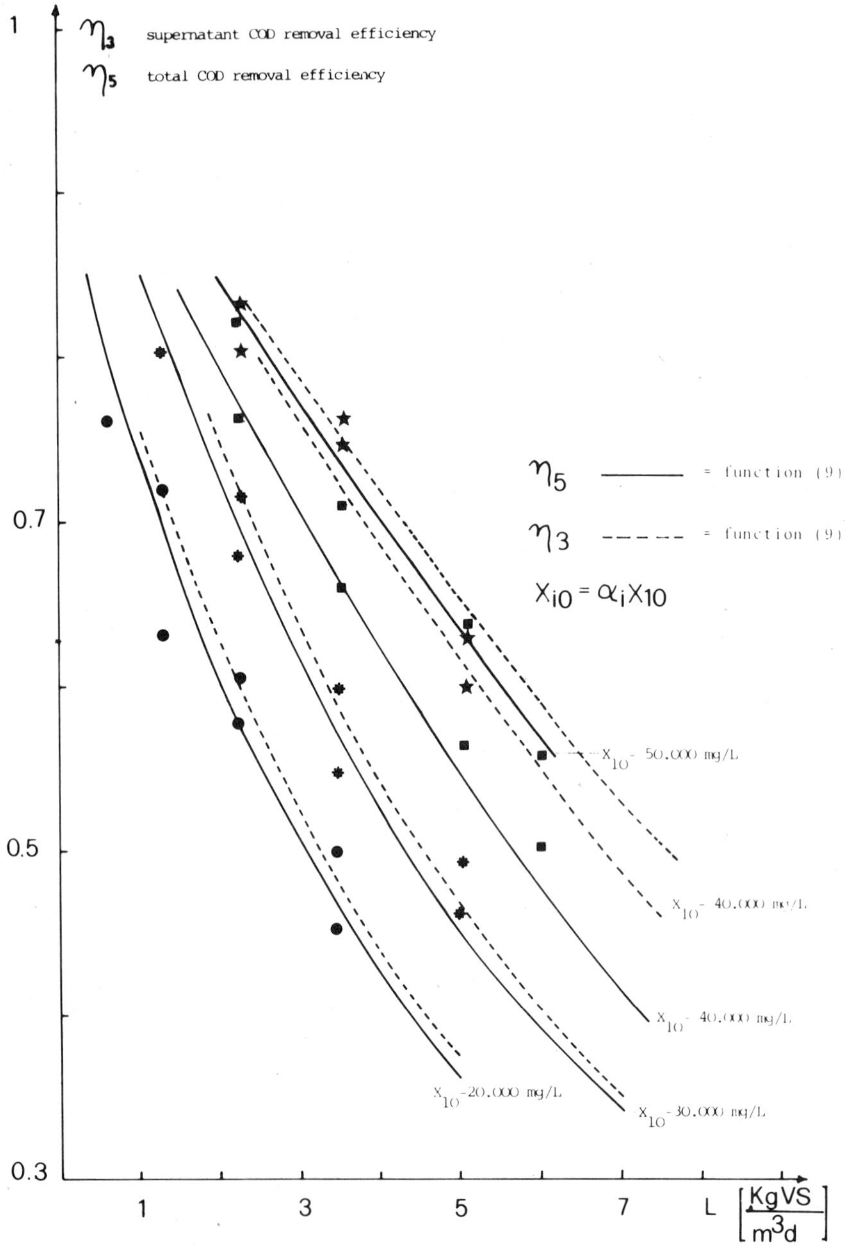

FIG. 6—*Total and supernatant COD removal efficiencies (η_3, η_5) versus loading rate at constant X_{10}.*

giving the maximum allowable flow rate and the minimum allowable inlet concentration of VS, respectively. The latter, derived from Eq 18, should be 53 000 mg/L, whereas the corresponding allowable loading rate should not exceed 4.8 kg VS per cubic metre per day (Eq 17).

According to Eq 18, a prethickening should be generally suitable. In fact, $X_{10,A}$ cannot exceed a given value, depending on practical factors, namely, the pumping and mixing-stirring requirements for the sludges.

The optimal allowable production factors are given by the following expressions

$$(P_{S,\mathrm{opt}})_A = \frac{1 - \dfrac{X_{1f}^*}{X_{10,A}}}{\gamma} \geqslant \frac{0,4}{\gamma} \tag{19}$$

$$(P_{R,\mathrm{opt}})_A = \frac{10^{-3} X_{1f} \mu_1 (X_{1f}^*)}{\gamma} = \frac{\eta_{1A} L_A}{\gamma} = \frac{4.8 \times 0.4}{\gamma} = \frac{1.92}{\gamma} \tag{20}$$

$$(P_{G,\mathrm{opt}})_A = 10^{-3} \frac{Q(X_{10,A} - X_{1f}^*)}{\gamma} = \frac{Q \cdot 21\ 206}{\gamma} \tag{21}$$

where the numerical values have been obtained by calculation of X_{1f}^* and $\mu_{1f}^*(X_{1f})$, according to the figures given in Table 2.

It is worth noting that, for any flow rate Q, the optimal global biogas production is expressed only as a function of the inlet allowable concentration of the VS. The former can be obviously increased by increasing the latter.

Finally, the optimal (minimal) volume of the reactor is given by

$$V_{r,\mathrm{opt}} = \frac{Q(X_{10,A} - X_{1f})}{X_{1f}^* \mu_1 (X_{1f}^*)} = 11.5Q \tag{22}$$

In Fig. 7, where the reactor productivity as a function of X_{10} and η_1 is shown according to Eq 14, the optimum values $(P_{R,\mathrm{opt}})_A$ are clearly evident. Once again, it seems worth noting how the experimental results mentioned in

TABLE 2—*Numerical values of the coefficients* X_{if}^*, $\mu_i^*(X_{if})$.

Coefficient	Value, mg/L	Coefficient	Value, 1/day
X_{1f}^*	31 794	$\mu_1^*(X_{1f})$	0.06
X_{2f}^*	1 343	$\mu_2^*(X_{2f})$	4.3×10^{-3}
X_{3f}^*	4 992	$\mu_3^*(X_{3f})$	0.031
X_{4f}^*	6 063	$\mu_4^*(X_{4f})$	0.0166
X_{5f}^*	45 000	$\mu_5^*(X_{5f})$	0.149

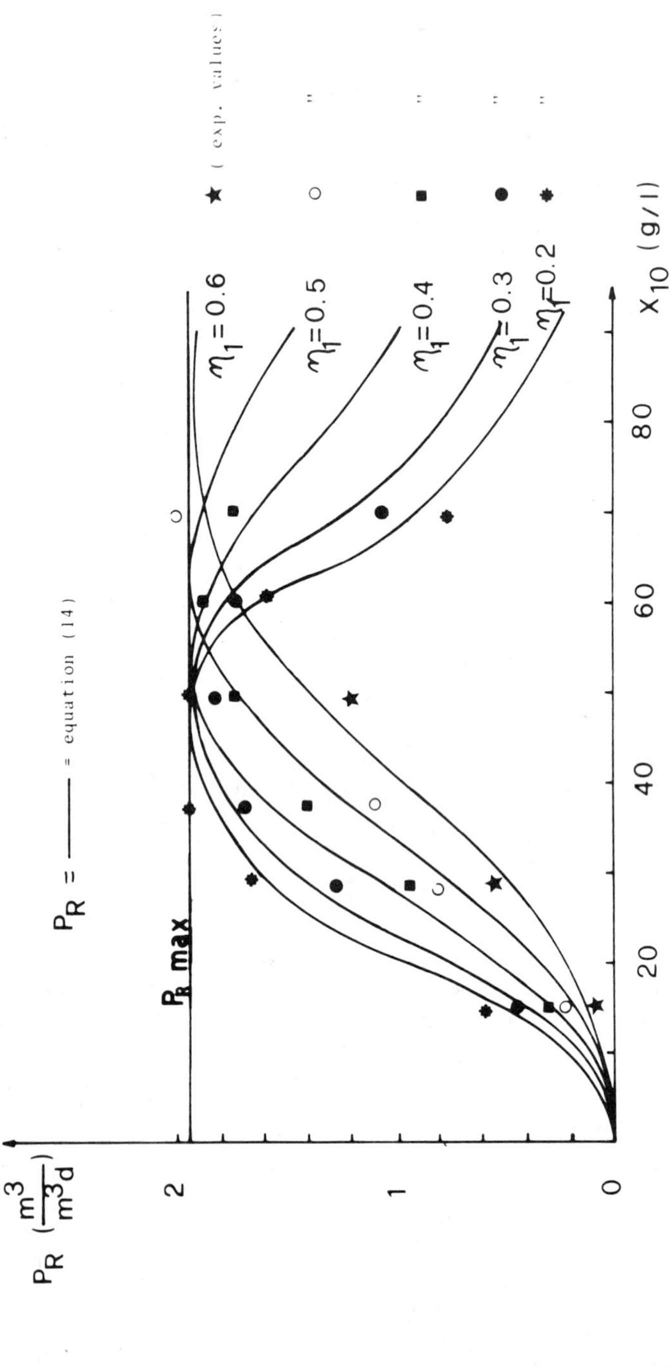

FIG. 7—*Reactor productivity according to Eq 14 (——) compared with experimental results.*

the Experimental Procedures section are in good agreement with the theoretical expression.

As a general procedure, to quantify the process variables, $X_{10,A}$ and L_A are introduced into Eq 8 where η_1 now stands for $(\eta_{1,\max})_A$. Once this is done, the efficiencies $\eta_2 \ldots \eta_5$ can be calculated for any $(\eta_{1,\max})_A$, and through them also the concentrations $X_{2f}, \ldots X_{5f}$ are given by means of Eq 5.

As already shown, the concentrations X_{i0} in Eq 8 are related to X_{10} through empirical Relationship 4, so that all the process variables can finally be determined only as functions of $(\eta_{1,\max})_A$. This provides the optimal solution under the ecological constraint due to sludge stabilization. As the global biogas production increases, so does the desired $(\eta_{i\max})_A$.

Triple Ecological Constraint

This is the case when, in addition to sludge stabilization, the digester might be required to provide an outlet supernatant-clarified liquor complying with enforced limits, A_2 and A_3 as to the BOD_5 and COD concentrations, X_{2f}, X_{3f}.

From Eqs 4, 5, and 8, the following relationships can be written

$$1 - \frac{X_{1f}}{X_{10}} > \frac{X_{1f}\mu_1(X_{1f})}{L_1} 10^{-3} \tag{23}$$

$$1 - \frac{A_2}{\alpha_2 X_{10}} > \frac{X_{1f}\mu_2(A_2)}{\alpha_2 L_1} 10^{-3} \tag{24}$$

$$1 - \frac{A_3}{\alpha_3 X_{10}} > \frac{X_{1f}\mu_3(A_3)}{\alpha_3 L_1} 10^{-3} \tag{25}$$

The solution of these equations gives $(X_{10})_A$, $(X_{1f})_A$, and L_{1A}; then, if the ecological constraint given by sludge stabilization, $(1 - X_{1f}/X_{10} \geqslant 0.4)$, is satisfied, the triple ecological constraint can be achieved by a single digester. Otherwise, one of the three constraints (namely, A_2 or A_3) must be given up and will have to be fulfilled in a further anaerobic or aerobic stage.

Let us consider two examples concerning the following Italian regulations [*11*]:

(*a*) $A_2 = 250$ mg/L and $A_3 = 500$ mg/L, after 1 September 1982, and
(*b*) $A_2 = 40$ mg/L and $A_3 = 160$ mg/L, after 1 March 1986.

By substituting the values of the first instance into Eqs 23, 24, and 25, one obtains the following value:

$$X_{10} \geqslant 1745$$

which is not acceptable because $\alpha_2 X_{10} < A_2$ and $\alpha_3 X_{10} < A_3$.

Similarly, in the second instance, the value $X_{10} \geqslant 790$ again is not acceptable because $\alpha_2 X_{10} < A_2$ and $\alpha_3 X_{10} < A_3$. This proves the impossibility of meeting the constraints enforced by regulation because the inlet BOD_5 and COD concentrations must be lower than those in the outlet-purified liquor.

To achieve a consistent solution, one should increase the values A_2 and A_3 to a limit far higher than the allowable concentration enforced by Italian law. Therefore, the triple ecological constraint cannot be fulfilled by means of a single digester. To increase the sludge stabilization and the biogas production, the sludge obtained from the preliminary digestion should be thickened and then treated in a second digester, according to the procedure of the previous case of a single ecological constraint.

An alternative can be represented by giving up, for instance, the ecological constraint of the limitation of concentration A_3.

From the condition enforced by sludge stabilization and from Eqs 23 and 24, one obtains the following:

$$(X_{1f})_A \leqslant 18\ 000 \text{ mg/L}$$

$$(X_{10})_A \leqslant 30\ 000 \text{ mg/L}$$

$$(X_{1f})_A \geqslant 35\ 000 \text{ mg/L}$$

$$(X_{10})_A \geqslant 58\ 000 \text{ mg/L}$$

so that $\eta > 0.4$.

In this alternative, the further limit, A_3, should be achieved through an anaerobic stage applied only to the supernatant-clarified liquor coming out of the single anaerobic digester.

In fact, the choice between the alternatives discussed here depends upon peculiar economic considerations and assessments, which are beyond the purposes of this paper.

Conclusions

The kinetic equations previously derived and discussed allow one to analyze and predict the role of the main operating variables in the performance of the anaerobic digestion process in order to achieve a proper design project for the whole system.

In particular, the analytical procedure achieved through the aforesaid experimental results leads to the prediction and optimization of both sludge stabilization and biogas production as a function of the "minimum allowable removal efficiency" of VS in the digester, so as to ensure effective stabilization without neglecting a possible source of energy. Correspondingly, a minimum allowable concentration of VS in the digester feed is also established.

When the digester must also provide a purified supernatant liquor, the optimization of the production factors is no longer achievable, since in this instance the reactor would work as a purifier rather than a biogas production unit.

This means that the triple ecological constraint represented by stabilization of the sludges and purification of the outlet supernatant-clarified liquor, within allowable limits, is very unlikely to be achieved in a single digester. Its feasibility depends on the severity of the allowed pollutant concentrations. Usually, two stages should be required: an anaerobic digester, followed by a second anaerobic or aerobic unit [*12–14*].

References

[*1*] Ferraiolo, G., Del Borghi, M., and Solisio, C., "Product Recovery from Solid Wastes, Sludges and Residues. The prospect in Italy," Keynote paper, Environmental Biotechnology Workshop, European Biotechnology Federation and Society of the Chemical Industry, Versailles, France, 7–8 October 1982.

[*2*] Eckenfelder, W. W., Jr., and Santhanam, C. J., *"Sludge Treatment,"* Marcel Dekker, New York, 1981, pp. 240–243.

[*3*] Del Borghi, M., Solisio, C., and Ferraiolo, G., "Activated Sludges Stabilization and Biogas Production Through Anaerobic Digestion. An Experimental Approach," Submitted for publication in *Water Research*.

[*4*] Forecasting and Assessment in Science and Technology, Environmental Biotechnology Workshop, European Biotechnology Federation and Society of the Chemical Industry, Versailles, France, 7–8 October 1982.

[*5*] Del Borghi, M., Fumarola, G., Rubino, L., Solisio, C., and Ferraiolo, G., Fourth Miami International Conference on Alternative Energy Sources, 14–16 December, Miami Beach, Fla., 1981.

[*6*] *Standard Methods for the Examination of Water and Wastewater*, 13th ed., American Public Health Association, New York, 1971.

[*7*] Hobson, P. N. and Shaw, G. B., *Water Research*, Vol. 7, 1973, p. 437.

[*8*] Grignaschi, G. C., Tombetti, F., Del Borghi, M., and Ferraiolo, G., *Studies in Environmental Science*, Vol. 9, 1981, p. 119.

[*9*] Aiba, S., Humphrey, A. E., and Millis, N. F., *Biochemical Engineering*, Academic Press, New York, 1965.

[*10*] Senes, B., Sechi, S., and Vaccaro, C., "Kinetics of Fermentation in Anaerobic Digestion," Thesis, Istituto di Scienze e Tecnologie dell'Ingegneria Chimica, Genoa University, Genoa, Italy, 1981.

[*11*] Italian Regulation, Law, 24 Dec. 1979, No. 650.

[*12*] Clausen, E. C. and Gaddy, J. L., *American Institute of Chemical Engineers Symposium Series*, Vol. 181, No. 74, 1978, p. 56.

[*13*] Gosh, S., Conrad, J. R., Kolass, D. L., *Journal of the Water Pollution Control Federation*, Vol. 47, No. 1, 1975, p. 30.

[*14*] Solisio, C., "Separate Stage Fermentation of Wastewater Sludges," Thesis, Istituto di Scienze e Tecnologie dell'Ingegneria Chimica, Genoa University, Genoa, Italy, 1981.

Brian A. Bolto,[1] *Marek Kotowski,*[2] *and Lucjan Pawlowski*[2]

Recovery of Chromium from Plating Wastes

REFERENCE: Bolto, B. A., Kotowski, M., and Pawlowski, L., **"Recovery of Chromium from Plating Wastes,"** *Hazardous and Industrial Waste Management and Testing: Third Symposium, ASTM STP 851*, Larry P. Jackson, Alan R. Rohlik, and Richard A. Conway, Eds., American Society for Testing and Materials, Philadelphia, 1984, pp. 278–288.

ABSTRACT: Recovery of chromium from a plating sludge is discussed. The process consists of two parts: roasting the sludge with sodium carbonate (Na_2CO_3) and calcium oxide (CaO), followed by leaching. The optimal parameters for roasting are a temperature range of 950 to 1000°C and Na_2CO_3 and CaO contents of 2.4 and 2.8 mol per mole of chromium oxide (Cr_2O_3) respectively.

Demineralized water is recommended for the leaching of sodium chromate (Na_2CrO_4) from the agglomerate formed during roasting. The leached water is recycled through a strongly acidic cation exchanger followed by a weakly basic anion exchanger. A solution of chromate or chromic acid having a concentration up to 60 kg Cr/m^3 is obtained.

KEY WORDS: chromium, recovery, recycling, plating, sludge, ion exchange, leaching, solid wastes, hazardous solid wastes

Heavy metals in their ionic forms, present in surface water and groundwater, belong to the category of pollutants that can be acutely toxic to humans, animals, and plants. They can also induce chronic illness by accumulating in the tissues of the organism, and, being nonbiodegradable, they may be concentrated along the food chain, producing toxic effects at points often far removed from the source of pollution.

One such metal is chromium, which is widely used in the metal finishing and tanning industries. As these industries are often widely dispersed, supervision and control of chromium waste disposal is often difficult.

Among the methods that have been employed in dealing with chromium-bearing wastes, the precipitation of its ionic form in an insoluble form and its

[1] Group leader, CSIRO Division of Chemical Technology, Clayton 3168, Victoria, Australia.
[2] Assistant and department head, respectively, Department of Environmental Chemistry and Technology, Technical University of Lublin (Politechnika Lubelska), 20-618 Lublin, Poland.

storage in special areas has been widely practiced. Frequently, the soluble ionic form of metal is precipitated as a hydroxide and stored in dump sites. However, doubts have been raised about possible leakage of soluble chromium from these sites which may enable it to reenter the biosphere. In some countries, for example, Poland, more rigorous regulations for chromium waste disposal have been implemented, and these have substantially increased handling and disposal costs.

Economically workable chromium ore bodies are not widely distributed, and some countries lack supplies of this essential metal. Thus, the recovery of the chromium salts from waste is environmentally desirable and may be economically feasible.

The two main types of chromium reclamation are (1) direct recycling from wastewater and (2) transformation of chromium from sludges into reusable compounds.

The reclamation of chromium from plating sludges will be discussed in this paper.

Theory

A well-known commercial process for manufacturing sodium chromate or dichromate is roasting the chromite ore [which consists of approximately 44% chromium oxide (Cr_2O_3)] with soda ash (Na_2CO_3) and lime (CaO) at 1100 to 1250°C. Atmospheric oxygen oxidizes Cr^{3+} to Cr^{6+}. The soda ash alkalizes the mixture, while lime increases the roasting rate appreciably and converts the alumina and silica in the ore to insoluble calcium aluminates and silicates. The leach solutions then require no additional processing. Since sludges bearing chromium are usually precipitated with lime, no lime addition is needed before sludge calcination, although soda ash is required.

The expected calcination reactions are given in Eqs 1 through 3. In these equations, M represents metals and can stand for calcium, copper, iron, magnesium, or zinc.

Primary reactions

$$2Cr_2O_3 + 4Na_2CO_3 + 3O_2 \xrightarrow{\text{Heat}} 4Na_2CrO_4 + 4CO_2 \quad (1)$$

$$M(OH)_{2n} \xrightarrow{\text{Heat}} MO_n + nH_2O \quad (2)$$

Competing reaction

$$2Cr_2O_3 + 4CaO + 3O_2 \xrightarrow{\text{Heat}} 4CaCrO_4 \quad (3)$$

The primary reaction results in the conversion and oxidation of chromium (Cr^{3+}) oxide to sodium chromate. The other metal hydroxides or hydrated oxides in the sludge are simply dehydrated by the calcination to form their corresponding oxides. During manufacturing of chromates from ore, only about 90% of the chromium contained in ore is extracted. A residual amount is stored as slag, which is harmful to the environment. Thus, the technology is not free of pollution.

The main reason for processing the plating sludge is to eliminate the hazard to the environment from storage of chromium-bearing plating sludges. In this case, complete removal of chromium compounds from the sludge is the main task from the viewpoint of protection of the environment. However, the calcium chromate formed in the competing reaction has a lower solubility in water than sodium chromate, so leaching with water is ineffective for complete recovery of the calcium chromate from the calcined sludge.

The use of acid leaching for recovering chromium from the calcined sludge results in a large amount of nonchromium compounds (mostly zinc and some ferric cations) also being dissolved.

Despite this difficulty, calcination is probably the most promising method of chromium recovery. There are two problems to be solved: optimization of calcination and improved methods for chromate leaching.

Calcination

The efficiency of chromium recovery from sludges by calcination depends greatly on the efficiency of oxidation. There are three basic parameters which influence the efficiency of the oxidation of Cr^{3+} to Cr^{6+}—the composition of the sludge admixed with CaO, Na_2CO_3, and sodium hydroxide (NaOH); the temperature; and the time of calcination. Most experiments have been carried out at temperatures between 850 and 900°C, [*1,2*]. Deacon et al, and Bahensky and Kubankova [*2-4*] have shown by thermal analysis studies that soduim chromate formation occurs at about 860°C. Machida and Nakagawa [*5*], while studying the calcination of exhausted catalyst, found that heating a mixture of catalyst containing 32.5% Cr^{3+} with Na_2CO_3 at 769°C for 4 h caused almost full oxidation of Cr^{3+} to Cr^{6+}. Yasuda et al [*6*] have found that calcination of plating sludge with a mixture of Na_2CO_3 and $CaCO_3$ at 550 to 730°C for 2 h also allows the conversion of Cr^{3+} to Cr^{6+}.

From Reaction 1 the proportion of chromium oxide (Cr_2O_3) to Na_2O_3 is in the molar ratio of 1:2. However, the ratio of Na_2O_3 to Cr_2O_3 may be lowered, as some amount of Cr_2O_3 may react in accordance with Reaction 3.

Leaching

The efficiency of leaching depends on the particle size of the agglomerate. Therefore, before leaching took place, the agglomerate was milled to obtain a powdered sintered sludge, which had the following distribution of particle sizes.

Particle Diameter, mm	%
<0.1	2.7
0.1 to 0.3	20.4
0.3 to 0.5	73.5
0.5 to 0.75	2.1
0.75 to 1.0	1.3

Leaching with water gives good results when the calcined sludges do not contain calcium, as in exhausted catalysts. If calcium is present in the pretreated sludge, a slightly soluble calcium chromate ($CaCrO_4$) is formed during calcination. Dissolution of the $CaCrO_4$ requires acidification, which also solubilizes some other metals. We have devised a special leaching system based on an ion-exchange process, which produces a regeneration effluent that is a concentrate of the chromate.

Experimental

The sludge of a typical plating plant was used in this investigation. The aim was to find the optimal levels of the following:

(a) the Na_2CO_3 to Cr_2O_3 ratio,
(b) the temperature of roasting, and
(c) the time of roasting.

The investigated sludge, after drying at a temperature of 105°C for 4 h, had the following composition:

Chromium (21.0%)— as hydrous chromium oxide ($Cr_2O_3 \cdot nH_2O$)
Calcium (11.3%)— as a mixture of calcium sulfate ($CaSO_4 \cdot 2H_2O$), calcium phosphate [$Ca_3(PO_4)_2$], and calcium hydroxide [$Ca(OH)_2$]
Zinc-(4.7%)—as zinc hydroxide [$Zn(OH)_2$]
Iron (2.7%)—as hydrous ferric oxide ($Fe_2O_3 \cdot nH_2O$)

This means that the molar ratio of Cr_2O_3 to CaO is 1:1.4.

An electric furnace was used to roast the dried sludge. An analysis of the sludges, after adjustment of their composition with Na_2CO_3 and CaO, is shown in Table 1. The roasting was carried out at temperatures of 700, 800, 900, 950, and 1000°C.

TABLE 1—*Analysis of stock used for roasting.*

Total Amount of Plating Sludge, g	Mole Ratios		
	Cr_2O_3 to	CaO to	Na_2CO_3
724	1	1.4	2.0
	1	1.4	2.4
	1	1.4	2.8
724	1	1.4	2.4
	1	1.8	2.4
	1	2.2	2.4
	1	2.6	2.4
	1	2.8	2.4

After roasting, the resultant agglomerates were ground in a ball-mill, and the chromate formed during roasting was quantitatively leached out with 0.1 *N* H_2SO_4 and analyzed, using standard methods of X-ray fluorescence (XRF) technique with the VRA-2 Carl Zeiss Jena. A column (Fig. 1) was used for the leaching operation.

When the necessary parameters for the calcination had been determined, another system for leaching was investigated (Fig. 2). The leaching solution [0.1 *N* sulfuric acid (H_2SO_4) or demineralized water] was circulated through a strongly acidic cation exchanger (Amberlite 200), which adsorbed all the ca-

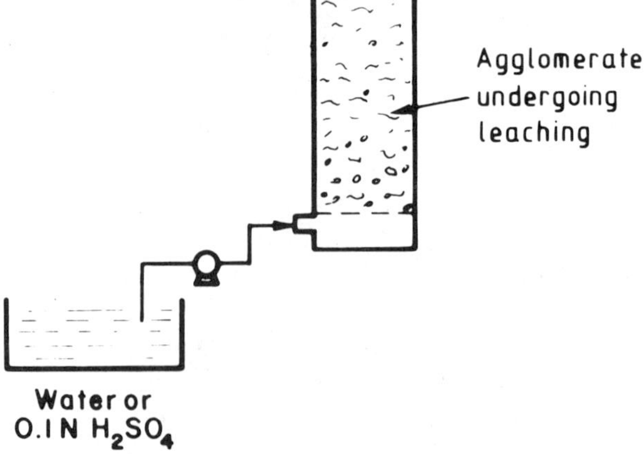

FIG. 1.—*Leaching column used to leach Na_2Cr_4 from the calcinated sludge.*

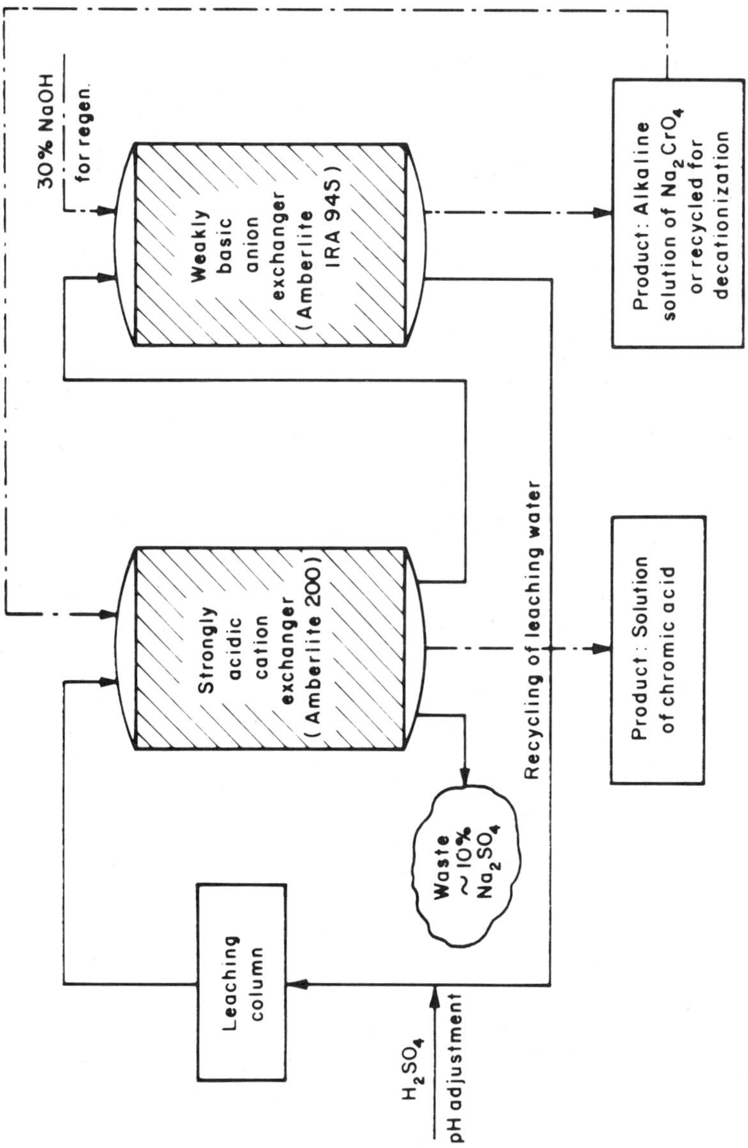

FIG. 2.—*Ion exchange system for leaching of Na_2CrO_4 from agglomerates.*

tions. The effluent from the cation exchanger, containing pure chromic acid, was percolated through the weakly basic anion exchanger, which adsorbed the chromate (plus sulfate when 0.1 *N* H_2SO_4 was used for leaching.)

A solution of 10% H_2SO_4 was used to regenerate the cation exchanger, and solutions of 20 to 45% NaOH were used for the anion exchanger. Recovery of the chromium was achieved either by directly collecting the regeneration effluent as an alkaline solution of chromate or by percolating the regeneration effluent through a cation exchanger bed where the chromate converted into chromic acid.

Results

Calcination

The influence of the amount of Na_2CO_3 added to the plating sludge is illustrated in Fig. 3. The roasting was carried out for 3 h to ensure completion of the reaction. The yield of the oxidation reaction of Cr^{3+} to Cr^{6+} increased significantly when the amount of Na_2CO_3 added was increased from the stoichiometric level (1:2) to a 20% excess (1:2.4). Since a further increase of 20% (1:2.8) increased the yield only slightly, a 20% excess of Na_2CO_3 (1:2.4) was considered the optimal amount.

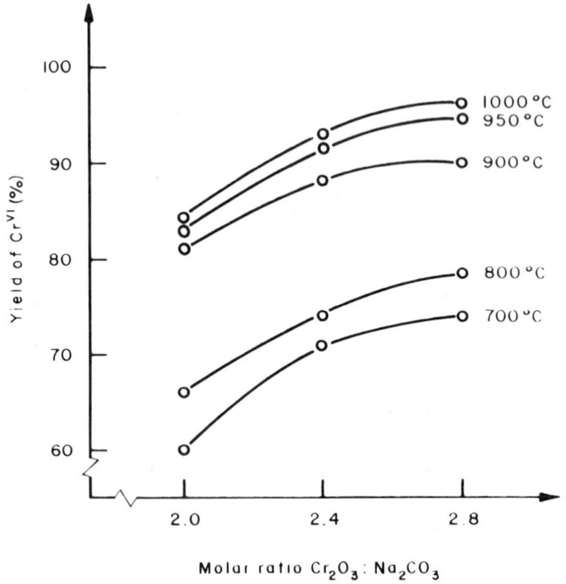

FIG. 3.—*The influence of molar ratios of Cr_2O_3/Na_2CO_3 on the yield of the oxidation reaction of Cr^{3+} to Cr^{6+}. The ratio of Cr_2O_3 to CaO is the same at 1:1.4. The roasting time is 3 h.*

An increase of temperature from 700 to 900°C led to a rapid rise in the yield; however, a further increase to 950°C caused only a slight increase in the yield. Since there was no difference in the yield for temperatures between 950 and 1000°C, the optimal temperature for this sludge was probably about 950°C. Although the optimal temperature varies from one composition of sludge to another [*1–6*], our data should be typical for most plating sludges, which have similar compositions.

The influence of roasting time on yield for a temperature of 950°C and a Cr_2O_3 to Na_2CO_3 molar ratio of 1:2.4 is shown in Fig. 4, when the roasting time was increased from 1 to 2 h, the oxidation yield rose significantly. A further increase from 2 to 3 h led to only a slightly improved yield. In another series of experiments, the effect of lime content on the oxidation yield was evaluated (Fig. 5). An increase in lime content from a Cr_2O to CaO molar ratio of 1:1.4 for the raw sludge to 1:1.8 (by the addition of 0.4 mol of CaO per mole of Cr_2O_3) increased the yield significantly (Fig. 5). However, a further increase in lime content did not improve the efficiency of the roasting. Thus, the efficient recovery of chromates from chromium plating wastes may require the following optimal parameter:

Temperature: 950 to 1000°C
Time of roasting: 2 h
Composition ratio of main components: $Cr_2O_3/Na_2CO_3/CaO = 1:2.4:1.8$

Leaching

Because leaching is the most difficult part of the process, several systems were evaluated.

For direct leaching by 0.1 *N* H_2SO_4, 8.10^{-3} m^3 of the acid was required to

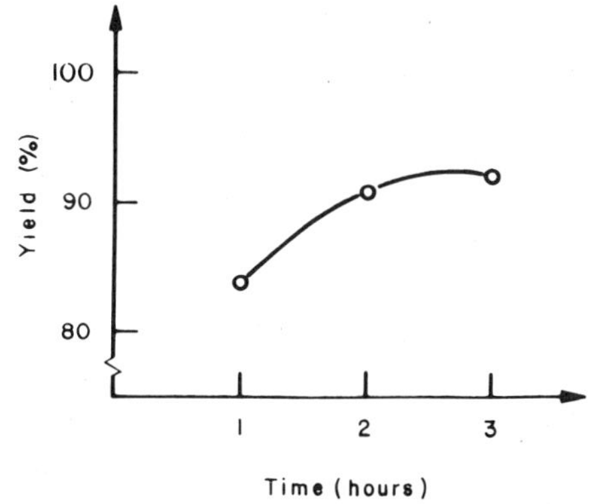

FIG. 4—*The influence of roasting time on the yield of the oxidation reaction of Cr^{3+} to Cr^{6+}. The roasting temperature was 950°C, and the molar ratio of $Cr_2O_3/Na_2CO_3/CaO$ was 1:2.4:1.4.*

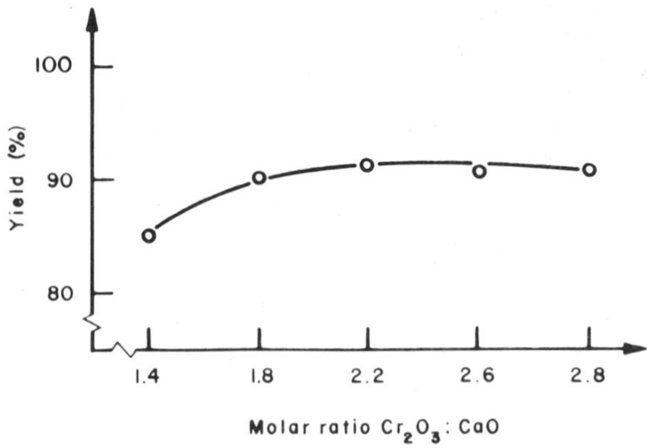

FIG. 5—*The influence of CaO content on the yield of the oxidation reaction of Cr^{3+} to Cr^{6+}. The roasting time was 2 h, and the temperature was 950°C. The ratio of Cr_2O_3/Na_2CO_3 was kept at 1:2.4.*

wash out the Na_2CrO_4 from 1 kg of agglomerate. The solution obtained contained about 15 kg Cr^{6+}/m^3 2 kg H_2SO_4/m^3, and 1 kg/m^3 of Zn^{2+}, Fe^{3+}, and Ca^{2+}.

When demineralized water was used for direct leaching, 0.03 m^3 was required to wash out the Na_2CrO_4 from 1 kg of agglomerate. The resultant solution contained pure Na_2CrO_4 with only low levels of $CaCrO_4$ (10^{-2} to 4.10^{-2} kg $CaCrO_4/m^3$). However, the concentration of Cr^{6+} was low (4 kg Cr^{6+}/m^3).

The best results of leaching were obtained with the system shown in Fig. 2.

A role of the ion exchange unit is to purify the chromate content in a leachate from metal cations and to increase concentration of the chromate. During leaching not only are chromates solubilized but also the calcium, zinc, and ferric cations are dissolved. These cations are taken up by the strongly acidic cation exchanger. The effluent leaving the cation exchanger contains a pure chromic acid, at an average concentration of about 4 kg Cr^{6+}/m^3. A role of the weakly basic anion exchanger is to concentrate chromates up to 60 kg Cr^{6+}/m^3. To obtain such a concentration, a special procedure of the regeneration is required. It was found that use of highly concentrated sodium hydroxide for regeneration leads to an increase in concentration of the regeneration effluent [7].

It was found during this investigation that an optimal concentration of sodium hydroxide in the regenerant is ~30% (Fig. 6). Using such a regenerant, it is possible to obtain a regeneration effluent having an average concentration of about 60 kg Cr^{6+}/m^3 as the sodium chromate is contaminated with 0.5 kg Zn^{2+}/m^3, 0.4 kg Ca^{2+}/m^3, and 0.1 kg Fe^{3+}/m^3.

The efficiency of leaching is shown in Fig. 7. The amount of the residual chromate in the agglomerate decreases sharply in the first 6 bed volumes of

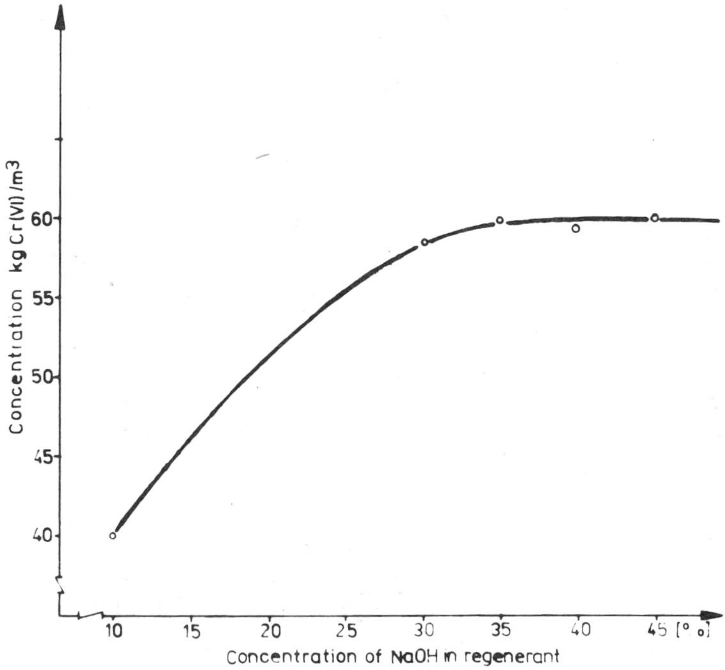

FIG. 6—*The influence of the concentration of NaOH in a regenerant on the concentration of Cr^{6+} in the regeneration effluent.*

the leaching water. Complete removal of the chromates is obtained when 16 bed volumes have passed through the agglomerate bed. The leaching efficiency in the last 10 bed volumes and the concentration of Cr^{6+} in the leachate are low.

However, it is important to completely remove Cr^{6+} from the agglomerate. Use of the ion exchange unit is therefore very important because it allows recycling of the leaching water; during the investigation, only 8 bed volumes of the leaching water recycled through the system, with a flow rate of 5 bed volumes per hour.

Summary

Chromium from plating sludge can be recovered by roasting the sludge with Na_2CO_3 and CaO at a temperature of 950 to 1000°C for 2 h. A reasonably concentrated and pure sodium chromate or a chromic acid solution can be obtained by using demineralized water for leaching; the leached water is then recirculated through the ion exchanger beds.

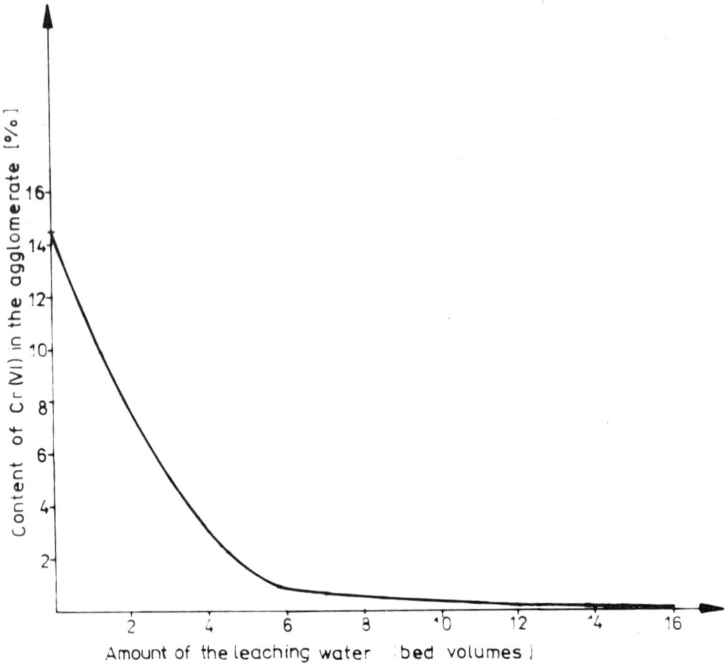

FIG. 7—*The influence of the water volume used for leaching on the efficiency of leaching.*

References

[*1*] Hallowell, J. B., Barrlett, E. S., and Cherry, R. H., "Ammonium-Carbonate Leaching of Metal Values from Wastewater-Treatment Sludges," U.S. NTIS PB Report, PB-271014, National Technical Information Service, Springfield, Va., 1977.

[*2*] Deacon, L. E., Holland, M. E., and Kaplan, R. I. "Chromium Recovery from Sludges Produced from Chromate-Inhibited Cooling Water," Report 1975, GAT-T-2374, National Technical Information Service, Springfield, Va., 1975.

[*3*] Deacon, L. E., Holland, M. E., and Kaplan, I. R. "Chromium Recovery from Sludges Produced from Chromate-Inhibited Cooling Water," *Proceedings, Environmental Third Protection Conference*, Vol. 2, 1975, pp. 511-526.

[*4*] Bahensky, V. and Kubankova, E., "Removal of Sludge from Purification Plants," *Koroze A. Ochrana Materialu* (Translation: *Corrosion and Material Protection*), Vol. 17, No. 4, 1973, pp. 81-83.

[*5*] Machida, T. and Nakagawa, M. "Recovering of Copper Oxide and Alkali Metal Chromate from Copper Chromate Catalysts." British Patent G.B. 1,223,734, 1968.

[*6*] Yasuda, K., Komoda, S., Natsukawa, K. and Yamazaki, K., "Recovery of Chromium in Electroplating Sludges," *Kagaku Kojo*, Vol. 21, No. 2, 1977, pp. 71-74.

[*7*] Pawlowski, L., Klepacka, B., and Zalewski, R., "A New Method of Regeneration of Anion Exchangers Used for Recovering Chromates from Wastewater." *Water Research*, Vol. 15, 1981, pp. 1153-1156.

National Perspectives in Waste Management

B. B. Sundaresan,[1] P. V. R. Subrahmanyam,[1] and A. D. Bhide[1]

The Hazardous Waste Scene in India

REFERENCE: Sundaresan, B. B., Subrahmanyam, P. V. R., and Bhide, A. D., **"The Hazardous Waste Scene in India,"** *Hazardous and Industrial Waste Management and Testing; Third Symposium, ASTM STP 851,* Larry P. Jackson, Alan R. Rohlik, and Richard A. Conway, Eds., American Society for Testing and Materials, Philadelphia, 1984, pp. 291–304.

ABSTRACT: India has made significant advances in the manufacture of basic organic chemicals, dyes, fertilizers, pesticides, drugs, and so forth during the last three decades, resulting in increased generation of hazardous wastes. Presently, these wastes are being indiscriminately disposed of into fallow land in the public domain. Legislation to control air and water pollution has not covered hazardous waste disposal. The magnitude of hazardous waste generation in general and the problems posed by such wastes from pesticide, dyes, and other industries are identified, and available data are presented and discussed.

KEY WORDS: hazardous wastes, industrial wastes, phosphatic fertilizer industry, nitrogenous fertilizer industry, by-product phosphogypsum, fly ash, toxic metals, trace metals, leachate, pesticides, dyes, pigments, chromates, basic organic chemicals, coke oven, steel mill, metallurgical industry, caustic chlorine industry, pollution, legislation

Industrial production in India has been on the increase during the last three decades (Fig. 1), which has resulted in the production of larger amounts of gaseous, liquid, and solid wastes. These are traditionally discharged into the air, water bodies including the sea, or the land, leading to considerable environmental degradation. Although the present level of industrialization has not reached as high a level as in industrialized countries, compounds hitherto unknown are being detected in an ever-increasing number of water supplies even in developing countries because of indiscriminate discharge of waste into the environment. So far the pollution problems are viewed on an *ad hoc* basis, and solutions are found to solve one problem at a time. Nonbiodegradable pollutants are phase transferred, resulting in solid and semisolid wastes, and are disposed of without regard for their effects on the total environment. During

[1]Director, scientist and head of Industrial Wastes Division, and scientist and head of Solid Waste Division, respectively, National Environmental Engineering Research Institute, Nagpur 440020, India.

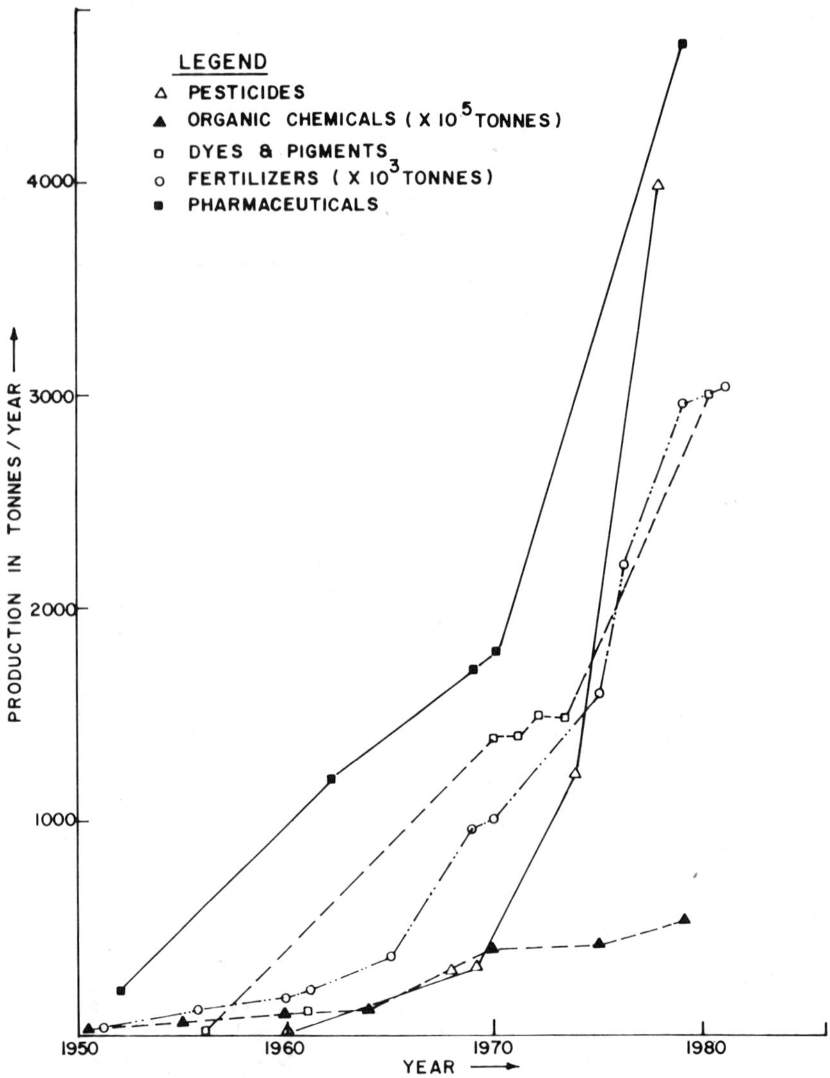

FIG. 1—*Growth pattern for selected industries in India.*

the past decade, many countries have taken regulatory measures to control indiscriminate dumping of chemicals such as spent cyanide from case hardening industry in the United Kingdom [*1*].

The first officially recorded case in India of water pollution due to hazardous industrial waste was the blaze in the river Ganges near Monghyr, Bihar, in 1968. The discharge of large quantities of oil along with the effluent from the Barauni oil refinery set the river aflame. The pollution of the water resulted in suspension of the water supply to the town and closure of the refinery for a short period [*2*].

A large-scale fish kill in the estuary near Goa was reported to be due to discharge of wastewater containing ammonia and arsenic from a nitrogenous fertilizer factory. The Damodar River is the main source of domestic and industrial water in the highly industrialized Asansol–Durgapur region. As early as 1969, a study [*3*] showed that the river water had a biochemical oxygen demand (BOD) of 30 mg/L, 3 mg/L phenols, 8 mg/L iron, and 0.3 mg/L cyanide because of waste discharges from coal-based industries.

Deaths of animals due to consumption of water receiving toxic wastes have been reported in a number of places from time to time. Untreated wastewaters from dye manufacturing units discharged into the Kalu River, Kalyan, lowered the pH of the river water to 4.0 [*4*].

Large-scale use of chemicals to protect crops from pests and other dangers in developing countries, to increase food production, has begun to cause serious concern about chronic toxic effects in human beings. The symptoms of poisoning due to accidents or occupational hazards have been documented in such episodes of poisoning as those in the Cochin area in 1955 and the Sitapur-Hardoi area in 1977 [*5*].

Legislation

Recognizing the danger of pollution, the government of India has enacted the Water (Prevention and Control of Pollution) Act (1974) and the Air Act (1981). Water pollution control boards have been functioning at the central and state government levels, for prevention and control of pollution due to discharge of domestic and industrial wastewaters and gaseous emissions. The Water Act also includes solid substances as pollutants for discharge into water courses, but no special provision exists to control the discharge of hazardous wastes onto land or other media. Most industries discharge semisolid wastes into wastewaters.

Hazardous Waste-Generating Industries

An attempt has been made to outline the magnitude of the hazardous wastes problem in India. Some of the industries known to be generating these wastes have been identified, and their locations are shown in Fig. 2. The quantity and composition of the wastes, in liquid, solid, and semisolid states, from these industries along with the methods adopted for their processing and disposal are described briefly. These industries are grouped into two categories:

(*a*) conventional solid wastes, which are produced in large quantities and stored near the factory, and

(*b*) unconventional semisolid and solid residues, normally in small quanti-

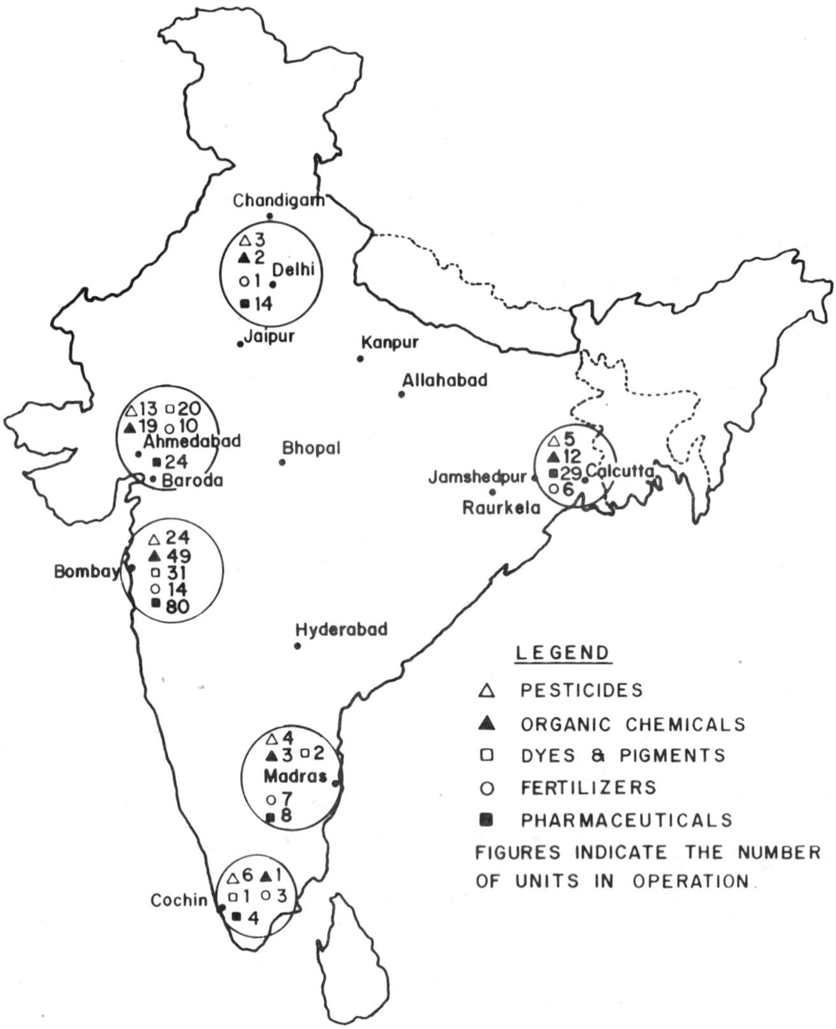

FIG. 2—*Major areas generating toxic and hazardous wastes in India.*

ties generated from the manufacture of a variety of organic and inorganic chemicals.

Conventional Industries

The phosphatic fertilizer industry and thermal power plants and a few others fall into the conventional industry group.

Phosphatic Fertilizer Industry—In India, there are a total of eleven major phosphoric acid and diammonium phosphate plants. About 4.5 to 5.0 tonnes of wet by-product phosphogypsum (BPG) are released per tonne of 100% phosphoric oxide (P_2O_5). During 1980, the total quantity of BPG generated was around 3.5 × 10^6 tonnes, containing 10 to 20% moisture [*6*]. Out of this, 20% is being used for ammonium sulfate manufacture and, to a small extent, for consumptive use in mixed fertilizers. The remaining BPG is being disposed of by landfilling and lagooning without any precautions, resulting in seepage of leachate into groundwater or overflow into surface water. BPG is observed to contain a number of toxic heavy metals in addition to fluoride, such as cadmium, chromium, copper, lead, and manganese, as well as radioactivity (29 pCi/g), as shown in Table 1 [*7*].

Leaching studies with tap water indicated that heavy metals and fluorides leach out from BPG and their concentration tends to decrease as leaching progresses. The heavy metals and fluoride expressed as milligrams per kilogram of BPG leached in the first two days were cadmium, 0.31; chromium, 0.24; copper, 0.71; lead, 0.01; manganese, 0.14; and fluoride, 3110 [*7*]. The National Environmental Engineering Research Institute (NEERI) has initiated work on the impact of BPG dumping on the surface water and groundwater quality. Utilization of BPG for plasterboard manufacture has been found feasible. The radioactivity in the boards has been measured to be 13.6 pCi/g.

Thermal Power Plant—Coal-based thermal power plants generate 57% of the total electric power in India [*8*]. These plants produce about 30 to 40% fly ash and soot per tonne of coal used. About 16.6 × 10^6 tonnes of fly ash were reported to be generated during 1980 from these plants, with an installed capac-

TABLE 1—*Composition of BPG produced in India.*

Constituents	Dry Weight
Nontoxic, %	
$CaSO_4 \cdot 2H_2O$	90 to 95
Calcium (CaO)	30 to 32
Sulfate (SO_4)	50 to 54
Water of crystallization	19 to 20
Total phosphate (P_2O_5)	0.6 to 1.2
Total fluoride	0.5 to 1.9
Sodium (Na_2O) + potassium (K_2O)	0.5 to 1.0
Silica (SiO_2) and other insolubles	2.0 to 3.0
Organic matter	0.4 to 0.8
Toxic, mg/kg (except radioactivity)	
Cadmium	1.25 to 3.22
Chromium	1.47 to 1.72
Copper	4.00 to 4.76
Lead	6.37 to 6.47
Manganese	5.48 to 10.31
Fluorides	0.51 to 2.41
Radioactivity	29 pCi/g

ity of 16 470 MW. The capacity is being increased by adding a few superthermal plants. The chemical composition of the fly ash from a thermal power plant is given in Table 2.

It can be seen that heavy metals such as zinc, barium, and copper are released in significant quantities. Fly ash dumps are a common sight near thermal power stations. The leachate from such sites tends to pollute both surface and groundwater. The extent of leaching depends on the form in which the toxic metals are present, the pH of the water, and the soil strata through which it flows. Use of fly ash in the manufacture of fly ash cement and in building construction are being tried.

Unconventional Industries

The industries known to generate hazardous wastes include (1) pesticides, (2) dyes and pigments, (3) fertilizers, (4) chromates, (5) organic chemicals, (6) acrylonitrile, (7) steel, (8) engineering works, (9) metallurgy, and (10) caustic chlorine.

Pesticide Industry

Dichlorodiphenyltrichloroethane (DDT)—Chlorinated pesticides like DDT and hexachlorocyclohexane (BHC) are the most widely used in agriculture and public health besides organophosphorus pesticides and carbamates, in limited quantities. The present annual production of DDT stands at 14 000 tonnes. The process wastewater generated is about 50 to 60 m^3/tonne of DDT [*9,10*]. The wastewater is corrosive (pH 1.0 to 1.7) and also toxic. For every tonne of DDT produced, 185 to 480 kg sulfuric acid, 3.3 to 8.5 kg suspended solids, 134 to 265 kg chemical oxygen demand (COD), and 0.91 to 1.2 kg DDT are dis-

TABLE 2—*Trace metal analysis of typical fly ash.*[a]

Metals	%	10^4 Tonnes/Year
Aluminum	3.73	61.92
Phosphorus	0.07	1.16
Sulfur	0.24	3.98
Chloride	0.05	0.83
Potassium	2.01	33.37
Thallium	1.02	16.93
Vanadium	0.08	1.33
Manganese	0.23	3.82
Iron	4.12	68.39
Copper	1.30	21.58
Zinc	6.00	99.60
Arsenic	0.02	0.33
Strontium	0.28	4.65
Molybdenum	0.06	1.00
Barium	12.20	202.52
Wolfram	0.21	3.49

[a] Unpublished internal report.

charged in the wastewater [9, 10]. The wastewater is treated with lime, and the treated effluent contains about 1.4 mg DDT/L.

The lime sludge produced is about twice the quantity of DDT [11] and is reported to contain 0.05% DDT on a dry basis, and thus 28 000 tonnes of sludge containing 14 tonnes of DDT will be produced annually. The present method of sludge disposal on land or use in agriculture is not satisfactory as it would leach out DDT and other associated organics into ground and surface waters.

Hexachlorocyclohexane (BHC)—The production of BHC is 41 000 tonnes/year, with 13% gama fraction. The process wastewater, 30 m^3/tonne of BHC made, is corrosive (pH, 2.6 to 3.0) and toxic. Further, the wastewater contains about 3.8 kg sulfuric acid, 1.2 kg suspended solids, 4.7 kg COD, and 1.2 kg BHC per tonne of product [10].

Lime neutralization is observed to remove about 90% of the BHC in the wastewater and results in about a 55% reduction in COD [10]. Since 17 tonnes of BHC will be lost in the treated effluent and the sludge retains about 15 tonnes per year, adequate measures need be taken for treatment of the effluent and the disposal of sludge to prevent BHC leaching into the environment.

Carbamates—About 30 m^3 of wastewater is produced per tonne of 1-naphthyl-*N*-methylcarbamate (Sevin)[2] made. The waste has an acidic pH and a foul smell of mercaptans. The wastewater is toxic because of the presence of carbamates and organics such as alpha-naphthol and methyl isocyanate, which are not readily biodegradable. Hence it is neutralized with lime and solar evaporated in ponds lined with polythene sheets all around. The sludge from the lime neutralization is about 50 kg/tonne of the product and is disposed of as landfill. As the sludge contains carbamates, adequate safeguards are required. The spillages and the contaminated packages are burnt along with other combustible trash in single-cell incinerators of conventional design with no precautions for safe disposal of the ash and the gaseous emissions.

Organophosphorus Pesticides—Organophosphorus pesticides like quinolphos, malathion, and parathion are made, using mainly indigenous technology. The process wastewaters are lime neutralized, and after drying, the sludge is disposed of on land. The effluent is lagooned in solar evaporation ponds, and the collected sludge, along with other combustible wastes, is incinerated. The semisolid waste [12] produced at a rate of 350 kg/tonne quinolphos has a flash point of 68°C and mainly contains organic impurities (58%) and quinolphos (2.2%) on a dry basis, in addition to potassium salts.

It is reported that some pesticide manufacturing units collect the concentrated and toxic semisolid and solid sludges in containers and store them for final disposal in areas earmarked for this purpose.

Dyes and Pigment Industry—The Indian dyestuff industry produced 18 500 tonnes of dyes and pigments during 1975. The production is estimated to increase to 46 306 tonnes/year by 1988–1989 [13]. About 139 organic chemicals,

[2] Union Carbide India, Ltd., Bhopal, India, personal communication, 30 November 1982.

heavy metals, and their salts (zinc, lead, chromium, copper, magnesium, molybdenum, and barium), acids, and alkalies used in the process are released in small quantities in wastewaters, filter cake, and sludges. Some of the organic compounds and intermediates are known to be highly toxic, and the incidence of bladder cancer, particularly due to aromatic amines, among workers in dyestuff industry has been documented [*14*].

Lime neutralization is the common method of treatment. The effluent still containing toxic substances is discharged without any treatment. The settled sludge is often carted out along with other sludges and filter cakes for disposal on fallow land in the public domain. Sometimes the organic sludges are mixed with township garbage and burnt in open pits. No data on the effects of such disposal methods are available.

A factory producing 500 tonnes of organic dyes and 50 tonnes of pigments per year generates 800 m^3/day wastewater [*15*]. The characteristics of the combined wastewater are given in Table 3.

The wastewater is highly toxic and the 96-h median lethal concentration (LC_{50}) value is 0.3% (volume/volume) using *Lebistes reticulatus* as the test fish. The wastewater is subjected to lime treatment, and the sludge amounting to 160 m^3/ day, after drying on a sand bed, is disposed of as landfill.

Another dye factory producing 250 kg daily of vat dyes and 450 kg of cyanine dyes produces 220 m^3 wastewater per day [*16*]. The combined wastewater contained 1540 mg zinc per litre and 1.2 mg pyridine per litre, and exerted a COD of 1940 mg/L.

Inorganic pigment—Titanium dioxide (TiO_2) pigment is produced by chloride and sulfate routes using low-grade ilmenite. A recently constructed 60-tonne/day plant using the chloride route is expected to generate 60 m^3 of wastewater per tonne of TiO_2. The simulated wastewater has a low pH (2.1) with high total solids (2.25%) and suspended solids (0.6%). Iron, to the extent of 3700 mg/L, 40 mg/L Cr^{3+} and 1.8 mg/L zinc are also present [*17*].

The wastewater treated with lime produces about 47.5 m^3 of sludge per tonne of TiO_2, containing mostly iron and other metallic hydroxides (Cr^{3+}, zinc, and

TABLE 3—*Combined wastewater characteristics from organic dyes and pigments manufacture.*

Parameters	Samples		
	1	2	3
Color	dark blue	dark blue	blue
pH	8.7	1.6	1.4
COD, mg/L	2200	1350	1120
Suspended solids, mg/L	1700	150	360
Lead, mg/L	11.8	12	20
Zinc, mg/L	1.2	2.0	1.2
Copper, mg/L	8.0	5.8	4.8

other heavy metals), depending on the raw material used. The waste sludge having 1 to 2% solids requires treatment and careful disposal. The sludge is proposed to be disposed of in lagoons or by filling the exhausted ilmenite mines. The other plant, which makes TiO_2 through the sulfate route, situated near a seacoast also produces highly acidic wastewater, which after lime neutralization is discharged as such into the sea. The neutralized wastewater will contain high concentrations of iron as well as all the heavy metal components in either soluble or insoluble form, depending on the pH of the neutralized wastewater.

Fertilizer Industry—A good number of nitrogenous fertilizer plants use the Giammarco Vetrocoke process, which employs a scrubber solution consisting of 20% potassium carbonate (K_2CO_3) and 15% arsenic oxide (As_2O_3) for separating out CO_2. Even though the process is a closed-loop system, spillages occur, and one factory making 1000 tonnes of urea discharges nearly 40 kg of arsenic oxide as complex potassium ferric arsenite. The spillages are collected and solar evaporated. The sludge containing about 12% weight/weight arsenic is stored in steel drums, sealed, and kept on the factory premises. A safe method of disposal is yet to be established. A very small concentration of arsenic (1 to 2 mg/L) is present in the wastewater into which Vetrocoke section waste is discharged.

Chromates—A large number of industries use water for cooling with recycling. Chromate, zinc, and phosphates are the preferred corrosion inhibitors. In a composite urea plant, the total volume of cooling wastewater (bleed-off) will be about 5.2 m^3/tonne urea. For a 1000-tonnes urea per day plant, the total quantity of chromate discharged in the cooling bleed-off at 10 mg/L CrO_4 will work out to be 52 kg/day or 23.4 kg/day as Cr^{6+}. The corresponding phosphate and zinc losses will be 104 kg/day and 11.7 kg/day, respectively [*18*].

The total sludge generated by using ferrous sulfate ($FeSO_4$) and lime to reduce Cr^{6+} and to precipitate chromic hydroxide [$Cr(OH)_3$] will be 678 kg dry solids per day with 3.5% Cr^{3+}. Current practice is to dispose of the sludge by landfilling or along with other wastewaters without settling it into lagoons. Several other industries such as refineries use Cr^{6+} in cooling water systems, and thus the total quantity of Cr^{3+} discharged into the environment is very high. Although efforts are being made to replace Cr^{6+}, no effort is being made to recover the chromium or collect the sludge containing Cr^{3+} for safe disposal. The tannery industry is another industry which discharges Cr^{3+} in large quantities along with other solid sludges on land, and no study has been made to evaluate the impact by leaching and surface runoff from dumping sites.

Basic Organic Chemicals Industry—The organic chemical industry generates a variety of semisolid and tarry residues (back ends) from the distillation columns. Data collected by NEERI indicate that one such unit making a variety of nitro and chloronitro products, aniline, and *m*-aminophenol generates about 5 tonnes of back ends per day. The back ends from an aniline plant contain phenylcyclohexamine, which is highly toxic. Similarly, nitrotoluene (NT) plant back ends contain mostly dinitrotoluene (DNT) and *p*-nitrotoluene (PNT) and a few unidentified organic compounds. DNT is explosive and hence a hazardous waste. Nitrochlorobenzene (NCB) distilling column residue, on gas chromatographic analysis, revealed the presence of highly toxic dinitrochlorobenzenes as the major component, in addition to small quantities of nitrobenzene, *m*-NCB, and *p*-NCB. Similarly, *m*-aminophenol back ends, which are brittle, contain polymerized organics and are reported to burn like charcoal. Until recently these wastes, along with huge quantities of calcium sulfate sludge from lime neutralization of acid wastewaters, have been carted out for disposal on fallow land. The factory has now established an incinerator to burn all the back ends along with other combustible solid residues. The sulfite sludge generated by the industry, containing 10 to 20% sulfite, is disposed of on land. This sludge can leach out the sulfite into surface water, causing depletion of dissolved oxygen.

Acrylonitrile—Using propylene, ammonia, and air, a plant produces acrylonitrile, acetonitrile, and hydrogen cyanide (HCN). Hydrogen cyanide and acetonitrile are recovered along with acrylonitrile. In the two wastewater streams, the flow of the concentrated wastewater is reported to be 1.3 m^3/tonne of acrylonitrile with a pH between 4.0 and 5.0, having 5.9 kg HCN, and 0.33 kg acetonitrile and acrylonitrile combined. Further, this wastewater, with 12% total dissolved solids, contains 9.6 kg of nitrogen as ammonia (NH_3). During abnormal conditions in the plant operation, the concentration of HCN, acrylonitriles, and acetonitriles may be as high as 18.0, 26.0, and 4.0 kg/tonne, respectively. This wastewater is being incinerated in a furnace along with fuel oil at 800°C. The exhaust gases are reported to contain less than 5 ppm of HCN, which is considered to be safe. The capital cost of the incinerator is $1.5 million, and the running cost is about $5000 per day.[3]

Steel Industry—About 1.1 tonnes of coke is used per tonne of steel made. Ammoniacal liquor from a coke oven is taken through tar catch pits to separate the heavy tar fraction, which is about 0.15 kg tar/tonne of coke [*19*]. This material, because of the presence of phenolics and tar acids, is toxic and flammable, but no specific data on the chemical composition are available. It is currently being disposed of on land, and information on specific damage caused by this disposal method is also not available. Based on the total steel production during 1979–1980, the tar slurry generated in India amounted to about 1 million tonnes.

Other toxic wastes include those from intensive coolers of benzol and napthalene plants. Although their volume is small, they contain highly toxic pollutants [*20*]. Waste from the benzol plant contains 56 000 mg/L benzene extractables, 60 mg/L steam volatile phenolics, and a COD of 5200 mg/L. The correspond-

[3]Indian Petrochemicals Corp., Ltd., Baroda, India, personal communication, 29 December 1982.

ing values from the napthalene plant are 2500 mg/L, 1450 mg/L, and 13 160 mg/L, respectively.

Steel mills recover ferrous sulfate from spent pickle liquor. However, the dilute liquor containing ferrous sulfate and acid is corrosive, and about 3000 m^3/day of this wastewater, containing 4.5 tonnes of total iron, is generated in a 1-million-tonne/year steel plant. The wastewater, after neutralization with sodium carbonate, is discharged into a drain. If not properly neutralized (pH, 2 to 4.6), the wastewater causes considerable corrosion.

Recovery of benzene, toluene, and solvent naptha generates a small volume of acid tar (1.4 m^3/tonne solvents), which is extremely corrosive (pH, 0.5) with a very strong acrid odor. The sludge is toxic because of the presence of sulfonated hydrocarbons and heterocyclic compounds and other organics which exert a COD of 134 000 mg/L. Alkaline sludge (0.03 m^3/tonne of solvents) is thick with a pH of 14.0, contains 3300 mg/L phenols, and exerts a COD of 2×10^6 mg/L [*21*].

Normally, the acid tar sludge is collected and carted away for land disposal. Since the volume of alkaline sludge is small, it is discharged along with other factory wastewaters. Because these sludges contain highly toxic chemicals, special treatment is needed before their discharge.

Engineering Industry—Production of taper rollers and of ball bearings utilize producer gas for heating the metal parts. They also utilize different varieties of oils. The producer gas plant generates about 8.2 m^3 of wastewater per tonne of coal gasified [*22*]. The major pollutants in the wastewater are 60.8 kg of tar oils, 2 kg of phenolics, a COD of 42.6 kg, a BOD of 11.8 kg, and 2.9 kg of soluble tar per tonne of coal gasified. Settled tar from tar catch pits is collected and disposed of on land or sold for by-product recovery. The liquid wastes, even after removal of tar, are highly toxic but can be detoxified by biological treatment. It was reported that the untreated wastewater, when discharged into a dry natural channel, polluted the wells in and around the factory, making the well water unsuitable for use.

Over and above the oil recycled, waste oil to the extent of 1420 L/day is discharged from the industry as liquid waste consisting of cutting and quenching oils, turpentine, and hydraulic oils. The industry recovers the spent oil and disposes of it by burning.

Metallurgical Industries

Lead—Lead metallurgy produces 1.55 tonnes of slag per tonne of refined lead produced. The lead lost in slag amounts to 1.45 tonnes/100 tonnes of refined lead. Slag contains about 1% lead in addition to 10.7% zinc. The slag is being dumped on land with no metal recovery [*23*].

Copper—Four tonnes of ore concentrate are needed to produce 1 tonne of copper metal [*24*]. Data on the quantity of slag generated per tonne of copper produced are not available. The slag obtained in smelting contains silicon, iron, copper, and magnesium oxides as well as 2 to 8% zinc, 0.5% manganese,

0.4% copper, and 1.69% sulfur. The spent acid from electrolytes is reused but sometimes is discharged and may contain copper, iron, nickel, cobalt, zinc, arsenic, and antimony as sulfates and is a hazardous waste [25]. Data on the present method of treatment and disposal are not readily available.

Zinc—Zinc ores invariably contain lead, copper, and cadmium, and the concentrates are processed to produce metals. Cadmium present in zinc concentrate is recovered. In India, the total capacity in zinc smelters is around 68 000 tonnes/year. Liquid wastes are produced during ore dressing and leaching. The settled tailing water [23] from zinc floatation contains 928 mg/L total solids, 30 mg/L of cresols, 2.2 mg/L of copper, and 0.02 mg/L of cyanide. Normal practice is to settle the tailings in ponds, and allow the clear water to overflow into the nearest stream either as is or after pH correction. Characteristics of wastewater [25] from one of the zinc processing plants (50 tonnes of zinc per day) are given in Table 4.

The wastewater is taken through three settling ponds in series and discharged into a river at a distance of 1.5 km. River water analysis at 1.5 km downstream of the discharge point showed a pH of 3.1 to 3.6, 17 to 23 mg/L zinc, 1.6 to 2.0 mg/L copper, 7 to 10 mg/L iron, and 13 to 17 mg/L fluoride [25].

Caustic Chlorine Industry—Caustic chlorine plants using mercury cells discharge mercury in wastewater and sludges. It is reported that 0.23 kg mercury is consumed per tonne of chlorine produced. Part of it is lost into the air, water, and sludge produced during brine purification. A plant producing 60 tonnes of alkali per day is reported to discharge 1400 m^3 of wastewater having a pH of 10.9 to 11.2 and 3 to 5 mg/L of mercury [26]. Lime sludge from a brine purification plant of this capacity will be 8 tonnes per day with 35% moisture. The wet sludge before filtration is estimated to contain 10 to 20 mg/L mercury [27], which compares with the value of 18 to 20 mg/L reported in brine sludges from other plants in India [28]. Commonly, the wet sludge is disposed of in lagoons or, after filtration, dumped into low-lying areas.

TABLE 4—*Characteristics of wastewaters from a zinc processing plant.*

Parameters	Minimum	Maximum	Average
pH	2.2	2.7	. . .
Total dissolved solids, mg/L	3700	4100	3952
Phosphate (PO_4),[a] mg/L	430	575	501
Iron, mg/L	35	48	39
Zinc, mg/L	120	160	146
Copper, mg/L	1.7	4.2	2.8
Fluoride[a]	85	110	96

[a] The fluoride and PO_4 must have originated from a superphosphate plant.

Summary and Conclusions

An attempt has been made to identify the industries in India that generate hazardous wastes, and the current methods followed for treatment and disposal are outlined. The data presented are based on the work carried out by NEERI and other published information.

1. Presently, the industries are mostly concerned with wastewater treatment to meet the requirements of water pollution control boards. However, the sludges produced, whether hazardous or otherwise, are rarely processed and allowed to accumulate before they are carted out.
2. Such wastes are disposed of or disbursed on fallow land, which practice needs more effective regulation and enforcement.
3. Some industries burn the solid wastes along with other combustible materials either in open pits or in single-cell conventional incinerators.
4. Solid and semisolid wastes are also being disposed of without segregation from liquid wastes.
5. In a few cases, wastes are collected, sealed in drums, and stored on the premises, awaiting further disposal. However, with the passage of time the available land space will become limited, and suitable methods need to be considered for the safe disposal of such wastes.
6. Generation, quantification, and analysis of hazardous wastes from various industrial sources need to be assessed systematically to evolve suitable policies to be followed by the regulatory agencies.
7. The existing Air and Water Pollution Control Acts do not have specific provisions to deal with the collection, treatment, and disposal of hazardous wastes; such provisions should be suitably incorporated through legislative action.

Acknowledgments

The authors are indebted to S. K. Titus, Dr. V. K. Kondawar, A. D. Patil, R. V. Bhoyar, J. K. Bhattacharya, and K. V. R. Verma for analyzing some samples and collecting some of the information presented in this paper.

References

[*1*] Wilson, D. C., "Management of Hazardous Wastes," *Kemia-Kemi*, No. 9, 1980, pp. 531-536.
[*2*] "Ganges Water Pollution Inquiry Report—Parts 1 to 4," *The Civic Affairs*, Vol. 17, 1969-1970, pp. 34-37, 31-35, 35-38.
[*3*] "Damodar Valley Industrial and Water Pollution Surveys," Part 1—Durgapur Region; Part 2—Asansol Region, Reports, Central Public Health Engineering Research Institute, Nagpur, India, 1969, pp. 48 and 77.
[*4*] Varde, R. S. and Ajwani, S. H., "Industrial Pollution of Ulhas River," *Environmental Health*, Vol. 6, No. 1, 1964, pp. 13-17.
[*5*] "A Status Report of R & D Work Done in India," *Proceedings*, Indo-U.S. Workshop on Biodegradable Pesticides, Industrial Toxicology Research Centre, Lucknow, India, 1979, p. 69.

[6] Muthukumar, G. and Subrahmanyam, P. V. R., "Disposal and Reuse of By-product Gypsum from Wet Process Phosphoric Acid Industry," *Fertilizer News*, Vol. 25, No. 8, 1980, pp. 23–25.
[7] "Analysis, Assessment and Treatment of Toxic Discharges from Phosphatic Fertilizer and Basic Organic Chemical Industries," Indo–U.S. Project No. PR-1-501-1, Interim Report, National Environmental Engineering Research Institute, Nagpur, India, 1982, p. 110.
[8] *The Times of India Directory and Year Book Including Who's Who*, Times of India Press, Bombay, India, 1982, p. 101.
[9] "Characterisation of Wastewaters from DDT Plant," Hindustan Insecticides Ltd. Report, National Environmental Engineering Research Institute, Nagpur, India, 1979, p. 35.
[10] "Treatment of Wastewater from HIL, Udyogmandal," Report, National Environmental Engineering Research Institute, Nagpur, India, 1979, p. 26.
[11] Sharma, S. K. and Dave, J. M., "Efficiency of Lime Treatment of DDT Factory Effluent—A Field Study," *Journal of Institution of Engineers* (India), Vol. 62, EN3, 1982, pp. 111–114.
[12] "Treatment of Wastewaters from the Manufacture of Organic Phosphorus Pesticides," Report, National Environmental Engineering Research Institute, Nagpur, India, 1980, p. 46.
[13] Johary, N. S., "A 15 Year Perspective Plan for Dyestuff and Intermediate Industries," Paper presented at Indian Chemical Manufacturer's Association Seminar on Perspective Plan for Chemicals, Bombay, India, 1976, p. 42.
[14] Nigam, S. K., "Dyestuff Industry and Its Effect on Human Health," *Proceedings*, Seminar on Pollution Control in Dyestuff and Dye Intermediate Industry, Industrial Extension Bureau and Gujarat Water Pollution Control Board, Ahmedabad, India, 1980, p. 1.
[15] "Treatment and Disposal of Wastewater from M/s. Pigments and Dyestuffs Ltd., Kalyan, Bombay," Report, National Environmental Engineering Research Institute, Nagpur, India, 1979, p. 35.
[16] Mudri, S. S., Kulkarni, A. L., Subrahmanyam, P. V. R., and Mohanrao, G. J., "Treatment of Wastes from a Dye Factory," *Environmental Health*, Vol. 12, No. 3, 1970, pp. 201–217.
[17] "Treatment and Disposal of Wastewaters from the Titanium Dioxide Pigment Plant, Quilon," Report, National Environmental Engineering Research Institute, Nagpur, India, 1980, p. 65.
[18] Muthukumar, G., Badrinath, S. D., Subrahmanyam, P. V. R., and Sundaresan, B. B., "Cooling Water in Fertilizer Industry—Treatment, Pollutional Aspects and Control Strategies," *Proceedings*, No. R&D/2, Water Treatment (Raw and Cooling) in Fertilizer Industry, Fertilizer Association of India, Tuticorin, India, 1981, pp. 33–43.
[19] "Effluent Treatment at Rourkela Steel Plant," Report, National Environmental Engineering Research Institute, Nagpur, India, 1975, p. 94.
[20] "Physico-chemical and Biological Survey of Oxidation Lagoon at Rourkela Steel Plant," Report, National Environmental Engineering Research Institute, Nagpur, India, 1977, p. 136.
[21] "Treatment of Coke Oven Wastes from M/s. Tata Iron and Steel Co. Ltd., Jamshedpur," Report, National Environmental Engineering Research Institute, Nagpur, India, 1973, p. 56.
[22] Moghe, C. A., Gavane, A. G., Bakde, S. N., and Kathuria, A. K., "Steel Mills Waste Treatment and Disposal," Paper presented at National Workshop on Environmental Preservation, Rajasthan State Board for Prevention and Control of Pollution, Jaipur, India, 1982, p. 18.
[23] Sarkar, S., Ed., *Lead: The Science and Technology*, A. Mukharjee & Co. Pvt. Ltd., Calcutta, India, 1965.
[24] Mukherjee, S. N., "Ore Beneficiation at Zawar Mines," *Indian Mining Journal*, Vol. 5, 1957, p. 220.
[25] Bhandari, L. M., "Characteristics of Zinc Smelting Industrial Wastes and Simultaneous Removal of Toxic Elements and Phosphates from It," *Indian Journal of Environmental Health*, Vol. 15, No. 3, 1965, pp. 236–242.
[26] Hanumanulu, V. and Subrahmanyam, P. V. R., "Planning of Water Pollution Control Measures from an Integrated Kraft Pulp and Paper Mill," *IPPTA*, Vol. 14, 1976, p. 138.
[27] "Treatment of Wastewaters from Nagaland Paper Project," Report, National Environmental Engineering Research Institute, Nagpur, India, 1975, p. 61.
[28] Venkataraman, T. P., "Mercury Pollution with Respect to Chlor-Alkali Plants," *Proceedings*, Symposium on Recent Advances in Chlor-Alkali Industries, Alkali Manufacturer's Association of India, Bombay, India, 1974, p. 198.

Ahmed Hamza[1] *and James D. Gallup*[2]

Management of Industrial Solid Wastes in Alexandria, Egypt

REFERENCE: Hamza, A. and Gallup, J. D., **"Management of Industrial Solid Wastes in Alexandria, Egypt,"** *Hazardous and Industrial Waste Management and Testing: Third Symposium, ASTM STP 851*, Larry P. Jackson, Alan R. Rohlik, and Richard A. Conway, Eds., American Society for Testing and Materials, Philadelphia, 1984, pp. 305–319.

ABSTRACT: Developing nations must locate their manufacturing operations where available resources and facilities make them most efficient. The presence of more than one third of the Egyptian industries in urban Alexandria has placed an undue burden on natural resources and has aggravated pollution problems.

The U.S. Environmental Protection Agency (EPA) is conducting a study to identify existing and future industries in Alexandria, estimate the solid wastes produced by these industries, examine methods of reusing solid and other wastes, and assess the feasibility of these methods from the technical, economic, and environmental points of view.

This paper presents a summary of the first phase of the EPA project, which encompasses surveys of residues from industrial sources in Alexandria. Studies to date indicate that wastes from various industries can be recovered economically. Wastes such as tin cans, glass, wastepaper, and food residues from processing of fruits, starch, and beer are examples of reusable industrial wastes in Egypt.

The results of experimental studies for reuse of residues from oil refining, starch and yeast processing, and steel pickling are presented. Spent clay from edible oil refining is currently discarded, causing both handling and disposal problems. This clay contains as much as 40% oil; 90% can be recovered by extraction. The recovered oil can be successfully used in soap production, and the spent clay can be reused in oil bleaching. Other examples include starch and yeast wastes, which can be used for animal feed, and spent pickling liquor, which can be used as a coagulant aid for treating paper mill wastewater.

A centralized system for collecting and treating hazardous residues is proposed to permit economical recovery of valuable materials.

KEY WORDS: solid industrial waste, hazardous materials, residue management, Egypt, hazardous wastes, industrial wastes

The growth of the Egyptian population has created an increasing demand for goods produced by industry. In the past, the public attitude permitted ex-

[1]Associate professor of environmental health, Alexandria University, Alexandria, Egypt.
[2]Chief, Office of Quality Review, U.S. Environmental Protection Agency, Washington, D.C. 20460.

ploitation of resources at the expense of environmental quality. The expectation was that gains in material well-being would exceed losses in environmental quality. However, this attitude has changed.

Alexandria is the largest seaport in Egypt, a prime industrial center, and a major summer resort on the Mediterranean. The city is located in the northwest corner of the Nile Delta and at the eastern edge of the Western Desert. Industries in the Alexandria metropolitan area (AMA) are distributed (Fig. 1) in industrial complexes near Mohmoudia Canal, along the north and south shores of western Lake Maryut, in Amryia, along the coastal areas of El-Mex and Abu Kir, and near the city of Kafr El-Dawar.

The industrialization in the AMA has had a profound effect on public services, transportation, water supply, and waste disposal systems. Environmental quality and natural resources have been severely affected by the continued industrial development.

Industrial operations in the AMA generate toxic wastes, hazardous materials, and inert residues. Existing solid wastes management schemes encompass (*a*) the reuse or recycling of wastepaper, metals, and other recyclable byproducts, (*b*) on-site concentration of nonrecyclable wastes before disposal, and (*c*) pretreatment of wastes to reduce or eliminate harmful constituents before land disposal.

Identification of Major Industrial Residues

The AMA constitutes one of the heaviest concentrations of industry in the Middle East. Industrial activities range from food processing and oil refining to textile finishing and paper production.

During the past year a study cosponsored by Alexandria University and the U.S. Environmental Protection Agency (EPA) has characterized sources of industrial emissions and alternatives for treatment and disposal of residues within the AMA. The specific objectives of the study include the following:

(*a*) identification of existing and potential sources of industrial pollution,
(*b*) prioritization of major waste sources,
(*c*) performance of treatment studies for the pollution abatement program currently implemented by the General Organization for Industrialization (GOFI), and
(*d*) establishment of treatment and disposal options.

The initial list of industrial establishments in the AMA was compiled from the directory of industries published by the Egyptian Union and records of governmental agencies in Alexandria. The list included a total of 1243 industrial plants, branches, and production units. In order to direct the study toward significant sources of wastes, a screening program was developed. Indus-

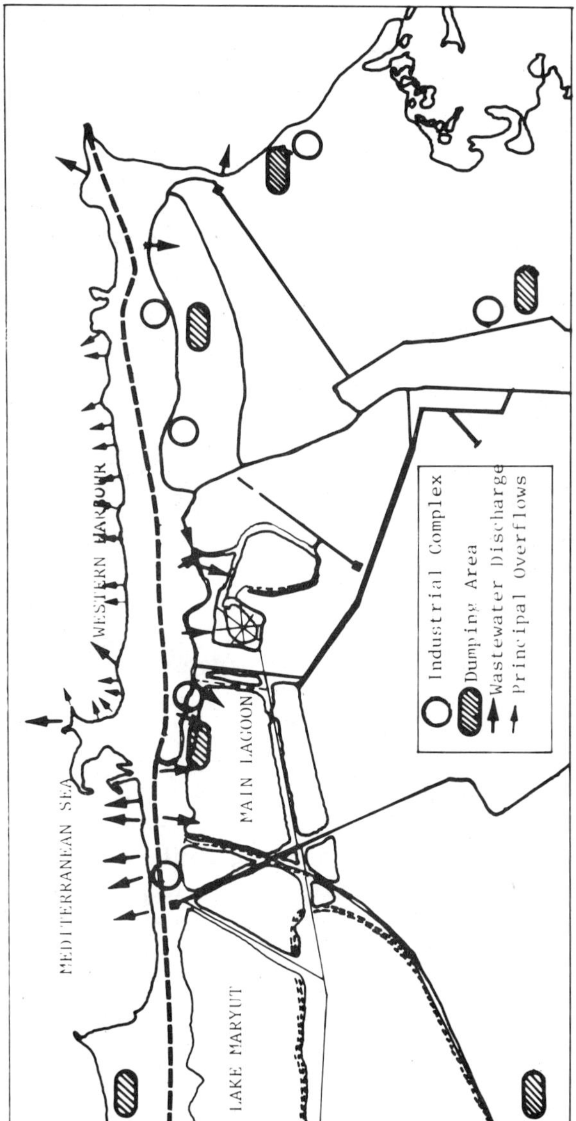

FIG. 1—*Major industrial complexes and dumping areas in Alexandria.*

tries that generate significant residues were selected based on one of the following characteristics:

(*a*) production of one metric ton or more of wastes per working day,
(*b*) production of one metric ton or more of wastes per working day resulting from expansion or planned manufacture of new products, or
(*c*) hazardous consistuents identified in the waste.

The preliminary screening survey involved compiling the following information:

(*a*) a brief description of the processing operations,
(*b*) present amounts of the wastes and future residue prediction,
(*c*) properties of the residues,
(*d*) present reuse of the wastes, and
(*e*) methods of ultimate disposal.

Most of the information was gathered from responses to questionnaires, field visits, and records of the local government. For some industrial establishments, no solid waste records were available. In these cases, residue generation was calculated based on the size and type of production. Based on these screening factors, 145 industries and establishments were identified as significant sources of wastes.

Residue Classifications and Amounts

The industrial residue classification scheme consists of four major classes divided into 28 subcategories, as shown in Table 1. Inorganic residues that are inert and nonhazardous account for 85.5% of the total industrial residues generated in the AMA. Organic and food residues represent 10.8 and 3.3% of the total residue, respectively, and metal and miscellaneous hazardous residues account for about 0.4% of the generated residues.

Chemical precipitation sludges and backwash effluents from water treatment plants are by far the major sources of wastes, representing about 74% of the industrial residues. However, this residue does not represent a direct hazard to the environment as it contains no known harmful constituents. Other major sources of residues are inorganic and organic chemical processing sludges, wastepaper, and residues from bakeries, canneries, and poultry plants.

The amount of domestic solid wastes in the Alexandria area is 1500 metric tons/day, or approximately one third of the contribution from the various industrial operations in the AMA.

The composition of the combined industry and domestic residue and the domestic solid waste are shown in Table 2 [*1,2*]. In comparison with those of the United Kingdom, the AMA domestic wastes contain larger amounts of putrescible food residues; the residues of glass, plastics, and paper are lower in the AMA domestic wastes because of extensive recycling of these materials.

Reclamation and Disposal Practices for Solid Wastes

The recovery of valuable materials from industrial residues is controlled by economic and technological factors. When assessing the economics of recovery schemes in industrialized countries, the higher labor cost required for recovering some materials in comparison with the relatively lower cost of primary production frequently hinders recycling of valuable materials. In developing countries, the situation is reversed, since most valuable industrial and household residues can be recovered. Despite the fact that the recycling agent has no concern for resource conservation or environmental damage, the economic incentive acts favorably for development of successful recovery systems. Although advanced reclamation technology is not applied to residue in the AMA, the level of technology currently employed is cost-effective. Examples are provided in the following sections.

Metal Recovery

Of all industrial residues, metals offer the greatest inducement for recycling. The importance of metal recycling in Egypt arises from the lack of indigenous resources, the high value of ferrous and nonferrous metals, the lower recovery cost in comparison with primary production, and the ease of recycling.

The Egyptian copper works produces 100 000 metric tons/year of steel, 20 000 metric tons/year of copper, and 15 000 metric tons/year of aluminum from scrap metals. Scrap iron and steel are reprocessed pyrometallurgically by open hearth and basic oxygen methods. Careful presorting of the scrap is essential to remove metals retained in molten iron such as tin and copper.

The sources of secondary aluminum used at the plant are residues of new scrap from manufacturing processes and old scrap arising from obsolete products. Aluminum scrap is melted in furnaces with flux for controlled removal of contaminants. Spent fluxes, drosses, and slags from aluminum and iron scrap reprocessing are stored for further processing or disposed of at an authorized dumping site.

Imported copper-bearing wastes, mostly high-grade electrolytic scrap, are melted and used for manufacture of electric cables and miscellaneous industrial products. The scrap is refined in converters in which zinc, lead, and tin are volatilized. The converted copper contains about 2% impurities, the majority of which are nickel, lead, tin, and antimony. The metal is further electrolytically refined to pure copper. The wastes from the electrolytic refining are exported to recover their content of precious metals. Methods of scrap reprocessing at the Egyptian copper works are illustrated in Fig. 2.

TABLE I—*Industrial solid wastes in Alexandria.*

Type of Waste	No. of Plants	Volume, metric tons/day	Major Constituents	Disposition
Inorganics				
Water treatment	6	3300	precipitation and backwash sludges	slurry discharge to drainage canals
Glass	3	2.6	cullet	recycle or reuse
		0.2	fiberglass	incineration
Cement and ceramics	2	22	dust	air emission
Iron and steel	1	30	slag	open dumping
Foundries	5	6	foundry dust	reclaim or dumping
Chemicals	1	420	salt refining and processing [$CaCO_3$, $Ca(OH)_2$,$CaSO_4$]	direct discharge to the sea
Subtotal	18	3810.8		
Organics				
Dyes and chemicals	3	260	organic residues and $CaSO_4$	controlled dumping
Tires	1	0.7	pastes and rejects scrap	reclamation
Textile	13	30.9	reject textile and waste fiber	reprocessing
Batteries	1	0.8	sludge (lead 6 to 9%)	dumping
Plastics	1	0.9	rejects	reprocessing
Paper	6	180	rejects and wastepaper	reprocessing
Tanneries	6	4	fleshing and rejects	glue manufacture
Subtotal	31	481.5		

Food				
Poultry	1	35	viscera and feathers	animal feed
Flour mills and bakeries	28	49	bread rejects and flour	animal feed
Beer	1	0.5	fermentation sludges	animal feed
Starch	2	10.4	processing residues	sewerage discharge
Yeast	1	3.6	fermentation residues	sewerage discharge
Canning	2	33	vegetables and fruit residues	animal feed
		3	fish residues	animal feed
Dairy	1	5	whey	sewerage discharge
Edible oils	8	6	spent bleaching earth	dumping
		0.3	spent sludge (nickel catalyst)	dumping
Subtotal	44	145.8		
Metal and Other hazardous wastes				
Iron	1	1.3	pickling wastes	neutralization and dumping
Metal cleaning	5	0.5	cleaning wastes, solvents	sewerage discharge
Pharmaceuticals	1	0		
Caustics	2	1.2		discharges to sea
Oil refining	2	3.6	tank bottoms (intermittent)	dumping
Municipal WWT[a]	2	2	sludges	drying beds
Service stations	39	12	crank case oil	reclamation
Subtotal	52	18.6		
Total	145	4456.7		

[a]WWT = wastewater treatment.

TABLE 2—*Composition of solid wastes used for land filling and open dumping in the AMA.*[a]

Type of Waste	Combined Industry and Domestic Residue, %	Domestic Residue, AMA, %	Domestic Residue, United Kingdom, %[b]
Putrescible	41.8	78.1	13.2
Paper	3.5	12.6	51.4
Rags	1.5	2.0	2.4
Construction materials	4.8	...	3.1
Plastics	1.3	0.4	1.1
Weeds	8.8	...	...
Bones	0.8	1.8	...
Glass	0.4	0.7	6.5
Metals	0.5	0.7	6.4
Inert (dust)	36.6	3.7	15.9
Total	100.0	100.0	100.0

[a]See Ref. *1*.
[b]See Ref. *2*.

Lead is mainly recovered from batteries after the containers are broken and discarded. The collected scrap lead is smelted at battery plants to yield crude soft lead and antimonial lead, which are further refined to remove arsenic and tin by caustic treatment and reused in the manufacture of automobile batteries.

Wastepaper Recycling

The escalating prices of imported pulp in Egypt and the concomitant increased demand for paper have contributed to recent increases in wastepaper reprocessing. The output of the four wastepaper reprocessing plants in Alexandria is 180 metric tons/day.

The key to efficient paper recycling is careful segregation of wastepaper from other materials. The wastepaper and residual cuts represent 80% of the raw materials used for paper reprocessing, while rice straw comprises the other 20%.

Reprocessing of wastepaper at the AMA (Fig. 3) involves cutting of straw, separation of foreign matter, and semichemical digestion in alkaline medium. The wastepaper is sorted and subjected to a series of screening and homogenization processes with the digested straw before it is sent to the paper machine.

Paper manufacturers suggest that half the total of Egyptian paper is suitable for recycling; however, the current estimate of reprocessing is 20%. Development of efficient collection systems from household and trade sources will increase the stocks of wastepaper and permit expansion of the paper reprocessing industry.

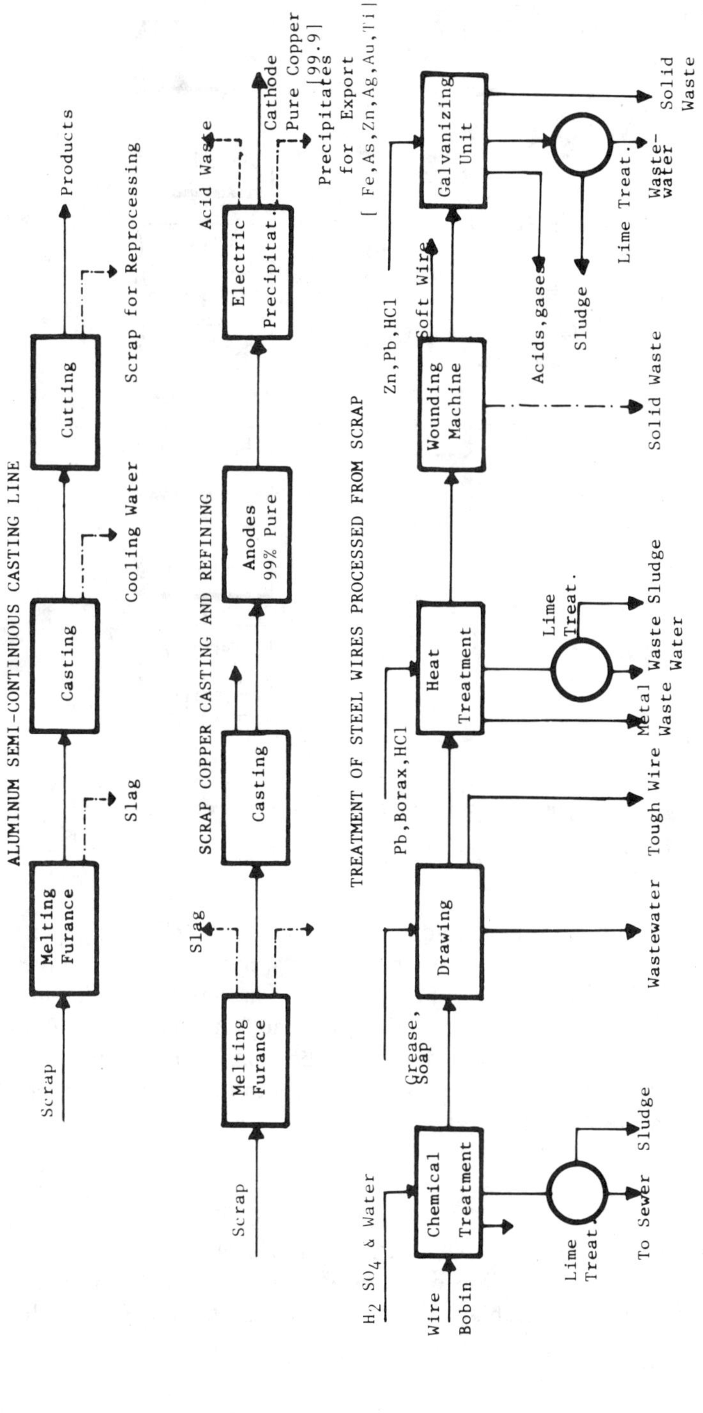

FIG. 2—*Metals recovery at the Egyptian copper works.*

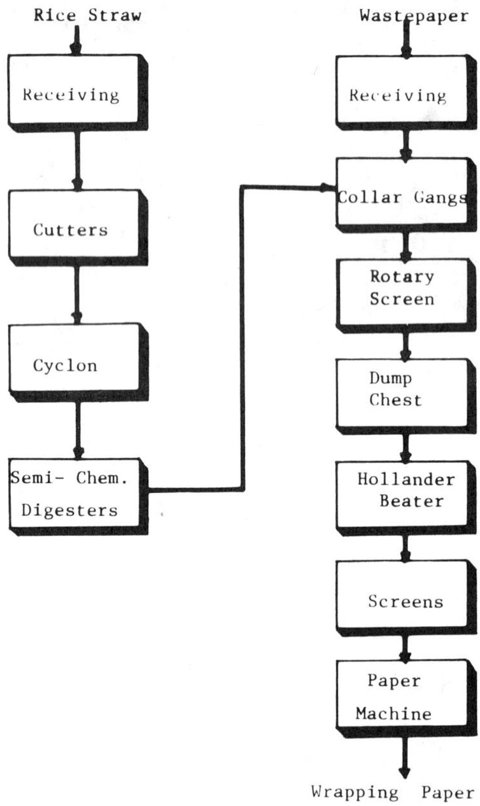

FIG. 3—*Schematic of wastepaper reprocessing.*

Rubber Reprocessing

Vehicle tires are scrapped at an estimated rate of 2 million every year. Two main products are obtained from scrapped tires, namely, reclaim and crumb. Reclaim is devulcanized rubber that is processed in sheet form for other industrial uses. Crumb is a granular material produced from the peripheral tire surface. The crumb is reincorporated into a rubber compound for manufacture of footwear, tubing, and cables. Retreading of scrap tires is widely practiced in Egypt and is a profitable operation.

Reuse of Textile Fibers

Textile manufacture is the oldest industry in Egypt and is the largest industrial activity in the AMA. Waste fabrics are sorted manually and reduced to fibers by a ragbulling machine. The used fibers are then blended with virgin fibers, carded, and spun into threads. Man-made fibers are not reprocessed because of their high recycling cost.

Food Processing Residues

Residues from food processing constitute a valuable source of animal foodstuffs. Analysis of residues from canning of vegetables and fruits is shown in Table 3.

These residues, together with residues from the fermentation and dairy industries, are currently used as animal feed. However, the lack of an organized collection and reuse system often results in putrefaction of food residues. Poultry processing residues are rendered instantaneously at the Alexandria poultry processing plant to produce chicken feed. In this case, prompt residue processing reduces losses. Residue from extracted cottonseed is the largest source of blended animal feed in Egypt. The AMA produces over half the national production of the blended feed.

Miscellaneous Wastes

Several other wastes are economically recycled, though on a smaller scale. Lubricating oils are collected and rerefined in small plants. Plastic, glass, and ceramic wastes are reprocessed in private establishments. The recovery technology is not advanced and the products are typically articles of lower quality, which can be easily marketed.

Potential for Reuse of Selected Industrial Residues

There is an industrial residue problem in the AMA. There are no controlled disposal sites within the metropolitan area for dumping hazardous wastes. Most of the industrial sludges and manufacturing residues that cannot be recycled are being shipped to disposal sites in the desert areas or being stored permanently or temporarily on site. Several residues are intentionally dumped into the sewer system with industrial wastewater. This unfortunate practice af-

TABLE 3—*Analysis of canning residues, %.*

Residues	Vegetable	Fruits	Seeds
Fat	1.4	4.4	41.5
Protein	6.8	2.0	12.0
Carbohydrates	13.5	20.5	9.5
Ash	7.6	2.4	7.0
Cellulosic matter	13.3	8.2	4.3
Moisture	57.4	62.5	25.7

fects the sewerage network, treatment works, and the receiving water. The following are examples of residues which could be reused.

Spent Clays

Diatomaceous earth is widely used as a filter aid and bleaching material in oil refining and preparing beverages. Spent clay is currently discarded; this practice poses environmental problems. Residual oil in the spent clay may reach 40%; however, no attempt has been made to commercially recover the retained oils and regenerate the clay. Studies for reactivation of spent clay [*3*] indicate that caustic treatment of the spent clay can recover about 70% of the entrained oil while extraction with a mixture of ethyl alcohol and benzene is capable of recovering 90% of the entrained oil. The spent clay is then burned under controlled conditions to produce reactivated clay with bleaching power similar to that of the new active clays. Application of this recovery system would be cost-effective because of the economic value of the recovered oil and the regenerated clay. Furthermore, this scheme would alleviate the disposal problems of the spent clay.

Spent Pickling Liquor

The pickling waste produced at the Egyptian copper works is neutralized in polyethylene-lined tanks, and the effluents are discharged into the sewerage network. The sludge is disposed of by dumping in a nearby site. The pickling waste is very effective as a coagulant aid for chemical treatment of paper mill wastewater [*4*]. This reuse scheme would provide significant environmental benefits.

Starch Wastes

The average protein content from steeping and centrifugation of starch is approximately 10.6 g/L as total Kjeldahl nitrogen (TKN). The TKN of the acidified sludge (pH 4.7) from the starch processing effluent varies from 47 to 59 g/L [*5*]. This amounts to potential recovery of 2 800 to 3 400 metric tons/year of high-protein residues that could be used in animal feed production.

Fermentation Wastes

The high nutritive value of residues from the yeast and brewing industries permits recovery of valuable constituents. Pilot studies are being performed to recover waste yeast from fermentation effluents. The system involves sieving of the waste yeast, heating to 70°C to inactivate enzymes, evaporation in multiple effect evaporators, and drying. The economics of the process and possible uses of the recovered yeast are being evaluated. At the least, the residues should be used as animal feed.

Industrial Residue Management for Alexandria

Management of industrial residues is needed to protect the environmental quality of the AMA. Decisions must be made by the government in cooperation with industry concerning an integral residue management scheme, the costs for various alternatives, and incentives to induce recycling and reuse. Such decisions should be formulated within the context of the social, economic, and technological constraints in the AMA.

Problems of residue management in Alexandria are compounded by (*a*) increases in residues resulting from industrial growth, (*b*) improper storage of hazardous residues, (*c*) the difficulty of monitoring disposal sites, and (*d*) the reluctance of most industries to invest money for adequate disposal of materials no longer useful.

Enactment of Law No. 38/1967 and Executive Decision No. 134/1968 gave certain powers and responsibilities relating to control of solid wastes to municipalities and public health authorities. In Alexandria, the responsibility for liquid waste collection and treatment is vested in the general organization for sanitary drainage, while solid waste collection and disposal is assigned to the local municipality. The function and duties of the solid waste section of Alexandria municipality are (*a*) preparation of long-range plans for management of residues from domestic sources; (*b*) operation of services, coordination of services assigned to private contractors, licensing, and authorization; and (*c*) ensuring the adequacy of dumping and landfilling areas.

The existing program needs strategic planning for residue management, particularly for solid wastes generated from industrial sources. If direct charges were levied against industry, it would eventually be encouraged to install waste control systems and introduce new residue recycling technologies. A system of equitable charges has already been developed by the municipality to limit disposal of harmful residues. Presently, the Egyptian government (GOFI) is investing about $60 million for wastewater treatment and residue control in the edible oil, starch, yeast, inorganic chemical, textile, and organic chemical industries. The first phase of the project will be completed in 1984.

Centralized System for Treatment of Hazardous Residues

Most hazardous residues, such as solvents, organic chemicals, acids, alkalies, spent catalysts, and inflammable solids, are currently discharged into the sewerage network or disposed of through open dumping. The disadvantages of on-site treatment are high capital and operating costs and low-volume operating levels. Treatment of these wastes at a centralized plant would permit process efficiency and economy of scale.

A proposal has been made to establish six large collection points at the major industrial complexes of the AMA. These collection points would send residues to the centralized plant, which might include processing by solvent distillation, catalyst regeneration, detoxification, incineration, or a combination of incineration and one of the other processes. For reprocessing specific high-volume residues, such as spent bleaching clays, textile, and plastic material, central recovery plants are proposed. The specialized facilities would be located near the production sites to minimize transportation costs. In all cases, a fee would be collected from the industry, depending on the quantity of waste, the location, and the type of treatment needed. The centralized system is illustrated in Fig. 4.

Summary

In the Alexandria metropolitan area, several recycling schemes are operated economically using labor-intensive techniques in the areas of metal recovery,

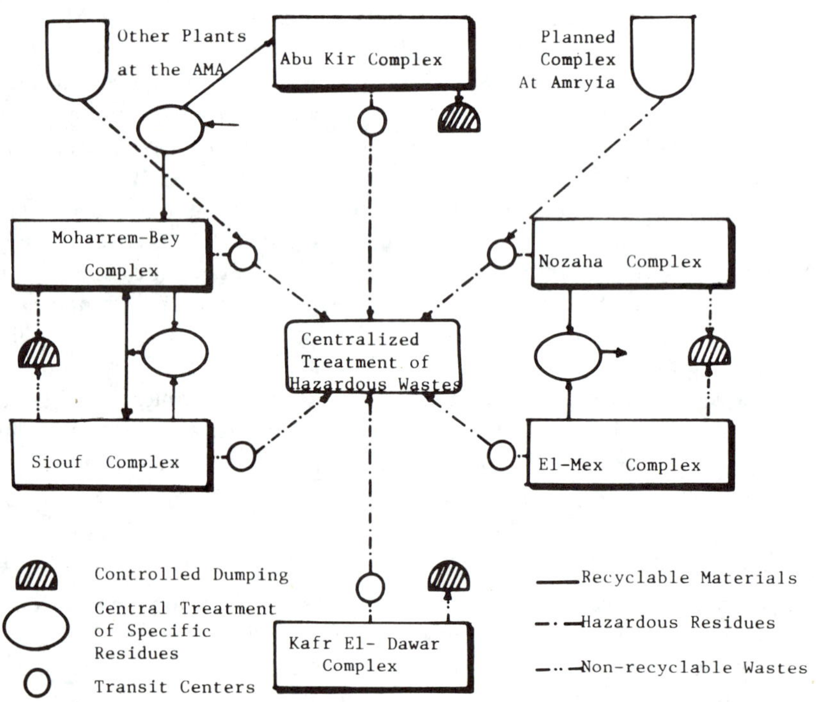

FIG. 4—*Proposed system for management of industrial residues in Alexandria.*

wastepaper reprocessing, rubber recycling, and processing of food residues for production of blended animal feed.

Several options for reuse of low-value residues are being investigated. Spent bleaching clay can be regenerated and entrained oil may be recovered. Spent pickling liquor from steel processing is used successfully as a coagulant aid for wastewater treatment. Starch waste presents an excellent protein source for animal feed production. Recovery of waste yeast from the fermentation industry is feasible. However, the economics and possible uses of the recovered yeast are still under investigation.

Waste management problems are compounded by lack of residue control legislation, improper dumping practices, continuous increase of industrial residues, and industry reluctance to invest in unprofitable residue management schemes. Establishment of a centralized system for treatment of wastes after collection in transit points at the six major industrial complexes in Alexandria has been proposed. The system would allow for recovery of valuable materials, reduce hazardous sources of pollution, and take advantage of the economy of scale.

References

[*1*] Hussein, A., "Field Study to Investigate the Problems of Solid Wastes in Alexandria City," MPH Sc. thesis, Institute of Public Health, Alexandria University, Alexandria, Egypt, 1979.

[*2*] Bridgewater, A., and Mumford, C., *Waste Recycling and Pollution Control Handbook*, George Godwin, London, 1979, p. 373.

[*3*] Baker, M., Abu Naser, A., and Hamza, A., "Suitability of Some Egyptian Clays for Bleaching of Cottonseed Oil: III—Regeneration of Spent Clays, *Indian Ceramics*, Vol. 13, No. 9, 1969, pp. 213–225.

[*4*] Hamza, A., "Biochemical Treatment of Paper Mill Wastewater," *The Bulletin of the High Institute of Public Health*, Vol. 11, No. 1, 1982.

[*5*] Hamza, A., "Biochemical Purification of Starch-Detergent Effluent," *Proceedings*, the Australian Water and Wastewater Tenth Federal Convention, April 1983, in press.

Norbert B. Schomaker[1]

Hazardous Waste Landfill Research

REFERENCE: Schomaker, N. B., **"Hazardous Waste Landfill Research,"** *Hazardous and Industrial Waste Management and Testing: Third Symposium, ASTM STP 851*, Larry P. Jackson, Alan R. Rohlik, and Richard A. Conway, Eds., American Society for Testing and Materials, Philadelphia, 1984, pp. 320–333.

ABSTRACT: The hazardous waste land disposal research program is collecting data necessary to support implementation of disposal guidelines mandated by the Resource Conservation and Recovery Act of 1976 (RCRA) (PL 94-580). This program, relating to the categorical areas of landfills, surface impoundments, and underground mines, encompasses state-of-the-art documents, laboratory analysis, economic assessment, bench and pilot studies, and full-scale field verification studies. Over the next five years the research will be reported as technical resource documents (TRDs) in support of the Permit Writers Guidance Manuals. These manuals will be used to provide guidance for conducting the review and evaluation of land disposal permit applications. This paper presents an overview of this program and reports the current status of work in the various categorical areas.

KEY WORDS: hazardous wastes, industrial wastes, pollution, environmental surveys, waste disposal, earthfills, environmental engineering, environmental surveys, environmental tests, waste treatment, decontamination, containment, hydrogeology, leaching, linings

Congress, through the Solid Waste Disposal Act [also known as the Resource Conservation and Recovery Act (RCRA), PL 94-580], directed the U.S. Environmental Protection Agency (EPA) to develop programs for the management of hazardous and nonhazardous solid wastes. To effect the degree of coordination required of such an endeavor, a committee was formed to prepare and update an agency-wide research strategy. The Solid and Hazardous Waste Research Committee coordinates the efforts of the Office of Research and Development (ORD) with those of the Office of Solid Waste (OSW), the Office of Waste Programs Enforcement (OWPE), the Office of Water Programs Operations (OWPO), the Office of Emergency and Remedial Response (OERR), and the EPA regional offices.

[1] Chief, Disposal Branch, Office of Research and Development, U.S. Environmental Protection Agency, Municipal Environmental Research Laboratory, Solid and Hazardous Waste Research Division, Cincinnati, Ohio 45268.

The fiscal year (FY) 1980 solid and hazardous waste research strategy was based on the anticipated needs of the EPA as it implemented the mandates of the RCRA. Major emphasis was placed on hazardous waste, with continuing efforts in the municipal solid waste area. The FY 1981–1983 strategies focused entirely on hazardous wastes.

Agency responsibilities under the RCRA are reasonably specific, and future activities to fulfill those responsibilities have been scheduled according to a step-by-step strategy. Within this hazardous waste strategy, four research categories, which are described in the following sections, have been established. The four categories are as follows:

(a) hazardous waste identification,
(b) hazardous waste environmental engineering technology,
(c) hazardous waste risk assessment, and
(d) energy and mineral wastes.

The land disposal research program is identified with the hazardous waste environmental engineering technology category.

Hazardous Waste Identification

Hazardous waste identification includes sampling, analysis, characterization, measurement, monitoring, and quality assurance of measurements. The development of valid identification protocols is vital to the separation of hazardous from nonhazardous wastes and to the subsequent "cradle-to-grave" management of hazardous wastes. At present, our knowledge of many of the factors in the effective management of hazardous wastes is far from complete.

The strategy for hazardous waste identification is based on the need to provide a complete capability for the sampling and analysis of wastes that are potentially hazardous. Initial efforts are developing engineering and chemical screening methods and protocols. Multimedia monitoring methods are equally important and occupy a high priority in this area. Quality assurance for the monitoring program, necessary to ensure the validity of all of the data to be collected for characterization and subsequent monitoring of hazardous waste sources, is addressed.

Hazardous Waste Environmental Engineering Technology

With the environmentally unsafe options for management of hazardous waste eliminated by the RCRA, generators must turn to the use of properly designed and constructed landfills or other alternatives. Improved treatment and disposal technology is necessary if hazardous materials are to be managed in an economically as well as environmentally sound manner.

The thorough study of all hazardous waste streams and potential treatment and disposal technologies is not feasible because of the variety of wastes considered hazardous; and funds would not likely be available to conduct such a mammoth undertaking. Immediate short-term efforts are focusing on those areas where the need is acute and industry research cannot be expected to be sufficient to meet needs. The land disposal strategy is presented in greater detail in following sections of this paper.

The EPA's treatment research strategy is directed toward three basic areas: preprocessing or predisposal treatment, thermal destruction, and detoxification treatment. Short-term research efforts will concentrate on the refinement of existing technology for each of these three areas. Investigations of advanced technology are being addressed in the area of thermal destruction.

Hazardous Waste Risk Assessment

The paramount objective of EPA in the management of solid and hazardous wastes is the protection of public health from the effects of these substances by direct contact or transmission through the environment. In order to accomplish this objective, it is essential that reliable tools be available to enable EPA to gage the potential impacts of various substances under various circumstances.

The research strategy for risk assessment includes the production of risk assessment documents for specific chemicals (scientific assessment), determination of toxic effects on humans likely to result from exposure to hazardous wastes, development of methods and data to define environmental transport and fate characteristics of hazardous wastes, determination of likely ecological effects of environmental insult from hazardous wastes, and development of socioeconomic criteria and model scenarios of the impacts to be expected for various waste management incidents or disposal methods.

Energy and Mineral Wastes

Certain industrial and mining operations yield large volumes of waste containing some hazardous constituents. These special wastes include uranium mining wastes; phosphate mining, milling, and processing wastes; certain other mining wastes; cement kiln dust; utility wastes (coal ash and flue gas desulfurization wastes); and gas and oil drilling muds and brines.

Specific efforts include studies of wastes from coal-fired electric generating plants, uranium mining, phosphate mining and beneficiation, and from selected ore mining industries. Lysimeter studies of oil shale leaching were also conducted.

Land Disposal

The previously mentioned four categories of research are being pursued by various research groups within the EPA. The hazardous waste land disposal research program is part of the hazardous waste environmental engineering technology category. Since the focus of this program was primarily hazardous waste land disposal research, it is this area which will be specifically addressed in this paper.

The program, encompassing state-of-the-art documents, laboratory analysis, bench and pilot studies, and full-scale field verification studies is at various stages of implementation.

The objective of this program is to compile a more adequate data base of information so that current hazardous waste disposal technology to the land may be upgraded by (*a*) developing proper site selection, design, and operational criteria for the establishment of new waste disposal sites and (*b*) developing proper control technology for upgrading existing waste disposal sites. The procedures thus developed will consider and describe those specific functions required to eliminate the pollution potential from landfills, surface impoundments, and underground mines. Specific functions include the ability to predict or provide guidance for determining leachate quality and quantity, quick indicator tests for the selection of lining materials, quick indicator tests for chemical fixation performance and compatibility with various generic waste streams, a methodology for minimizing moisture infiltration, alternatives to conventional landfills, and predictors of environmental and economic impacts.

Over the next five years the research will be reported as criteria and guidance documents for user communities. The land disposal research program is currently developing and compiling a data base for use in the development of guidelines and standards for waste residual disposal to the land as mandated by the Resource Conservation and Recovery Act of 1976 (RCRA). Permit Writer's Guidance Manuals, which provide guidance for conducting the review and evaluation of permit applications, are currently being prepared by the Office of Solid Waste in Washington, D.C. Technical resource documents (TRDs) in support of the guidance manuals are also being prepared by the Office of Research and Development in specific areas to provide the best engineering technologies and methods for evaluating the performance of the applicant's design. The information and guidance presented in these manuals will constitute a suggested approach for review and evaluation and should be useful references for design and operation.

The current hazardous waste land disposal research program has been divided into five general areas: (*a*) landfill, (*b*) surface impoundment, (*c*) underground mines, (*d*) economic assessment of hazardous waste disposal practices and alternatives, and (*e*) TRDs and guidance manuals.

Landfill

The landfill program examines components and unit operations of a hazardous waste landfill and optimizes their performance. The goal is to control or be able to predict the movement of liquids and gases in and around a landfill. The program develops basic information for each component and prepares reports of research and user manuals for permit writers, design engineers, and operators.

The components and unit operations of a landfill to be addressed in this research include waste leaching, pollutant movement, pollutant control, pollutant treatment, and waste modification.

Waste Leaching

The overall objective of this research activity is to provide techniques for leaching a waste sample to obtain leachate similar to that which will occur when the waste is landfilled. A leaching procedure is being developed for use in determining which pollutants will be of concern in movement of leachates out of the landfill and in evaluating leachate treatment schemes in order to better develop pollutant control technology. A TRD entitled "Solid Waste Leaching Procedure Manual" (SW-924), was completed in FY 1982 on a batch leaching procedure and is being submitted for public comment. Another comparative study of batch and column leaching methods will be implemented in FY 1983. The initial product will focus on leachate from well-defined mixtures or single waste streams.

Pollutant Movement

The objective of pollutant movement work is to develop techniques that can be used in the design and regulation of clay-lined disposal facilities. These techniques will be developed to achieve a minimized, predictable release into the environment.

Because the processes of pollutant retention by soils do not operate independently of one another, understanding the causes of pollutant movement and retention in a specific situation is difficult. Heterogeneity of soil characteristics and the usual lack of knowledge about the types and composition of the variety of wastes accepted at a general disposal facility are further complications. Consequently, it is not presently possible to use soil and waste characteristics to predict pollutant movement and retention in soils at most hazardous waste disposal facilities with sufficient precision and accuracy to include the action of the soil or earth material as a reliable part of the design for the facility. To enhance designs and predictive capabilities, future facilities over sensitive aquifers, if not prohibited, may be restricted to a single waste or limited wastes of known migration characteristics. The capacity of soils and earth materials below the facility liner to retard the movement of pollutants may be used in future designs, but most likely as a safety factor rather than as a part of the primary facility design. As a result, containment through use of impermeable membrane liners, facility location, and leachate collection are currently considered more important in protection of public health and the environment.

Current research is intended to develop tools for designing soil liners that limit pollutant release to a predictable minimum. The application is primarily for mono-waste landfills, but the information is also useful in designing soil liners for surface impoundments and in predicting the amounts of pollutants that will pass below the zone of incorporation in land treatment cells. The work is proceeding in two parallel phases. In the first, methods for predicting movement of liquid (leachate) are being developed and improved, and in the second, methods are being developed for predicting how well soil liners will remove pollutants from any leachate.

Work is in progress on empirical predictive techniques using samples of wastes and soils from the location of interest. This approach appears to be the most promising way to integrate the effect of waste and soil characteristics on pollutant retention processes. The precision and accuracy that will be achieved by this method remains to be determined.

In FY 1981, a TRD on "Landfill and Surface Impoundment Performance Evaluation" (SW-869) for use by permit writers was published for public comment. This TRD is for predicting how much leachate will be intercepted by a leachate collection system and how much will seep through the soil liner into underlying soils. Following analysis of public comment, this TRD will be published in FY 1983.[2] A related TRD on methods for measuring flow properties entitled "Soil Properties, Classification, and Hydraulic Conductivity Testing" (SW-925) will be issued for public comment in FY 1984. Research will be continued through FY 1985 to test the procedures in these two TRDs against field experience from the permit program for RCRA Section 3004. They will be updated as necessary.

One additional TRD on using batch techniques to predict pollutant removal from leachates by soil liners will be issued for peer technical review in FY 1984. Following public comment, final publication is planned for FY 1985. Field verification of this information will be achieved by a large-scale field project and by analysis of data collected under the hazardous waste permit program.

Pollutant Control

The overall objective of this research activity is to evaluate and assess technology that minimizes, contains, or eliminates pollutant release and leaching from waste disposed to the land. The studies are determining how various soils, flexible membranes, and admixed materials may be utilized as liners to contain and prevent leachates from migrating from landfill sites.

Procedures for selecting, placing, and predicting the life of liners in landfills and surface impoundments have been the main thrust of research efforts prior to FY 1981. Information on these procedures is contained in the TRD "Lining

[2]Copies of this document can be obtained from the U.S. Government Printing Office, Washington, D.C. (GPO Stock No. 055-000-00233-9).

of Waste Impoundments and Disposal Facilities" (SW-870). This TRD was revised and updated in FY 1982 based on public comments and additional research data. Final publication will be in FY 1983.[3]

In FY 1982 a research program was initiated to evaluate and assess advanced liner designs, including emerging flexible membrane liners and the use of nonpetroleum-based materials (such as sulfur/hydrocarbon compounds) in place of asphalt. FY 1983 studies will evaluate the seaming techniques necessary to adequately seam a variety of flexible membrane liner materials under various environmental conditions, the quality control necessary to ensure the integrity of the seams, and the beneficial aspects of placing soil or soil-like material over membrane liners. Studies will continue to ascertain the proper placement procedures and the use of nonsoil materials to assist in bedding requirements for synthetic membranes and leachate and gas collectors. Studies will be initiated to field validate nondestructive leak evaluation techniques of *in situ* materials. Repairs of damaged or leaking membrane liners will be evaluated on a quick patch, retrofitting, or complete replacement basis.

Another major effort in FY 1983 will be the identification of federal agencies actively involved in liner technology research for the purpose of data interpretation on a national scale. This information will be utilized to standardize tests and materials for industry use.

The research program will continue laboratory and pilot-scale operations and initiate field verification studies in early FY 1983. Laboratory studies will assess various approaches on predicting service life based on the chemistry of the liner and waste. Field conditions and their effects on liner materials will identify problems either as improper liner selection or placement and mismanagement of liner systems. The field surveys will also identify why certain systems continue to perform as based on original design concepts. As these data become available, the TRD will be updated. It is also envisioned that additional TRDs could be generated on specific liner technology achievements.

Pollutant control research is also addressing the issue of minimizing moisture infiltration into the site by constructing adequate cover systems. Two TRDs were developed in FY 1980 for public comment. SW 867, "Evaluating Cover Systems for Solid and Hazardous Waste," provides procedures for evaluating and designing cover systems utilizing existing site materials, bringing in additional soil types not found at the site, or using synthetic membranes to act as moisture barriers. "Hydrologic Simulation on Solid Waste Disposal Sites" (SW 868) is an interactive computer model to determine the moisture infiltration through a cover system. In FY 1982 this model was revised for two-dimensional considerations, and additional default values were added. Both TRDs were revised in late FY 1982 and were published in final form in FY 1983. Copies of these two documents can be obtained from the U.S. Govern-

[3] Copies of this document can be obtained from the GPO (Stock No. 055-000-00231-2).

ment Printing Office (GPO), in Washington, D.C.[4] As additional research information becomes available, the TRD will again be revised. Research initiated in FY 1982 includes a project addressing the quality control procedures which need to be applied to geotechnical parameters that are evaluated during land disposal site design and construction. In addition, a field-scale study designed to provide soil moisture data which would allow verification of drainage models applied to landfill cover systems was begun. This project includes an approach to measuring the impact of subsidence imposed on a small section of the constructed cover system. The FY 1983 emphasis will be directed toward field verification studied for validation of concepts and techniques developed by the research program.

Pollutant Treatment

The objective of pollutant treatment work is to identify the composition of leachates being generated from hazardous waste landfills and to assess the application of treatment technology for neutralizing or detoxifying the collected leachate.

The use of a leachate collection and removal system to minimize the seepage of leachate into underlying soils is an integral part of the landfill concept for hazardous waste disposal. Consequently, research has been initiated characterizing procedures for collecting and treating leachate to ensure that it is disposed of safely.

Leachate treatment schemes will likely include biological, physical, or chemical processes either singly or in combination, depending on the characteristics of the leachate involved. Similarities will exist between the combination of processes selected for leachates from codisposal and leachates from strictly municipal solid waste (MSW). The leachate characteristics from codisposal facilities may be expected to be very similar to municipal refuse leachate. Consequently, completed studies of MSW should be useful in selecting appropriate treatment schemes.

In FY 1981 a TRD, "Management of Hazardous Waste Leachate" (SW-871), was issued for public comment. Also, pilot-scale treatment studies with several waste leachates and contaminated groundwaters were initiated.

During FY 1982 the TRD was revised to incorporate public comment and interim results from treatment projects and was published in final form in FY 1983.[5] In FY 1983, further studies to develop physical, chemical, and biological approaches to treatment of leachates will be initiated. The research will focus on protocols for matching leachate quality with the appropriate combination of treatment processes. During this period additional field studies will

[4] Stock Nos. 055-000-00228-2 and 055-000-00225-8, respectively.
[5] GPO Stock No. 055-000-00224-0.

be initiated to validate the results obtained from bench and pilot-scale studies. By FY 1985 the TRD will be updated to take into account the results from the various research programs and experience with use of the TRD in the hazardous waste permit program.

Waste Modification

Chemical solidification and stabilization is achieved by incorporating the solid and liquid phases of a waste into a relatively inert matrix which exhibits increased physical strength and protects the components of the waste from dissolution by rainfall or by soil water. If this slows the rate of release of pollutants from the waste sufficiently and no serious stresses are exerted on the environment around the disposal site, then the wastes have been rendered essentially harmless and technical restrictions on siting will be minimal.

The objective of the waste modification work is to validate the performance of chemically solidified and stabilized wastes. This validation addresses four basic goals:

(*a*) to improve the handling and physical characteristics of waste materials,
(*b*) to decrease the surface area across which transfer or loss of contained pollutants can occur,
(*c*) to reduce solubility of any pollutant contained in the waste, and
(*d*) to detoxify the contained pollutants.

One of the options being investigated is alteration of the waste to minimize or prevent release of pollutants from the wastes. Fixation converts the waste to a physically stronger or less soluble form whereas encapsulation incorporates the waste into a strong, low-solubility matrix. Encapsulation, the more durable and expensive process, is designed for low-volume, high-toxicity wastes.

In FY 1981 a TRD entitled "Guide to the Disposal of Chemically Stabilized and Solidified Waste" (SW-872) was published for public comment. Although it contains some information on encapsulation, it deals mainly with fixation processes and gives procedures for evaluating the suitability of combinations of inorganic wastes and fixation processes. The TRD was revised in late FY 1982 based on public comments and additional research data and was published in final form in 1983.[6] A field survey of seven to ten-year-old sites containing chemically fixed inorganic wastes was undertaken to develop verification information for the TRD.

In FY 1983, information will be expanded on fixation processes for organic wastes and work initiated on development of procedures for evaluating fixation process suitability under long-term (100 years) worst-case scenarios. Studies of encapsulated waste products and production of state-of-the-art for encapsulation processes will be completed.

[6] GPO Stock No. 055-000-00226-6.

Surface Impoundments

The surface impoundment research program has been modified to provide a comprehensive understanding of the design, operation, and maintenance of surface impoundments as options for hazardous waste disposal. Information is needed on the use of natural soils as liners and the correlation of laboratory measurements with the construction standards achievable in the field. Of particular interest is the degree to which specification of construction techniques and inspection practice can influence uniformity and performance of the finished impoundment.

Containment

The use of natural soils as liners is attractive because on-site availability can lower construction costs. However, waste-soil interactions can pose problems in addition to those normally involved in earthen structures. The potential contact of soluble and liquid organic compounds with soil is much greater in surface impoundments than in other disposal facilities. Any resulting changes in soil structure or permeability are serious because of the large area of liquid/surface interface contact and the effect of the hydraulic head. Work is underway on predicting the effects of organic wastes on soil properties. This will be continued and opportunities for field verification developed. The emphasis is on identifying effective techniques for determining the compatibility of a specific waste and soil rather than on collecting comprehensive data concerning a variety of wastes and soils.

Work on the passage of pollutants through soil liners is being conducted as part of the work on landfills; work on the effects of organic wastes on the durability and other properties of soil and flexible membrane lining materials is also included. The information developed will be applied to surface impoundments as well. Procedures for evaluating the effects of wastes on natural and flexible membrane lining materials and earthen structures that are unique to surface impoundment—for example, wave action, aeration and mixing, sunlight, and solids settling on the liner in the impoundment—will be evaluated.

Because of the large ratio of exposed surface area to volume in surface impoundments, the potential for emission of volatile organic compounds into the ambient air is of particular concern. A procedure for predicting air emissions as a function of waste composition, impoundment design, and weather conditions is being developed. A field monitoring program will verify the predictive method. A final phase in this work will be the development of a TRD addressing the use of the predictive model and the field verification procedure.

Another area of concern is the siting and size problem. The socioeconomic aspects of this problem will be addressed in tandem with similar research effects for landfills and land treatment facilities. A complicating constraint that must be addressed is the economic disincentive to locate a surface impoundment at any great distance from the source generating the waste. The size problem is independent of location and requires development of methods for assessing the trade-off between increased potential for environmental damage (area seepage, dike failure) from larger impoundments and the economies of scale achievable as impoundment size is increased. Initial studies will examine the magnitude of these factors for a series of smaller impoundments containing the same total volume as a single large impoundment.

Closure of surface impoundments has been an initial concern because of the uncontrolled dump site problem. As such, a TRD entitled "Closure of Hazardous Waste Surface Impoundments" (SW-873) was published in FY 1981 for public comment and review. It has been revised and published in FY 1983.[7]

Performance Evaluation

Additional TRDs are being developed for conducting performance evaluations of surface impoundment under different designs. The source of information and data will be generated from field surveys where the advantages and disadvantages of specific designs, management procedures, geographic location, and waste contained are identified and evaluated. This data set is envisioned to produce information that can be modeled to predict the success or failure of future designs. This is expected to be available in early FY 1985.

Underground Mines

The use of underground salt mines for the long-term controlled storage of nonradioactive hazardous wastes has been shown to be promising as a result of studies by the EPA and others and by the successful commercial operation of such a facility in West Germany since 1973.

A research program on characterization and containment of hazardous wastes for mine disposal was completed in 1977. This effort included determination of suitable mine locations, determination of waste handling requirements, specification of operational techniques, and the preliminary design of a mine disposal system. This material was used to develop a feasibility report. Based upon this report, one or more candidate mines were selected for further evaluation and detailed economic design. An engineering plan for an initial pilot-scale operation was developed, and a full-scale demonstration was planned in cooperation with a private firm but never implemented. In FY 1983 an update of hazardous waste disposal into underground mines state-of-the-art will be initiated.

[7] GPO Stock No. 055-000-00227-4.

Economic Assessment

The objective of economic assessment work is to provide cost information and evaluation methods for land disposal and remedial action technologies.

Recently completed efforts include the development of a method to analyze the economic and social effects of alternative approaches to hazardous waste management, the determination of unit costs for 16 treatment and 5 disposal techniques applicable to various hazardous waste streams, and the analysis of unit operations cost data for long-term remedial action programs at uncontrolled hazardous waste sites. Ongoing efforts include the analysis of actual remedial action costs from case studies, the development of a standardized method for conducting cost-effectiveness analysis of remedial actions at uncontrolled sites, and the analysis of human safety and degree-of-hazard considerations on remedial action costs. Technology cost information is also being developed as part of many efforts involving land disposal and remedial action technology evaluation and development.

Technical Resource Documents

The TRDs mentioned in this paper consist of a series of documents which represent best engineering judgment (BEJ) for the design, operation, and closure of hazardous waste disposal facilities. These documents are a compilation of research efforts to date and are being developed to assist in the implementation of 40 Codified Federal Register, Parts 264, 265, and 267, regulations concerning hazardous waste disposal facilities (landfills, surface impoundments, and land treatment). These documents are technically oriented and not policy oriented. They are not regulatory by design, but are intended only to provide guidance in decision-making processes for use as a guide by permit writers for evaluating facility designs and potential performance of new waste disposal facilities. The documents will be updated continually as new information is developed. Some of the manuals provide specific design and operational information (such as closure and lining TRD), whereas others are intended as evaluation tools (for example, hydrologic simulation and performance evaluation). Eight TRDs have been completed to date. Six of these documents are related to landfills, and one document is related to surface impoundments and one to land treatment. Research on land treatment of hazardous wastes has been transferred to the Robert S. Kerr Environmental Research Laboratory, Ada, Oklahoma. However, the initial version of the "Hazardous Waste Land Treatment" (SW-874) TRD is being published by the Solid and Hazardous Waste Research Division and is therefore included in this report.[8] A brief description of each TRD document is listed here:

[8]GPO Stock No. 055-000-00232-1.

TRD 1, "Evaluating Cover Systems for Solid and Hazardous Waste" (SW-867): This document presents a procedure for evaluating cover systems for solid and hazardous waste landfills. It contains a checklist, with supporting documentation, that a permit writer can use to evaluate a proposed cover system.

TRD 2, "Hydrologic Simulation on Solid Waste Disposal Sites" (SW-868): This document provides a computer package to aid planners and designers by simulating hydrologic characteristics of landfill covers to predict percolation as a function or cover design and climate.

TRD 3, "Landfill and Surface Impoundment Performance Evaluation" (SW-869): This document describes how to evaluate leachate collection systems using compacted clay or flexible membrane liners to determine how much leachate will be collected and how much will seep through the liner into underlying soils. The adequacy of sand and gravel drain layers, slope, and pipe spacing are also covered.

TRD 4, "Lining of Waste Impoundment and Disposal Facilities" (SW-870): This document provides information and guidance on liner systems. It discusses waste, liner types, compatibility, liner selection, specifications, design of leachate collection systems, and case study analysis methodology. It also includes a glossary of liner system related terms.

TRD 5, "Management of Hazardous Waste Leachate" (SW-871): This document discusses leachate composition, leachate generation, selected management options, available treatment technologies, unit-treatment options, and system economics. It presents management options that a permit writer or hazardous waste landfill operator may consider in controlling and treating leachate.

TRD 6, "Guide to the Disposal of Chemically Stabilized and Solidified Wastes" (SW-872): This document provides basic information on stabilization and solidification of industrial wastes and examines the regulatory considerations, current and other options, involved in disposal systems using stabilization and solidification of wastes. A summary of the major physical and chemical properties of treated wastes is presented. A listing of major suppliers of stabilization and solidification technology and a summary of each process are included.

TRD 7, "Closure of Hazardous Waste Surface Impoundments" (SW-873): This document discusses and references the methods, tests, and procedures involved in closing a surface impoundment in a manner that (*a*) minimizes the need for further maintenance and (*b*) controls, minimizes, or eliminates, to the extent necessary to protect human health and the environment, postclosure escape of hazardous waste, hazardous waste composition, leachate, contaminated rainfall, and waste decomposition products to groundwater, surface waters, or the atmosphere. Problems that have been overlooked in abandoned impoundments and have caused environmental degradation are discussed. The document discusses closing an impoundment either by removing the hazardous wastes or by consolidating the waste on site and securing the site as a landfill. Relevant literature or specific procedures are documented to provide greater detail.

TRD 8, "Hazardous Waste Land Treatment" (SW-874): This document discusses the land treatment disposal concept, characteristics of hazardous waste, allowable waste loadings, facility planning for land treatment, land treatment permit process, closure, and items to be considered during permit evaluation.

Conclusions

This paper represents an overview of the hazardous waste land disposal research program conducted by the EPA. Specific areas discussed in order of priority were landfills, surface impoundments, and underground mines. All activities in these areas are being summarized into technical resource documents to provide guidance in regulatory decision-making processes. These documents will be updated regularly to reflect the latest research findings; the documents will eventually be condensed into guidance manuals. A guidance manual will be prepared for landfills and surface impoundments. More information about specific aspects of the hazardous waste land disposal research program can be obtained by contacting the author.

Brian D. Gillen[1] *and Wayne K. Tusa*[2]

Assessment of Hazardous Waste Management Practices for Potential Damage to the Environment in EPA Region III

REFERENCE: Gillen, B. D. and Tusa, W. K., **"Assessment of Hazardous Waste Management Practices for Potential Damage to the Environment in EPA Region III,"** *Hazardous and Industrial Waste Management and Testing: Third Symposium, ASTM STP 851*, Larry P. Jackson, Alan R. Rohlik, and Richard A. Conway, Eds., American Society for Testing and Materials, Philadelphia, 1984, pp. 334-345.

ABSTRACT: This paper evaluates damage case histories associated with land-based and non-land-based hazardous waste facilities in U.S. Environmental Protection Agency (EPA) Region III. This information was extrapolated from a nationwide study authorized by the EPA Office of Solid Waste. It provides an outline of the kinds of environmental damage associated with certain types of facilities and contamination events and measures the overall extent and damage resulting from mismanagement of hazardous wastes in the region.

Approximately 93% of the sites evaluated had evidence of suspected or documented contamination. Damage loss was suspected or documented at 73% of the sites involving contamination, with 38% of the sites evaluated having documented evidence of damage to human health or the environment. The data compiled in this study were used with similar information compiled from the other EPA regions, as part of the agency's ongoing Regulatory Impact Analysis (RIA) program and for use in preparing regulations tailored for specific facility types.

KEY WORDS: contamination, damage, documented damage, suspected damage, facility types, landfills, surface impoundments, containers, tanks, EPA Region III, volatile nonhalogenated organics, volatile halogenated organics, metals, hazardous wastes, industrial wastes

Under the Resource Conservation and Recovery Act (RCRA), the U.S. Environmental Protection Agency (EPA) Office of Solid Waste (OSW) is responsible for promulgating hazardous waste management facility regulations. In ad-

[1]Manager, engineering design group, Solid and Hazardous Waste Services, Fred C. Hart Associates, Inc., New York, N.Y. 10036.

[2]Associate and office manager, New York Office, Fred C. Hart Associates, Inc., New York, N.Y. 10036.

dition to developing the necessary data base to support these regulations, the OSW must also develop Regulatory Impact Analyses (RIAs), as required by Executive order 12291, issued 18 Feb. 1981. In order to achieve these objectives, the EPA felt it necessary to develop and compile a data base of damage case histories associated with hazardous waste disposal facilities. This task was completed, along with a subsequent follow-up analysis of facility types located in over 1000 sites nationwide. This paper is based on the evaluation of the 164 sites in Region III taken from the nationwide study.

The information generated under that study will serve to provide the following data:

(a) a compilation of damage information on a large number of active and inactive disposal sites meeting certain criteria,

(b) information concerning different kinds of environmental damage associated with certain types of contamination, and

(c) a measure of the overall extent of contamination and damage resulting from the mismanagement of hazardous wastes.

The OSW intends to submit the compiled information for incorporation into the administrative record prior to finalization of the RIA process. In addition, the data will be useful in preparing regulations tailored to specific facility types, in evaluating alternative regulatory scenarios, and for use as background information in resolving a variety of other technical issues.

Procedure

Outline Work Approach

A number of specific tasks were completed in order to accomplish the overall project objectives. These included the following:

1. Existing potential sources of information concerning damage from hazardous waste disposal sites were identified, reviewed, and assessed. These sources included the EPA's Site Tracking System (STS), the Superfund Notification System (NOTIS), the Hazardous Waste Data Management System (HWDMS), the Surface Impoundment Assessment (SIA) files, the Center for Disease Control files, the Eckhardt Report, EPA Regional Office files, and the Regional Field Investigation Team (FIT) files.

2. The FIT and EPA Regional Office files were selected as having the data most suitable and readily available for the intended purpose.

3. Site selection criteria were developed to meet the technical information requirements of the EPA at minimal cost. The criteria to be evaluated in this study included the preferential selection of the following sites:

(*a*) all sites having available sampling and analytical data,

(*b*) all sites scored using the Mitre Hazard Ranking System (HRS) model (the interim list of 175 sites, rescored under the direction of EPA in September and October of 1981), and

(*c*) any site associated with waste storage.

4. Approximately 1000 sites were selected for preliminary file review using these criteria. (Additional technical detail on the rationales behind the criteria is discussed in the next subsection.)

5. A Damage Incident Summary Form (DISF) to record the data collected from individual site files was developed.

6. Review criteria were developed to ensure uniformity of DISF responses regarding the identification of contamination and damage events, the rating of damage severity, and the determination of the amount of file documentation required to support given responses, and so on.

7. The selected damage incident cases at each of the FIT and EPA regional offices were reviewed to complete the DISF survey form.

8. The data were analyzed and a report was prepared entitled "Assessment of Hazardous Waste Mismanagement Damage Case Histories" (September 1982), which is the report on which this paper is based.

Case Selection Criteria

Table 1 summarizes the final six criteria developed to select damage case histories contained in the regional FIT and EPA files. Files or sites conforming with these criteria were identified as the most suitable in meeting the project goals. Since site selection was not a random process, the reader is cautioned against making generalizations based on the findings of this analysis.

Criteria 1 and 4 identify files associated with sites for which sampling and analytical data were available. These files generally were those sites inspected, investigated, and sampled by FIT or the EPA survey teams, or both. Sites for

TABLE 1—*Summary of case selection criteria for evaluated sites for FIT and EPA files.*

	Criterion Number	Criterion Description
Field Investigation Team (FIT) Files	1	files having sampling data
	2	files associated with waste storage
	3	files associated with Mitre (HRS) scored sites[a]
EPA Survey and Analysis Division Files	4	files having sampling data
	5	files associated with waste storage for which there is evidence of damage
	6	files associated with Mitre (HRS) scored sites[a]

[a]The interim list of 175 Mitre HRS scored sites, rescored during September and October 1981 under the direction of EPA.

which sampling data were not available usually were not sufficiently detailed to support damage case assessments. Environmental sampling results reported in approximately 40 to 50% of the files evaluated were based on analysis of samples collected by either the EPA or FIT field survey teams. While verification of the sampling and analysis protocols was beyond the scope of the report, it is assumed that samples collected and analyzed by the EPA and FIT teams conformed in most cases with the EPA "Test Methods for Evaluating Solid Waste, Physical and Chemical Methods" (May 1980) (SW 846) and *Standard Methods for the Examination of Water and Wastewater.*[3]

Criteria 2 and 5 identified sites associated with hazardous waste storage, such as tank and container facilities. Criterion 5 was further refined to include only those sites for which there was preliminary evidence of damage in order to generate a data base consistent with project goals and resources.

Criteria 3 and 6 targeted the 175 highest scoring sites on EPA's interim Superfund list identified under the Superfund program developed by the Mitre HRS model. These sites were included in the survey based on the assumption that environmental damage could readily be documented at these locations.

Evaluation Procedures

Files in each region were evaluated by a study team consisting of a project director, team leader, and four to five technical assistants. Guidelines, definitions, and criteria used by the study team in making the interpretations and judgments needed to complete the DISFs are discussed in detail in the project report.

In summary, the evaluation procedure consisted of a two-phase effort. Phase I consisted of visiting regional FIT and Survey and Analysis offices, screening files according to the selection criteria, and transferring the appropriate information to the DISFs. This effort was accomplished over a noncontinuous nine-week period beginning in November 1981 and ending in February 1982. The second phase consisted of reviewing the completed DISFs for consistency, format, and editorial standards; tabulating the conformed DISFs; and summarizing the information in the report. This effort was accomplished over a period of several weeks, beginning in late March 1982 and ending with the submission of the report.

The DISF was used to assess damage case histories and associated site characteristics. After the study team reviewed the file information, Appropriate responses were entered in Sections I through XII of the DISF. The case then was summarized in a brief narrative, Section XIII, and was attached to the DISF form. Sections I through XIII of the DISF are (I) Site Identification, (II) Site Description, (III) Date of Incident/Discovery, (IV) Status of Operations, (V) Exposed Media, (VI) Affected Areas, (VII) Epidemiological Studies, (VIII)

[3]Fourteenth edition, American Public Health Association, Washington, D.C., 1976.

Event Causing Incident, (IX) Waste Characterization, (X) Status of Response, (XI) Source of Information, (XII) Severity of Damage, and (XIII) General Comments.

Evaluation Criteria

The DISF responses for Sections I, II, III, IV, VII, X, and XI were prepared from information available in the files according to the definitions and instructions accompanying the DISF. The DISF responses for Sections V, VI, VIII, IX, and XII required value judgments based on evaluation criteria developed for each value judgment question. For example, the study team was frequently required to assess whether contamination had occurred, what media had been exposed, what event had caused the incident, and what waste resulted in contamination. Finally, the study team was required to assess the severity of damage to human health or the environment, or both. In order to ensure that the study team rated sites uniformly, specific evaluation criteria were developed for use as guidance in (1) identifying contamination and damage events, (2) rating the severity of damage, and (3) determining what file documentation required to support a given response (that is, documented versus suspected).

In this study, contamination is defined as the presence of pollutants in groundwater, surface water, soil, or air, as identified by standard sampling and analysis techniques. Identification is defined as positive contaminant verification at concentrations above the detection limits of the sampling and analysis techniques used. Pollutants are defined as substances not naturally found in the site-specific environment, which may interfere with the best use of, or cause environmental harm to, the affected resource. The verifiable concentration levels varied but in most cases were in the parts per billion (ppb) range.

Contamination was considered to be documented if the event was substantiated by a direct investigative action by a regulatory office or other recognized agency. File information required to support documentation included the following material:

(*a*) sampling and analysis data,
(*b*) relevant documents (engineering reports, environmental impact statements, NPDES and RCRA permits, enforcement actions, and so on, and
(*c*) professional evaluations, expert witness testimony, and so forth.

Damage was defined as the presence of pollutants at concentrations causing interference with, loss in quality of, or harm to human health, drinking water, the food chain, flora, fauna, or property. Damage was considered to be documented according to the same evaluation criteria discussed previously, along with these additional criteria:

1. The DISF responses indicating documented damage to human health were to be based on authoritative references in the file linking sickness, injury,

or death to contamination events occurring at the site. These references would typically include hospital reports, Occupational Safety and Health Agency (OSHA) citations, regulatory agency reports, facility operating reports and, in some cases, epidemiological data.

2. The DISF responses indicating documented damage to drinking water were to be based on authoritative references in the file correlating excessive contaminant concentration levels in the water supply with contamination events occurring at the site. These concentration levels were defined as exceeding EPA National Interim Primary or Secondary Drinking Water Standards or EPA Human Health Criteria for Maximum Contaminant levels (MCLs) in water supplies.

3. The DISF responses indicating damage to the food chain and flora were to be based on authoritative references linking visible vegetative stress to contamination events occurring at the site.

4. The DISF responses indicating documented damage to fauna were to be based on authoritative references, usually bioassay studies, linking damage to fish and wildlife to contamination events occurring at the site.

5. The DISF responses indicating documented property damage were to be based on authoritative references linking property damage to contamination events occurring at the site. These references generally include insurance claims, regulatory reports, OSHA citations, and enforcement actions restricting the use of residential property, drinking water, or other site or facility usage.

Damage was considered to be "suspected" if the responses in Section VI were based only on citizen allegations, newspaper reports, or inconclusive scientific studies.

The study team also rated each site according to the severity of human health and environmental damage. Table 2 outlines broad guidelines developed to rate site severity of damage. As noted in this table, high human health damage ratings were assigned to sites where incidents resulted in deaths, whereas low damage ratings were associated with minor, short-term injuries. High environmental damage ratings were typically associated with sites associated with substantial fish or animal kills, or with groundwater contamination incidents in which contaminant concentrations exceeded ten times the drinking water standards discussed previously. Low environmental damage ratings were usually associated with sites where soil or vegetation contamination was limited to small areas.

Results

The study team evaluated and completed DISFs for a total of 164 sites in Region III. Many of the sites contained multiple facilities. A total of 317 individual facility types were used in describing the 164 sites in the this region. Of the

TABLE 2—*Summary of guidelines used in rating severity of damage at evaluated sites.*

Category	Severity		
	High	Medium	Low
Human health			
	Damage incident to at least one person resulting in:		
	death	severe injury	minor injury
	Contamination of groundwater resulting in closure or restriction of		
		drinking water in a community water supply	single private well
Environmental			
Groundwater, surface water and air	Contamination incident in which sampling indicates the presence of pollutants in concentrations:		
	at levels *greater* than *10 times* applicable standards	at levels *equal* to applicable standards	at detectable levels, but less than applicable standards
Flora and food chain	Contamination incident resulting in stress to vegetated or food crop area:		
	greater than 1 acre	between ½ and 1 acre	in limited areas only
Fauna	Damage incident confirmed by:		
	massive kills	limited kills	bioassay studies confirming tissue contamination
Soil	[a]	[a]	contamination incident confirmed by sampling data

[a] Higher levels of damage were typically identified by using evidence in the other categories.

317 facility types evaluated, 27% were landfills, 19% were containers, 14% were surface impoundments, and 9% were tanks. The remaining 31% were described by other categories (see Fig. 1).

Discussion

Contamination, either documented or suspected, was identified in 151 sites, or at 92% of the sites evaluated. At 93 of the sites, or 57%, contamination was documented. Figure 2 indicates that 38% of the contamination incidents involved groundwater, with the remaining contamination occurring in soil (28%), surface water (27%), and air (7%). Of the 366 responses originally indicating contamination, only 126 (34%) could be documented using the evaluation criteria previously developed.

Each site was also evaluated for damage occurring to life, property, and various natural resources. As previously explained, this evaluation focused on six potentially affected areas, including drinking water, the food chain, flora, fauna, human health, and property. Damage (either documented or suspected) was identified in 119 sites, or 73% of the sites evaluated. Figure 3 (extreme left bar graph) compares the total number of evaluated sites with the

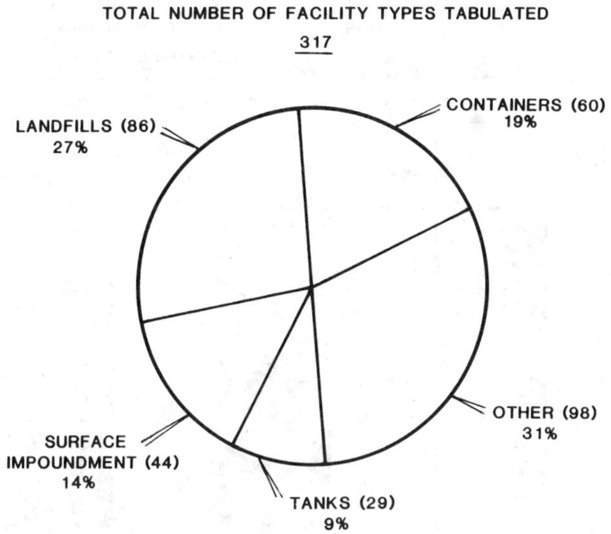

FIG. 1—*DISF summary of evaluated hazardous waste sites in U.S. Environmental Protection Agency Region III: facility types.*

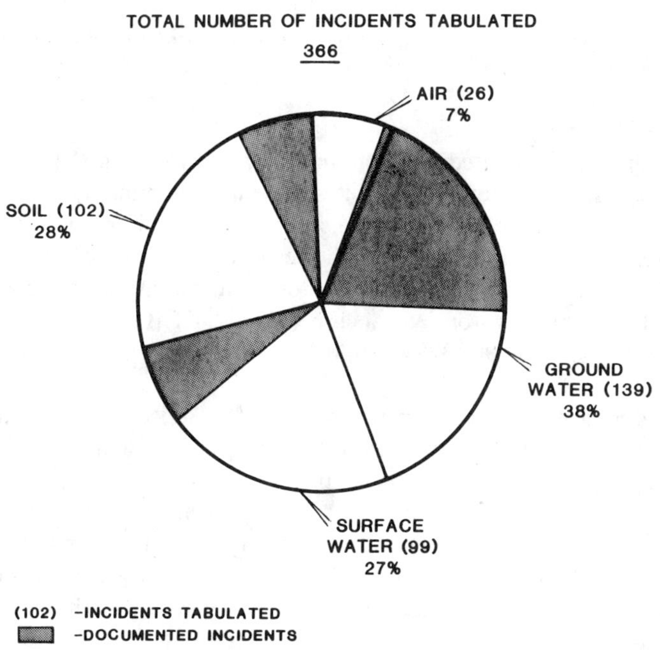

FIG. 2—*DISF summary of evaluated hazardous waste sites in EPA Region III: media contaminated.*

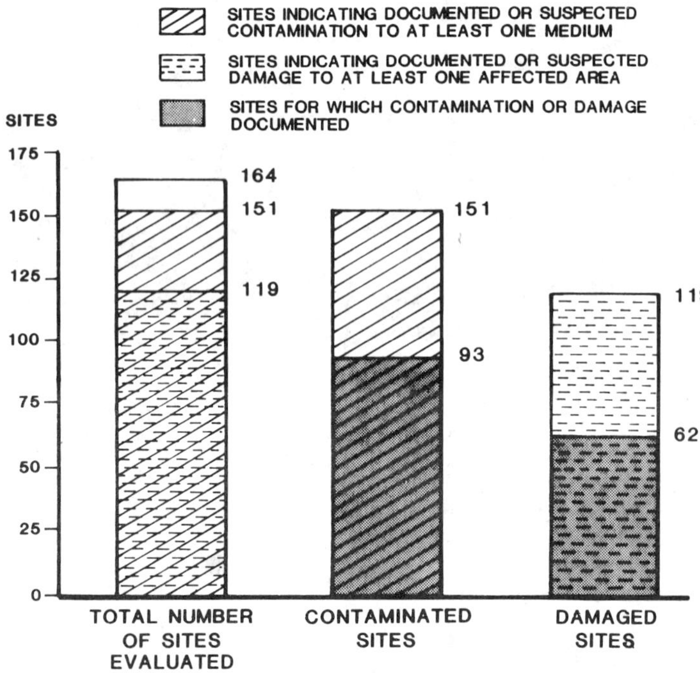

FIG. 3—*DISF summary of evaluated hazardous waste sites in EPA Region III: tabulation of sites contaminated and damaged.*

total number of sites rated as "contaminated" or "damaged." ("Contaminated sites" are here interpreted as sites causing contamination to be at least one medium, "damaged sites" are those sites resulting in damage to one affected area.) Figure 3 also compares the respective fractions of contaminated sites (middle bar graph) and damaged sites (right bar graph) associated with the files having the appropriate documentation. Of the 301 affected areas showing damage, only 96 (32%) could be documented using the evaluation criteria.

Figure 4 indicates that approximately 44% of the documented incidents involved damage to drinking water, with the remaining incidents including damage to property (28%), flora (10%), fauna (7%), human health (7%), and the food chain (4%). Figure 5 indicates that 76% of the incidents causing the damage or contamination just described were the result of leachate (33%), leaks (18%), spills (17%), erosion (8%), and other causes. These incidents involved contamination caused by volatile halogenated organics, volatile nonhalogenated organics, or metals, in 64% of the incidents tabulated.

The text of the report provides additional data for different facility types, as well as evaluation of individual EPA regions (Regions I through X). Specific in-

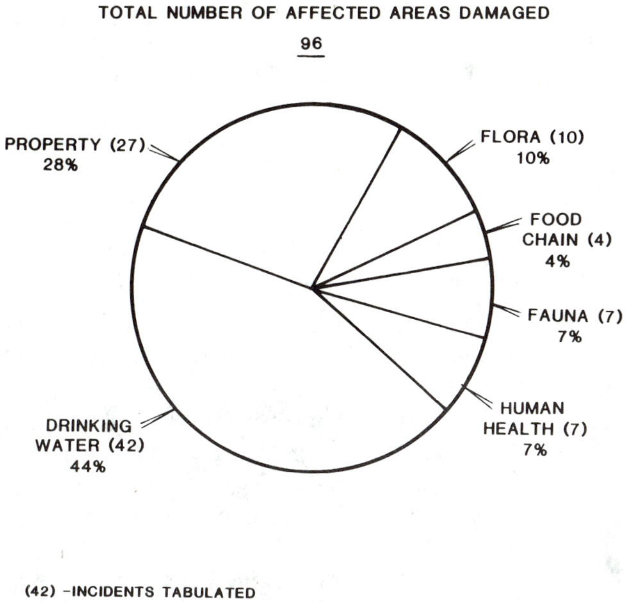

FIG. 4—*DISF summary of evaluated hazardous waste sites in EPA Region III: documented cases of affected areas damaged.*

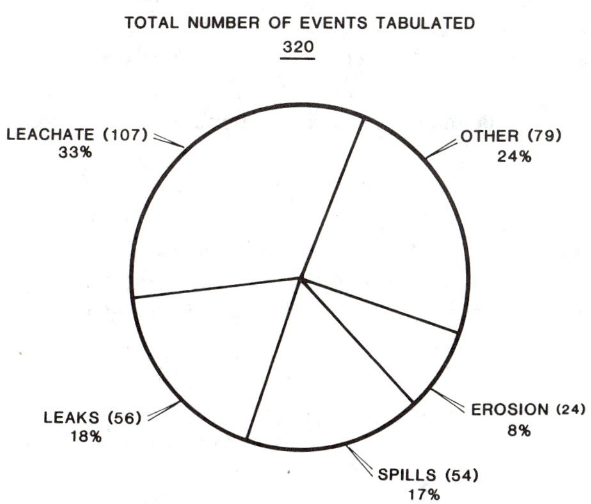

FIG. 5—*DISF summary of evaluated hazardous waste sites in EPA Region III: events causing contamination.*

formation is also provided on the sources of the data, profiles of the facilities, the events causing contamination, the types and ranges of concentrations of chemicals recorded in the files, and the status of remedial responses indicated in the files.

Findings and Conclusions

As a result of the data base review, the site history reviews, and the subsequent data analysis, the study team reached the following overall conclusions:

1. Of the data bases examined in this study, the FIT and regional EPA files contain the most readily accessible data on potential damages from hazardous waste disposal facilities.
2. For those sites that met the selection criteria, the facility types most commonly identified with potential contamination or damage included landfills, containers, surface impoundments, and tanks.
3. Approximately 92% of these sites had evidence of suspected or documented contamination.
4. Approximately 62% of the sites indicating the potential presence of contamination had documented evidence of contamination.
5. Groundwater was the medium that was most often contaminated, followed by soil and surface water, which were contaminated at approximately the same number of sites.
6. The events most often associated with contamination included leachate, leaks, spills, and erosion.
7. The most commonly identified contaminants included volatile halogenated organics, volatile nonhalogenated organics, and metals.
8. Damage was suspected or documented at 73% of the sites evaluated, or at 79% of the sites involving contamination; in total, approximately 38% of the sites evaluated had documented evidence of damage to human health or the environment.
9. The suspected damage most often reported was damage to human health, drinking water, flora, and fauna; the documented damage most often reported was damage to drinking water and property.
10. While remedial programs varied on a case-by-case basis, various legal actions or remedial activities, or both, have been initiated at a significant number of sites evaluated in this study. For example, legal or enforcement activities have occurred at 17% of the sites, while 46% of the site have had or are currently undergoing additional environmental investigations. At approximately 26% of the sites, remedial activities of some type have been initiated.
11. It is particularly important for the reader, throughout this analysis, to be cognizant of the fact that the 164 sites evaluated were selected on the basis of specific criteria. These criteria included preselection of active and inactive hazardous waste disposal sites associated with cases for which there were sampling

data, cases associated with waste storage, and Mitre scored sites. In view of this preselection process, it should be noted that these cases are not necessarily representative of all damage cases on file at the EPA or in actual existence. As a consequence, it is difficult to apply the findings of this analysis to other data bases on hazardous waste facilities, abandoned sites, and so forth, or to develop more generalized conclusions based on the efforts of this study.

Acknowledgments

The authors wish to express their appreciation to John Perry and Kent Anderson of the U.S. EPA Office of Solid Waste for their guidance in completing this project.

Harvey F. Ludwig,[1] *J. Warren Evans,*[1] *and Saksit Tridech*[2]

Industrial Solid Waste Management for the Eastern Seaboard—Thailand's First Systematic Approach to the Problem

REFERENCE: Ludwig, H. F., Evans, J. W., and Tridech, S., **"Industrial Solid Waste Management for the Eastern Seaboard—Thailand's First Systematic Approach to the Problem,"** *Hazardous and Industrial Waste Management and Testing: Third Symposium, ASTM STP 851*, Larry P. Jackson, Alan R. Rohlik, and Richard A. Conway, Eds., American Society for Testing and Materials, Philadelphia, 1984, pp. 346–360.

ABSTRACT: Accelerating population growth, urbanization, and industrialization in Thailand are creating unprecedented problems requiring government intervention for achieving systematic planning and growth of existing and new urban centers. The first such regional planning project in Thailand, carried out under the sponsorship of the National Economic and Social Development Board in 1981–1982, was focused on the Eastern Seaboard, the coastal zone on the Gulf of Thailand to the southeast of Bangkok. Because of the recent development of major natural gas fields in the gulf, the Eastern Seaboard region seems destined to become a new regional urban-industrial center. The planning project resulted in preparation of a comprehensive plan for guiding continuing development in the region, with particular attention to industrial development and environmental pollution control. Preliminary concepts were developed for management of industrial solid wastes at the various urban growth centers in the region, including bulk or nonhazardous wastes and hazardous wastes. Both conventional and innovative practices, adapted to the Eastern Seaboard situation, are utilized.

KEY WORDS: Thailand, solid wastes, industrial wastes, sanitary landfills, land reclamation, solid waste grinding, groundwater pollution, hazardous wastes, environmental pollution, marine ecology evaluation

The growth of cities in Thailand has taken place essentially without the intervention of government planning agencies, simply because the traditional role of government has not included this function. The government philosophy has largely been *laissez-faire* in this respect, but the pattern is now changing

[1]Chairman and consulting engineer, respectively, Seatec International Consulting Engineers, Bangkok, Thailand.

[2]Chief, Water Quality Section, National Environment Board of Thailand, Bangkok, Thailand 10112.

because of the pressures of accelerating population growth, urbanization, and industrialization. Whereas in the past, most of the population growth was in farm families, as usual, now the farm areas are mostly filled and the new population will be urbanites. With the total population expected to increase from the present 46 million to some 88 million before it stabilizes in the next several decades, this means an increase in urban population from its present level of about 6 million to more than 40 million. Thus far, the trend has been for the new population to move to the capital region at Bangkok, and in response to this situation, the government is initiating a role of intervention to try to induce the new population to move to other regional centers and, moreover, to guide the development of such centers through use of planning mechanics.

Eastern Seaboard Project

The first major government effort in this direction, carried out under sponsorship of the National Economic and Social Development Board (NESDB) in 1981–1982, was the Eastern Seaboard Regional Planning Project for preparing an appropriate plan for development of the coastal area along the Gulf of Thailand to the southeast of Bangkok (see Fig. 1). This Eastern Seaboard (ESB) region is of interest as an emerging regional center because of the recent development of large-scale natural gas fields in the gulf and because the gas reaches the mainland in the ESB region in the vicinity of Map Ta Phut, as shown in Fig. 1. Thus the stage is set for very significant development of new industrial complexes in the ESB, not only at Map Ta Phut but also at various locations along the coast stretching from Rayong in the south to Chonburi in the north.

The purpose of the ESB project was to develop a recommended growth plan by which the new industries can be wisely accommodated, including associated new communities and infrastructure, in harmony with the existing communities and infrastructure in the region. The ESB Regional Planning Project was completed in 1982 with the publication of a series of reports presenting the findings, conclusions, and recommendations of the study [*1*]. This represents the government's first effort in preparing a comprehensive plan for guiding systematic development of a regional center, and hopefully it will encourage continuing and increasing use of this approach. Needless to say, its implementation will involve quite a few changes in the traditional ways of doing things in Thailand.

The total investment proposed for the ESB development over the next decade includes about $2.94 billion for industrial plants and $0.86 billion for public infrastructure.

Environmental Aspects

The ESB project included an environmental analysis, for evaluation of all significant impacts of the expected developments on environmental resources

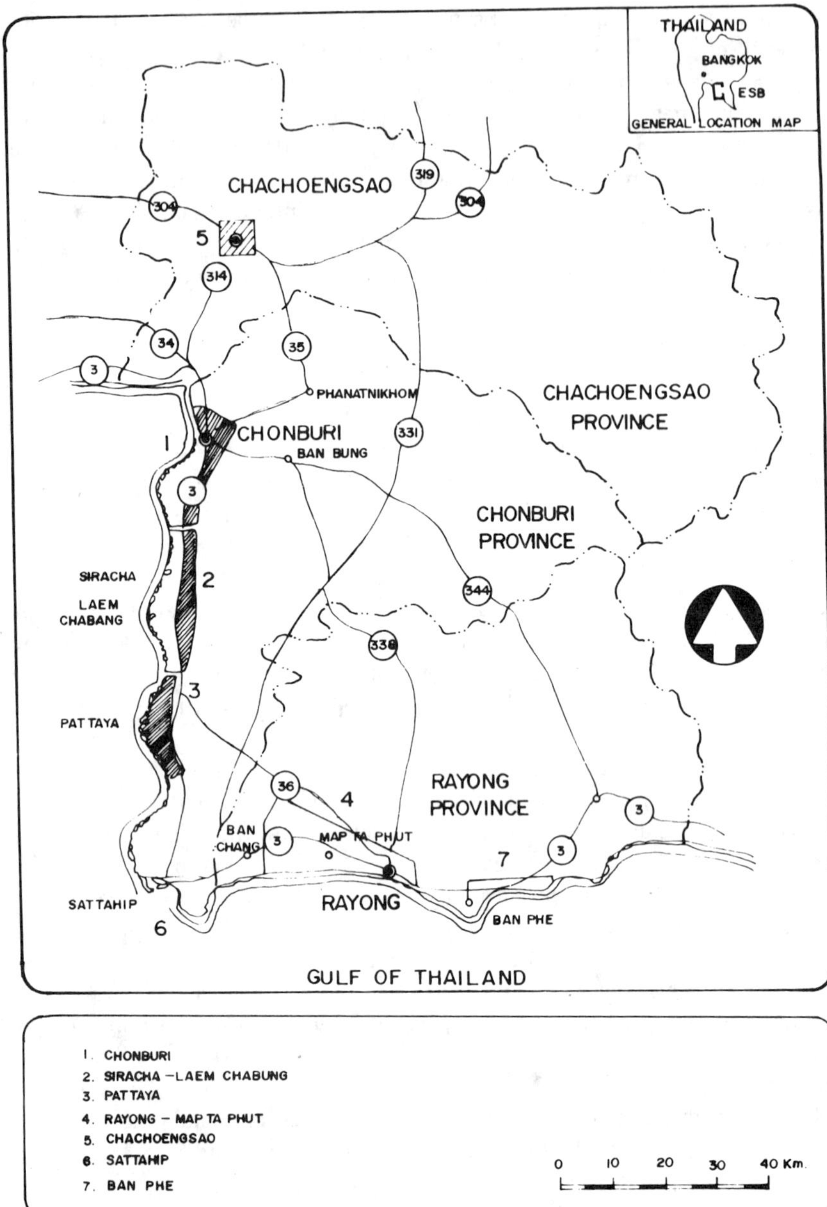

FIG. 1—*Proposed urbanizing growth centers for the eastern seaboard of Thailand.*

and values in the ESB region. Particular attention was given to environmental pollution control because of serious concern by officials of existing ESB communities that the new developments would degrade their community environments. This was especially so for (1) the municipality of Pattaya, an international beach and tourist resort city with many first-class hotels, which is heavily utilized by visitors to Bangkok and is in the vicinity of the proposed new major harbor project at Laem Chabang, and (2) Rayong, which is in the vicinity of the proposed new major industrial center at Map Ta Phut. Hence careful attention was given to environmental pollution control, for both sanitary and industrial sewerages, and to air pollution and solid wastes, as well as liquid wastes. For air pollution control, preliminary design analyses were made for evaluating the problem and for determining the appropriate solution on an industry-by-industry basis. For liquid wastes, the proposed system of disposal for most coastal communities involves the use of marine outfalls serving both communities and industries, with industries being required to install in-plant systems for achieving special preliminary treatment, when needed, prior to discharge into the public system. For municipal solid waste management, the study proposes primary dependence on the use of sanitary landfilling, including land reclamation along the coast.

All of the proposed solutions represent preliminary concepts to be accepted in principle on a conceptual basis for use in proceeding with implementation of the comprehensive ESB project. These concepts are to be developed in detail as part of the final planning and design of each significant project component. An important initial step in such further planning and in implementation of components will be preparation of an appropriate environmental impact assessment (EIA) for each such component. Also, an integral part of a proposed overall ESB management plan would be a continuing program of environmental monitoring of the entire ESB region, especially the coastal zone, to be carried out under the aegis of the National Environment Board (NEB) [*2,3*].

Industrial Solid Wastes

Existing and Projected Industries

Figure 1, taken from the ESB Planning Project report, shows the ESB region and the seven urbanizing growth centers included in the study. It shows that practically the entire coastal zone will become urbanized, which poses requirements for careful planning for preserving existing valuable environmental resources including Pattaya, Ban Phe, other beach resort areas, and the valuable coastal fisheries.

Figure 2 shows the same ESB region with specific information on the locations of new major industries. Also shown are the locations of (1) submarine outfalls for the proposed municipal sewage disposal systems (plus the inland treatment plant at Chachoengsao), (2) coastal mud flat zones, (3) proposed in-

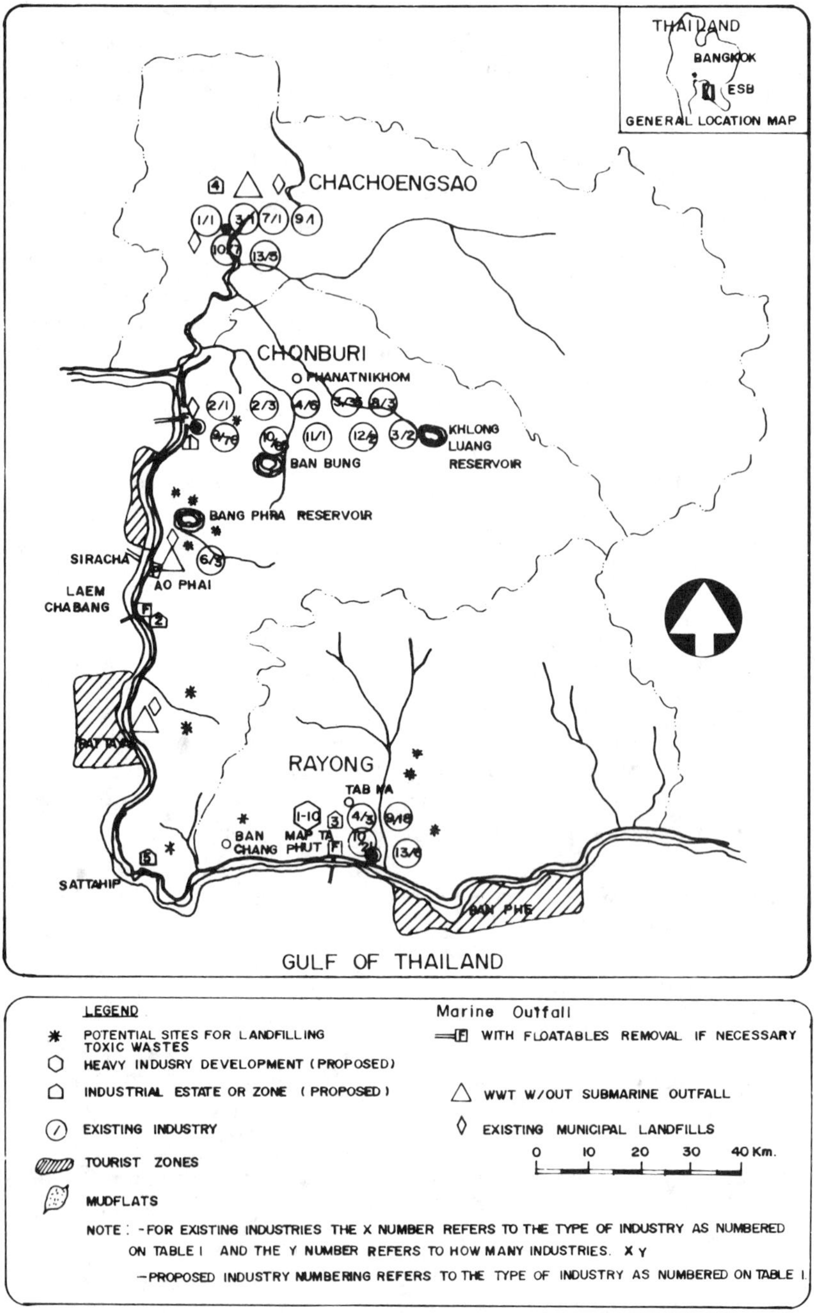

FIG. 2—*Background information relating to the disposal of industrial solid wastes in the eastern seaboard of Thailand.*

dustrial estates, (4) existing solid waste landfilling operations (essentially open dumps), and (5) preliminary selection of sites for inland landfilling where there are no significant groundwater resources.

Industrial Solid Waste Production

Table 1 lists the existing industries in the Eastern Seaboard, and Table 2 gives information on proposed new industries. With respect to existing industries, the only major sources of liquid and solid wastes are from petroleum refineries and from tapioca and sugarcane processing mills. Thus the ESB project will bring about a dramatic change in the regional complex. This will involve sizable expansions of the existing refineries plus establishment of various new major petrochemical industries, a major nitrogen fertilizer plant, a major soda ash industry, and others.

The ESB Planning Project focused major attention on management of industrial sewages and gaseous wastes but gave relatively little detailed attention to industrial solid wastes. However, Table 2 shows the types and amounts of solid wastes estimated to be produced for the major industrial projects included in the plan. These preliminary estimates are based on the various reports prepared by the proposed industries [for the Ministry of Industry or EIA reports for the National Environment Board (NEB)], on various publications of the U.S. Environmental Protection Agency (EPA), and on experience with existing industries elsewhere in Southeast Asia (for example, urea-producing plants in Indonesia). As indicated in Table 2, it is not now anticipated that any especially hazardous or toxic solid wastes will be produced, but it is recognized that such wastes will probably be produced sooner or later from plant processing add-ons or from numerous "downstream" industries yet to be defined.

TABLE 1—*Listing of major industries, by order of importance, in the Eastern Seaboard (existing and projected).*[a]

Existing Industries	Projected Industries	Projected Industrial Estates and Zones
1. Paper	1. vinylchloride monomer	1. Chonburi industrial estate (general purpose)
2. Wood processing	2. chemical fertilizer	2. Laem Chabang-Sri Racha industrial estate (general purpose and export processing)
3. Ceramic	3. gas separation	3. Rayong industrial estate (agro-processing)
4. Sugar refinery	4. steam cracker	4. Chachoengsao industrial estate (agro-processing)
5. Raw sugar processing	5. ethyl oxide ethylglycol	5. Sattahip export processing zone
6. Oil refinery	6. low-density polyethylene (recently started operation)	
7. Chemical fertilizer (NH_3, urea, acid plant)	7. high-density polyethylene	
8. Lime	8. soda ash	
9. Tapioca processing	9. polypropylene	
10. Tapioca pallet	10. methanol	
11. Coconut fiber		
12. Animal feed		

TABLE 2—*Estimates of amounts and characteristics of industrial solid wastes to be produced in the Eastern Seaboard.*

Source Industry	Production Capacity, tons/year	Solid Waste: Estimated Quantity, tons/year	Solid Waste: Source	Solid Waste: Hazardous Materials and Major Pollutants
Propylene[a]	73 000	500 to 600	process wastes, biological sludges	phenol, oil
Ethylene[a]	300 000	2 000 to 2 600	process wastes, biological sludges	phenol, oil
Vinylchloride[a]	80 000	600 to 700	process wastes, biological sludges	NA
Polypropylene[a]	70 000	500 to 600	process wastes, biological sludges	NA
High-density polyethylene[a]	110 000	1 000 to 2 000	process wastes, biological sludges	NA
Low-density polyethylene	73 500	1 200 to 1 300	process wastes, biological sludges	NA
Ethylene glycol[a]	50 000	400 to 500	process wastes, biological sludges	NA
Soda ash (AC process)[b]	400 000	1 200	water treatment sludge, wastewater treatment sludge	calcium and sodium salts, mercury, cadmium, lead, chromate, arsenic, polychlorinated biphenyls
Steam cracker	NA	NA	process wastes, boiler blowdown	acids, spent caustic phenols, tars, heavy oils, heavy metals, sulfite, sulfide

Glass[c]	NA	400[d]	coagulation/sedimentation sludge, spent diatomaceous earth, brine residue	salts
Fertilizer urea, ammonia[e]	NA	NA	processing waste, cooling tower blowdown	oils, grease, biocide
Petroleum refineries	63 000 120 000	600 1 200	process solids, wastewater treatment sludge, wastes associated with general plant activities	iron rust, iron sulfides, oil, inert solids, catalyst particles, hydrocarbons, calcium fluoride, bauxite, aluminum chloride, acid sludges, copper compounds, sulfides, spent caustics

[a] Estimate is based on reported average solid polymer waste generation of 0.4 to 0.8% of resin production for resin producers, except for the polyethylene resin producers, which reported that polymer wastes ranged from 0.5 to 4.0% of resin production and averaged 1.75%. Marynowski, C. W., "Disposal of Solid Polymer Wastes by Primary Producers and Fabricators," in *Proceedings*, National Industrial Solid Wastes Management Conference, University of Houston, Houston, Texas, 1970.

[b] Based on Japanese experience in using the activated carbon process described in the project description.

[c] Lacy, W. J. and Cywin, A. "Industrial Sludge Disposal Practices," Paper presented at U.S./U.S.S.R. Handling Treatment and Disposal of Sludges and Industrial Residual Symposium, Moscow, U.S.S.R., 5-8 May 1975.

[d] "Petroleum Refining—Point Source Category," EPA 440/1-74 014a, U.S. Environmental Protection Agency, Washington, D.C., April 1974.

[e] Ludwig, H. F., "Review of Water Pollution Control System at Kujang Fertilizer Plant," D. T. Pupuk Kujang, Cikampek, West Java," 1979.

In addition to the solid wastes shown in Table 2, sizable volumes will be produced from commercial shipping and harbor operations at Laem Chabang and Sattaheep harbors. The ESB planning requirements for these harbors include provision for management of these as part of the overall harbor sanitation systems (including provisions for water supply as well as management of sewage and gaseous wastes), to be managed by the municipalities.

Preliminary Concepts for Solid Waste Disposal

Disposal of Nonhazardous Wastes

As mentioned earlier, the ESB development and plan considers the overall project to comprise a group of individual urban and industrial centers at various locations in the ESB region (see Fig. 1). Each involves industrial project development, more or less, with the major industrial developments at Map Ta Phut and Laem Chabang. For each of these growth centers, a community solid waste management system is proposed based on disposal of the bulk of the solid wastes primarily by landfilling at one or more sites under municipal sponsorship. These would receive all sanitary solid wastes plus industrial solid wastes generated in the vicinity, except for hazardous wastes, which would require special handling. The landfilling operations would include, depending on the local situation, both inland sanitary landfills of a conventional nature and land reclamation projects along the coast where the environmental setting is suitable, such as mud flat zones and swampy areas or other low-lying zones.

Land Reclamation in Mud Flat Zones—Most of the upper portion of the Gulf of Thailand (the Inner Gulf) is a shallow estuarine zone with extensive muddy bottom sediments. Along the ESB coast, while some areas are sandy and suitable for beaches, much of the coastline consists of mud flats with a very low slope or gradient, so that the depth is only a few metres even 1 or 2 km offshore. These areas are shown in Fig. 2. These shallow-water mud flats are, of course, important ecologically, and in many estuaries (for example, San Francisco Bay) their extent is quite limited around the fringes of the estuary; hence, filling in such limited shallow zones should not be allowed. However, in the Inner Gulf these shallow areas are so extensive that filling in strips along the coast should not be ecologically significant. This has been done over the past decade at Chonburi where the municipality has converted some 14 km^2 of mud flats into valuable land.

Assuming coastal mud flat filling is pursued in the ESB region, a number of such operations could be established, as shown in Fig. 2. Each of these would be carefully designed to prevent pollution of marine ecology outside the fill area, using appropriate practicable technology. Fortunately, the necessary basic technical information to do this is now available as a result of the extensive in-depth analysis of the ecological effects of filling and dredging in coastal areas made by the U.S. Corps of Engineers over the period 1973 to 1978 [*4*]. This project, which represents a milestone in environmental engineering history, has produced a great deal of basic information, which can be adapted for use in guiding the design and operation of landfilling projects in developing countries. The adaptations suggested for use by the ESB are discussed in the following section.

Sanitary Landfilling—In addition to coastal land reclamation, sanitary landfills will be utilized where this alternative would be economically or financially more feasible. While at present there are no sanitary landfill operations in Thailand (there are only dumps), it would be quite feasible to use them, and it is believed this can be achieved in municipalities throughout the country simply by giving some serious attention to the problem. The technology is available and affordable, but interest in using it has been lacking. The situation is gradually improving as a result of activities of the government and international assistance agencies stemming from the United Nations Water and Sanitation Decade. One of these ongoing activities is a sanitation planning project for the urbanizing region of Chonburi sponsored by the World Health Organization (WHO) and the government of West Germany (German Agency for Technical Cooperation). This project is preparing a master plan for disposal of sanitary and industrial wastes in the Chonburi urbanizing zone, including both sewage and solid wastes.

One aspect of the proposed overall sanitation improvement program for Chonburi is the use of solid wastes, both sanitary and industrial, for filling in some very unsanitary areas under numerous homes built on stilts on mud flats along the coast. Originally these areas were flushed by tides, but with increasing development in the immediate vicinity they have become cut off from flushing; hence they have become sanitation messes of the worst type. The only solution is to fill them in, and the only affordable solution is use of the "swamp reclamation" technology discussed in the following section.

Swamp Reclamation—Another sanitation improvement measure now being evaluated by the Thai National Environment Board and the World Bank, with participation by the Asian Institute of Technology at Bangkok, is the use of solid wastes (both sanitary and industrial) for filling in low-lying swamp areas and paddy fields to accommodate expanding urbanization in communities and metropolitan areas such as Bangkok, which are located on low-lying areas formerly used as paddy fields. Filling in these areas (for Bangkok, a fill of from 1 to 1.5 m is usually needed), assuming imported dirt or sand is used; is a major expense. However, a far cheaper approach is to fill these areas with refuse, and, indeed, much of the built-up area of Bangkok today was filled in this manner, simply through the common practice of dumping solid wastes into the area.

In temperate zones this would hardly be considered appropriate because practice has required landfilling to be conducted above groundwater. In low-lying swamp areas in a tropical monsoon climate, however, the situation is vastly different because the deposited wastes are subjected to inundation or prolonged wetting, or both, throughout the monsoon season. Hence, unlike conventional sanitary landfills, where biological stabilization and physical consolidation can take many decades, landfills under the periodic inundation conditions just described can achieve both stabilization and consolidation in a relatively short space of time. While this process has been used traditionally, and very successfully, by the public without any planning or systems approach, its potential for use in a systematic engineered manner was first noted only in 1975 as part of the United Nations Development Program (UNDP)/WHO project for preparing a master plan for sewerage and sanitation in Jakarta [*6*].

The present interest in Thailand is based on use of this concept for creating new lands or large-scale housing projects and also for the special use noted earlier at Chonburi [*7*]. Based on observations of those areas of Bangkok that were filled in this manner, rather than by importing dirt or sand, it appears that such filled areas are suited for virtually all uses. Imported dirt or sand is required only for the final surface cover. For the special case at Chonburi, this appears to be the only affordable approach to filling in the low-lying unsanitary areas and thus solving the problem.

Disposal of Hazardous Wastes

The basic concept is to limit disposal of hazardous wastes, meaning any waste with serious toxicity potentials, to inland sanitary landfilling sites where there is no potential for groundwater pollution. The ESB region has quite limited surface water resources and very little groundwater resources, and, indeed, this shortage has been the main factor limiting growth in the region up to now. The situation at Pattaya, which has developed into a large-scale international beach resort over the past 15 years, is illustrative. This growth was made possible only by construction of impoundment reservoirs in the upland areas to store enough surface water to support the growth. The ESB Planning Project provides for further development of upland water impoundments sufficient to meet the foreseeable needs.

From the point of view of the impact on groundwater quality from landfilling of refuse, there are numerous potential sites (shown in Fig. 2) that could be used where there are no significant groundwater resources.

Proposed Solid Waste Disposal Plan for Eastern Seaboard

Municipal Wastes

As part of the detailed planning for each of the ESB urbanizing centers, engineering and economic analyses will be made to determine the optimal plan for disposal of municipal solid wastes. The alternatives will be (1) landfilling in mud flat areas to reclaim land and (2) use of sanitary landfills located at sites where there will be no significant groundwater pollution problem. Figure 2 shows the mud flat areas along the coast and a number of potential sanitary landfilling sites. The final choices will be governed mainly by economic analyses, but considering that coastal lowland areas suitable for development are relatively limited in the ESB region, the option for creating new land on mud flats may prove generally more attractive.

Industrial Wastes

In general, the municipal disposal facilities will be able to accommodate the bulk of the industrial solid wastes produced in the ESB region. Hazardous industrial wastes would be disposed of only in sanitary landfills at sites not posing groundwater pollution hazards. If necessary, special protection measures can be used to guard against possible contamination of groundwater from leachates from the fills. A variety of such protection measures have been developed in recent years, mostly under EPA sponsorship, such as those described in the "Proceedings of the Fifth Annual Research Symposium, Municipal Solid Waste, Land Disposal," published in 1979 [*8*]. However, it seems likely that the only protection measures needed will be for provision of dikes for diverting surface runoff away from the sites, plus the use of techniques to minimize infiltration of local rainfall into the site (such as the use of clay covers), and grading of surface slopes to facilitate runoff of this rainfall.

In addition to these two alternatives, a third alternative will be available for use by those industries discharging their liquid wastes into municipal sewerage systems that utilize submarine outfalls for marine disposal. The treatment to be provided by the municipality prior to discharge into the marine environment will be limited to the removal of floatables, with the vast seawaters of the gulf being depended upon to achieve biological stabilization and dispersion. Thus, it will be feasible for each industry to consider the option of grinding all or some of its solid wastes so that the ground-up materials can be discharged properly into the public sewers. Numerous studies in the United States, notably those of the Los Angeles County sanitation districts [*9*], have shown that sewers, when properly designed to maintain the flow velocities necessary for self-cleaning and prevention of sulfide generation, actually have great hydraulic carrying capacity, which has been only very partially utilized. Sewers can readily transport large amounts of suspended materials, such as solid waste reduced to particles by hammermills. Assuming these particulates do not include floatables (oils, greases, rubber particles, and so on), their discharge to the sewers would not add extra burdens on the municipal sewerage collecting and disposal system if eventual disposal is into unconfined offshore seawaters.

The choice of the best alternatives for each industry, or groups of industries in industrial estates, will again be mostly a matter of economics. In general, however, where the municipal sewerage systems employ a submarine outfall for final disposal, as is the case for most of the ESB urban growth centers, grinding solid wastes for discharge into the sewer will probably be the most economical method.

Monitoring Program

As mentioned earlier, a part of the ESB project implementation will be a continuing environmental monitoring program, especially of pollution control, to ensure competent attention to this problem. This will include monitoring of coastal and beach water quality as affected by sanitary and industrial wastes and of inland freshwater resources, both surface and groundwater. Detailed design of this regional monitoring program is now underway by NEB, with assistance from the World Bank [*3*]. Also, since its establishment in 1975 NEB has built up technical staff capability and physical resources, so that it is ready now to undertake the proposed program [*10*]. Again, this will be Thailand's first comprehensive regional monitoring program and will, of course, contribute a great deal to strengthening government awareness of the fragility of many environmental resources in the coastal zones and the need for continuing monitoring surveillance.

As envisioned at this time, with respect to the impacts of industrial solid wastes, the monitoring program will include two components, one on the coastal marine environment and one on groundwater. The coastal program will include (1) monitoring of near-shore ocean water quality at beaches (including Pattaya and the vicinity south of Rayong), especially for the presence of oily substances which might be transported by currents along the coast, (2) monitoring of water quality and of bottom muds in the vicinity of all marine outfalls and landfilling projects on coastal mud flats to check on possible transport of floatables, accumulation of heavy metals and toxic organics in the sediments, and effects on benthic fauna, and (3) monitoring of coastal fisheries (finfish and shellfish) and of coral beds to check on adverse effects of wastes. The NEB, working with the National Research Council and the Royal Thai Navy, has been engaged in monitoring the coastal water quality, including bottom sediments, in the Inner Gulf region under a program sponsored by the United Nations and initiated by the National Research Council in 1973 [*11*]. The technology being used for monitoring coastal waters and bottom sediments is that originally developed by the California State Water Resources Control Board in 1972 [*12*]. With respect to the monitoring of the coastal marine ecology, some valuable experience in the ESB region has already been gained by a program of detailed study sponsored by the Electrical Generating Authority of Thailand from 1973 to 1976 [*13*]. This study, done in connection with a proposed major power plant to be located at Ao Phai, is the first detailed marine ecological evaluation done in Southeast Asia.

The groundwater monitoring program will include checking impacts on water quality from inland landfilling operations. The parameters to be selected will include those toxic heavy metals or other toxic substances known or believed to be utilized by the industries making use of the landfill.

Conclusions

The Eastern Seaboard project represents Thailand's first approach to comprehensive planning for the development of an entire region, in order to realize a secondary urbanizing growth center away from the capital region of Bangkok. The comprehensive plan of development, completed only a few months ago, is now in the early stages of implementation under the lead of the NESDB. The planned implementation period is the next five to ten years, in order to make efficient use of the large natural gas energy supply discovered and developed in the Gulf of Thailand over the past decade.

Because the emphasis of the regional growth plan is on industrial development, and because the region's environmental resources include very valuable beach and tourist resort areas and a very valuable marine ecology, including fisheries and corals, the regional plan gives careful attention to management of pollutants of all types. For handling liquid wastes, a series of core municipal sewerage systems is planned for servicing both the population and industry, with utilization of submarine outfalls for final disposal for all but one of the growth centers. For handling solid wastes, a series of municipal disposal sites is proposed, including landfilling in coastal mud flat zones and the use of inland sanitary landfills, again serving both the population and industry. The availability of the marine disposal systems for discharging sewage effluents in open offshore seawaters will not only greatly simplify the problem of disposing of industrial sewage but also make it feasible for industries to consider another alternative for solid waste disposal, namely, grinding and discharge into the public sewers. Also, hazardous solid wastes can be disposed of at selected inland fill sites where there are no groundwater resources of value. The entire pollution control system for the region will be continually monitored to ascertain effects on the environment and to determine needed improvements. The overall objective is economic—with environmental—development.

All of this represents lots of "firsts" for Thailand, but growth pressures have reached such levels that NESDB is now exercising a much more vigorous role in development; hence the Eastern Seaboard expectations may not be unrealistic. All of the proposed environmental pollution control measures are practicable, affordable, and appropriate for use in developing countries. All that is needed is the will to use them, and it seems that this will is now shaping up in Thailand. With NESDB and the National Environment Board working together on environmental pollution control, the stage is set for competent management of industrial solid, liquid, and gaseous wastes. Assuming this will be achieved, it will set an example that will probably revolutionize basic sanitation concepts in the country.

References

[1] *Eastern Seaboard Study, Final Report* (series of volumes including one component on "Environmental Aspects"), Coopers & Lybrand Associates for the National Economic and Social Development Board of Thailand, Bangkok, 1982.

[2] Snidvongs, Kasem, "Environmentalism and the Eastern Seaboard," *Journal of the National Environment Board*, September 1982, pp. 85–98.

[3] "Terms of Reference for Establishing an Environmental Monitoring Programme for the Eastern Seaboard," in *Eastern Seaboard Study Final Report*, Coopers & Lybrand Associates for National Economic and Social Development Board of Thailand, Bangkok, 1982.

[4] "Synthesis of Research Results, Dredged Material Research Program, Executive Overview and Detailed Summary," Technical Report DS-78-22, U.S. Corps of Engineers, Vicksburg, Miss., December 1978.

[5] "Urban Sewerage and Excreta Disposal Planning for Chonburi and Thailand," report prepared for the Thai Department of Public Works and the World Health Organization by Seatec International, Bangkok, 1983.

[6] "Comprehensive Plan for Sewerage and Sanitation for Metropolitan Jakarta," report prepared for the United Nations Development Program, World Health Organization, and the Government of Indonesia, Nihon Suido, 1977.

[7] Ludwig, H. F., "Proposal for Research Project for Developing a Technical Manual for Swamp Reclamation by Refuse Landfilling in Tropical Countries," National Environment Board, Bangkok, 7 May 1976.

[8] "Proceedings of the Fifth Annual Research Symposium, Municipal Solid Waste, Land Disposal," EPA-600/9-27-023a, U.S. Environmental Protection Agency, Washington, D.C., August 1979.

[9] Reports on disposal of solid wastes by grinding and discharge to public sewers, Los Angeles County Sanitation Districts, Whittier, Calif.

[10] Ludwig, H. F., "Environmental Protection and the National Environment Board in Thailand," *Journal of the National Environment Board*, September 1978, pp. 27–43.

[11] Publications of the National Research Council (Subcommittee on Marine Resources) on water quality monitoring in the Inner Gulf of Thailand in Thai, Bangkok, 1974 to 1982.

[12] "Guidelines for the Preparation of Technical Reports on Waste Discharges to the Ocean and for Monitoring the Effects of Waste Discharges on the Ocean," California State Water Resources Control Board, Sacramento, Calif., 1972.

[13] Ludwig, H. F., "Final Report on Study of Marine Ecology Relating to Proposed Ao Phai Nuclear Power Plant," Seatec Thailand for Electrical Generating Authority of Thailand, Bangkok, October 1975.

Fatma El-Gohary[1] *and Mohamed S. El-Khouly*[2]

Management of Solid Wastes in the Iron and Steel Industry

REFERENCE: El-Gohary, F. and El-Khouly, M. S., "**Management of Solid Wastes in the Iron and Steel Industry,**" *Hazardous and Industrial Waste Management and Testing: Third Symposium, ASTM STP 851*, Larry P. Jackson, Alan R. Rohlik, and Richard A. Conway, Eds., American Society for Testing and Materials, Philadelphia, 1984, pp. 361-372.

ABSTRACT: Wastes from a local iron and steel factory are the subject of this study. The manufacturing operations in the plant under consideration are agglomeration of iron ore and sintering, pig iron manufacture, steel making, rolling mill operations, and pickling.

Liquid slag, produced in the blast furnace, is granulated in water and used as a concrete additive. Other wastes are directed separately to sedimentation tanks. The settleable solids are reused, and the treated effluents are pumped to a cooling tower for recycling.

As a result of the new manufacturing expansion, existing waste treatment facilities are not adequate, and it was found necessary to provide additional treatment techniques.

Departmental, as well as composite wastes were treated using plain sedimentation, centrifugal sedimentation, or chemical coagulation, or a combination of these methods.

The results obtained showed that the use of the hydrocyclone for solid-liquid separation is much more efficient than plain sedimentation. When this process was followed by coagulation, very promising results were obtained. The use of pickling liquor as a coagulant gave comparable results with alum and ferric chloride.

KEY WORDS: iron, steel, solid wastes, management, solid-liquid separation, hydrocyclone, coagulation, pickling liquor, hazardous wastes, industrial wastes

Ever-increasing amounts of hazardous waste materials are being generated as a result of Egypt's economic and population growth. The uncontrolled disposal of these wastes not only results in a potentially dangerous health and environmental threat but also wastes raw materials and energy. This situation is aggravated by the practice of one-time use of materials developed when raw materials and energy were in abundance.

[1]Research professor of environmental sciences, and head, Water Pollution Control Laboratories, National Research Center, Cairo, Egypt.

[2]Associate professor, Civil Engineering Department, Ain Shams University, Cairo, Egypt.

The source of these hazardous wastes is primarily the industrial sector. Steel mills are one of the major producers of solid wastes. In terms of tonnages, by far the largest component is the slag producted in iron blast furnaces and in steel-making operations. Modern blast furnace practice typically gives rise to about 320 kg slag per tonne of pig iron [*1*]. Additional amounts of solid wastes are being produced from gaseous and liquid environmental control operations. A large portion of the solid wastes is generated in the gas wash water in various types of furnaces. This water is always clarified to remove the dust. The clarified dust is thickened as a sludge to 80% dryness and sent to the sinter plant.

Sintering is a significant potential source of emissions. Emissions of particulate matter have been estimated to be about 11 kg/tonne of sinter [*2*], and the average particulate loading is about 1 gm/m^3 of gas [*3*].

The wastes from iron and steel factories in Egypt are the subject of this study. The manufacturing operations in the investigated plant are agglomeration of iron ore, pig iron manufacture, steel-making processes, rolling mill operations, and pickling.

In the initial design of this plant, wastes are segregated. Each waste receives only primary sedimentation. The effluents of primary sedimentation basins are pumped to a cooling tower, after which the effluents are reused. At the beginning, this treatment was sufficient. However, as a result of the new extensions and the increase in the volume of wastes requiring treatment, existing facilities are not adequate, and additional treatment techniques were found to be necessary.

This paper presents a trial management of by-products and wastes in the factory under investigation.

Procedure

Departmental as well as composite wastes were treated using plain sedimentation, centrifugal sedimentation, or chemical coagulation. The combined effects of two or more of these techniques were also investigated.

For centrifugal sedimentation experiments, a hydrocyclone was designed (Fig. 1). This consists of a tangential inlet nozzle, which separates coarse solids at the center of the vessel by centrifugal action.

For coagulation, the jar test procedure was used to determine optimum operating conditions. The coagulants examined were aluminum sulfate and ferric chloride. Pickling liquor as a coagulant was also investigated to reduce additional costs and to solve the problem of this most troublesome waste produced in the iron and steel industry. Calcium oxide (CaO) and Nalco (600 S) were used as coagulant aids.

Results

Blast Furnace Wastes

The blast furnace process consists of charging iron ore, limestone, and coke into the top of the furnace and blowing heated air into the bottom. The produc-

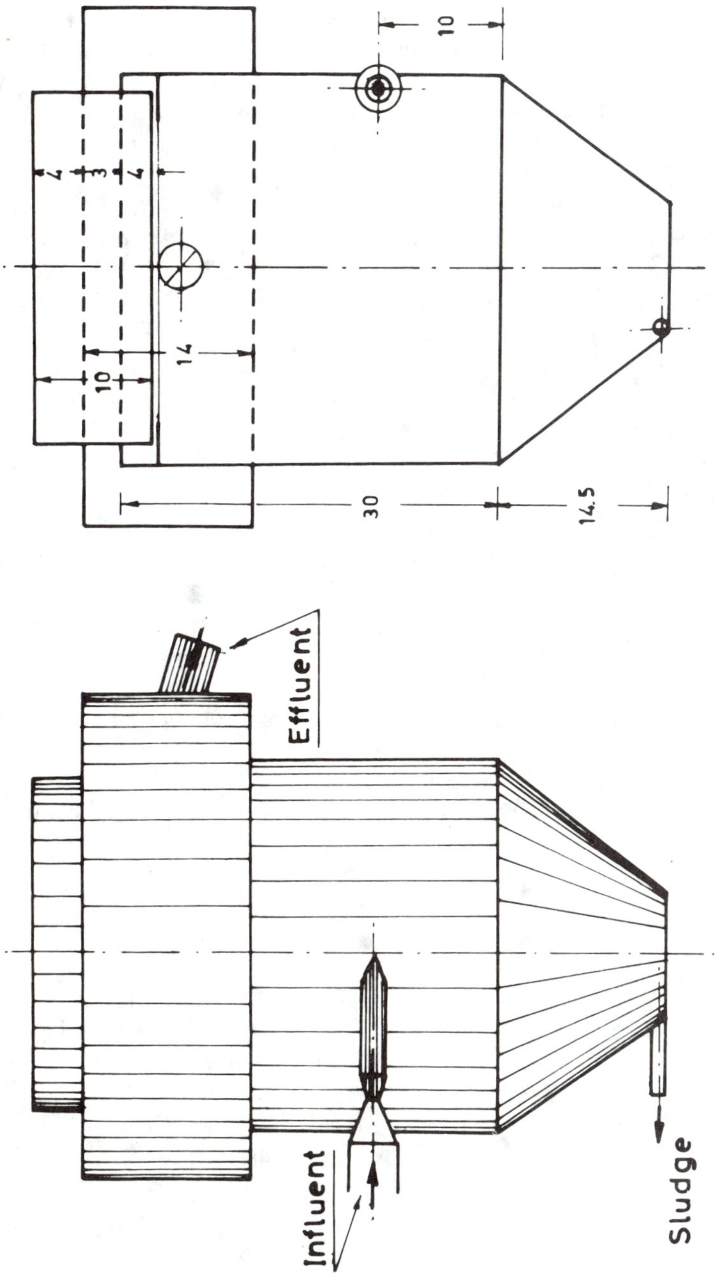

FIG. 1—*Schematic drawing of a hydrocyclone. All the measurements are in centimetres.*

tion of 1 tonne of iron gives rise to about 0.5 tonnes of slag and 5 tonnes of blast furnace gas containing the fines of the burden carried out by the blast.

Molten slag is granulated in water and used as concrete additive.

Wet scrubbers are used to entrain dust as the gas passes through the upper part of the furnace. This dust consists largely of iron oxides, alumina, silica, carbon, lime, and magnesia. About 40 tonnes of solids are produced per day.

The scrubber effluent is directed to a rectangular sedimentation tank for a retention period of 1 h. Average characteristics of the raw waste and the sedimented effluent are shown in Table 1. From these results it is clear that the effluent obtained is not suitable for recycling for cooling purposes.

Plain Sedimentation—Laboratory tests showed that about 80% of the suspended solids were removed after 1 h of sedimentation. Increasing the retention time up to 2 h raised the removal of suspended solids by only 8%.

Treatment with Hydrocyclone—The influent was introduced under different pressures—0.2, 0.4, and 0.8 kg/cm^2. Retention periods ranging from 5 to 25 min were used. The results obtained showed that the efficiency of sedimentation increases by increasing the influent pressure (Fig. 2). With respect to the retention period, it has been found that suspended solids and iron removal rates tend to be active during the first 15 min, after which a gradual decline is recorded.

Comparison of the quality of the effluent after plain sedimentation with that after centrifugal sedimentation indicates no marked differences. This may be because a large amount of the suspended solids are so fine that they tend to float and be discharged with the supernating effluent. Therefore, sedimentation is inadequate, and the waste must be chemically treated.

Coagulation—Consideration of the available results obtained for the treatment of blast furnace gas washing wastes, by means of different coagulants alone and in combination with coagulant aids, shows that over 99% removal of suspended solids can be achieved by the use of either 10 mg/L aluminum sulfate, 15 mg/L ferric chloride, or 20 mg/L ferrous sulfate (crystalized from pickling liquor), each aided by 1 to 1.5 mg/L Nalco.

Rolling Mill Wastes

Scale-bearing wastes orignate from various rolling mill operations. These wastes consist of the water used to dislodge scale and cool the rolled product, plus the water used to transport scale through the flumes beneath the mill line. The volume of water used is around 30 000 m^3/day. The mixture of scale and water is directed to a scale pit, which basically acts as a sump settling basin, having an average detention time of 5 min. In this pit, the heavy scale settles to the bottom and is removed by a clam shell bucket. The main overflow from the pit goes to the hot rolling sewer system, which flows by gravity, together with the cooling water, to the settler. The sedimented effluent, after 2 h of sedimentation, is pumped to the cooling tower to be reused. Analysis of effluents from the scale pits and from the main settler are represented in Table 1. The main

TABLE 1—*Average characteristics of raw and settled effluents.*

Parameters	Raw Blast Furnace Wastes	Settled Blast Furnace Wastes	Raw Rolling Mill Wastes	Settled Rolling Mill Wastes	Raw Agglomeration Wastes
Suspended solids (105°C), mg/L	1796.0	82.0	2928.0	100.4	3857.0
Suspended solids (550°C), mg/L	1535.0	63.5	2674.0	62.4	3646.0
Total residue (105°C), mg/L	3059.0	880.0	3426.0	650.8	4916.0
Total residue (550°C), mg/L	2509.0	753.7	3225.0	472.0	4281.0
Settleable solids, ml/L	10.0	...	1.7	...	8.8
Oil, mg/L	...	...	16.0	2.0	...

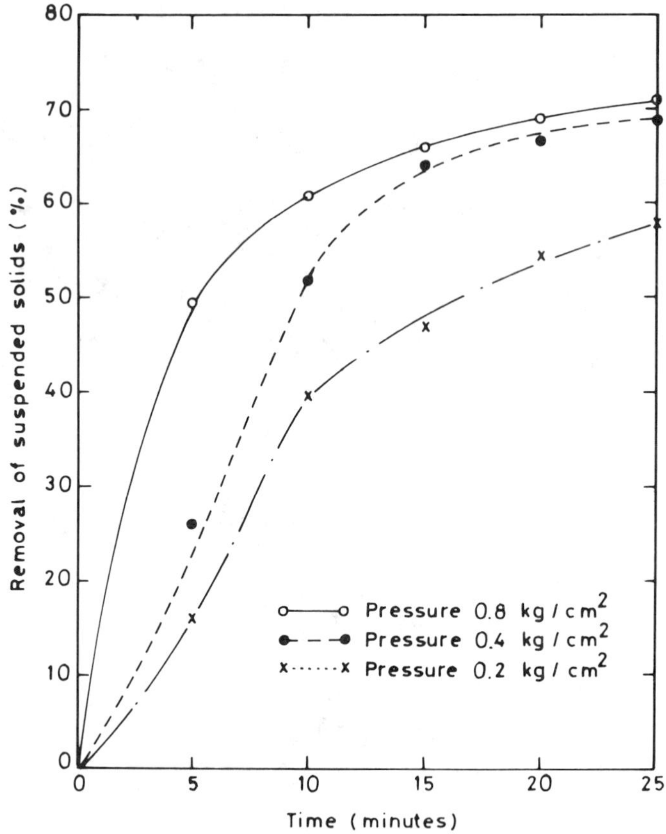

FIG. 2—*Effect of hydrocyclone treatment on suspended solids removal: blast furnace gas washing wastes.*

problem encountered in the primary treatment of hot mill wastes is the ineffectiveness of the initial scale pit.

Plain Sedimentation—The effect of sedimentation time on suspended solids removal was investigated. A retention time of 30 min produced an average removal of 67%. As the retention period was increased, a gradual improvement was observed. An average removal of 88.8% was recorded after 2 h of sedimentation.

Treatment with Hydrocyclone—The use of the hydrocyclone, even at the lowest investigated pressure at 0.2 kg/cm^2 and a detention period of 5 min achieved 91% oil reduction. At higher detention periods, 100% oil removal was obtained. The suspended solids removal values were 57%, 77.4%, and 84% at pressures of 0.2, 0.4, and 0.8 kg/cm^2, respectively (Fig. 3). The corresponding iron removal rates were 15.4%, 81.2%, and 83%.

Coagulation—Treatment of rolling mill wastes by coagulation revealed a

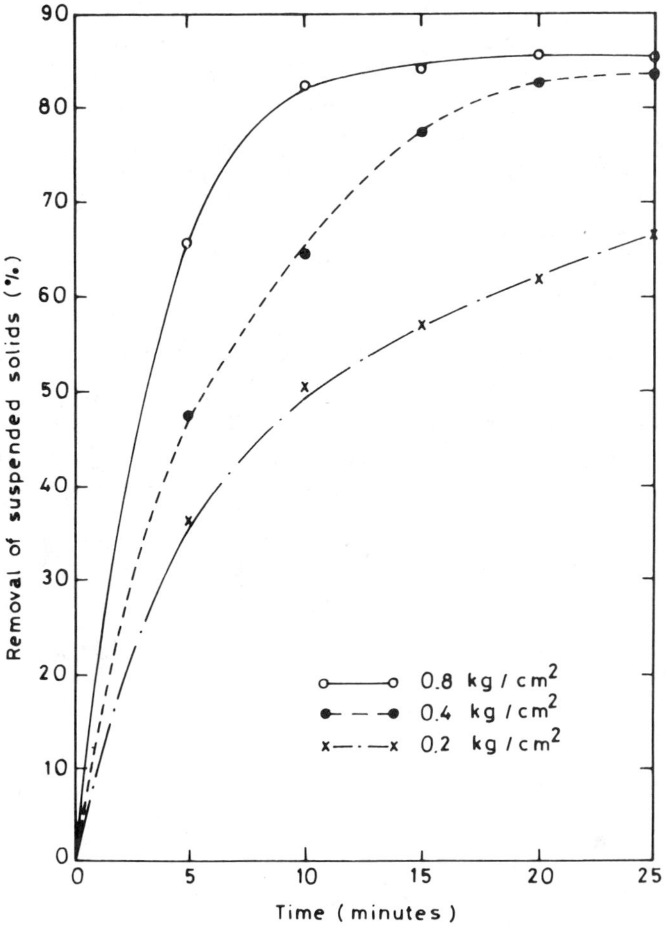

FIG. 3—*Effect of hydrocyclone treatment on suspended solids removal: scale-bearing wastes.*

number of facts. The most suitable results were obtained when 7.5 mg/L aluminum sulfate aided with 0.25 mg/L Nalco were used. Turbidity removal reached 98%, and the final effluent contained residual iron and suspended solids of about 1 mg/L.

The use of 0.1 ml/L pickling liquor waste in combination with 40 mg/L lime gave promising results. Residual iron and suspended solid levels were 2.5 and 9.5 mg/L, respectively.

Sintering Wastes

In the sintering process, fines of iron ore, lime, mill scale, coke, and recycled materials are mixed and granulated in drum mixers with added water. The mixture is charged on to the sintering strand and ignited. A great volume of water is usually used for cooling the exhaust blower bearings and in the drum mixer. Water consumption ranges from 3805 to 4025 m^3/day.

The wastewater from this department receives primary sedimentation in a rectangular basin. Settleable suspended solids collected at the bottom are removed from time to time by means of a clam shell bucket and used once more in the blast furnace. The effluent is usually reused for preparation of ores and sometimes disposed off into a drain. The average pollution characteristics of this waste are given in Table 1.

Plain Sedimentation—The results obtained showed that about 70% suspended solids removal may be attained after 30 min of sedimentation. Raising the detention period up to 2 h increased suspended solids removal by 15%.

Treatment with Hydrocyclone—From the results obtained (Fig. 4), it can be concluded that the hydrocyclone is much more efficient than plain sedimentation. It provides faster removal of suspended solids after a short retention time.

Chemical Treatment—Wastewater from the granulation processes usually

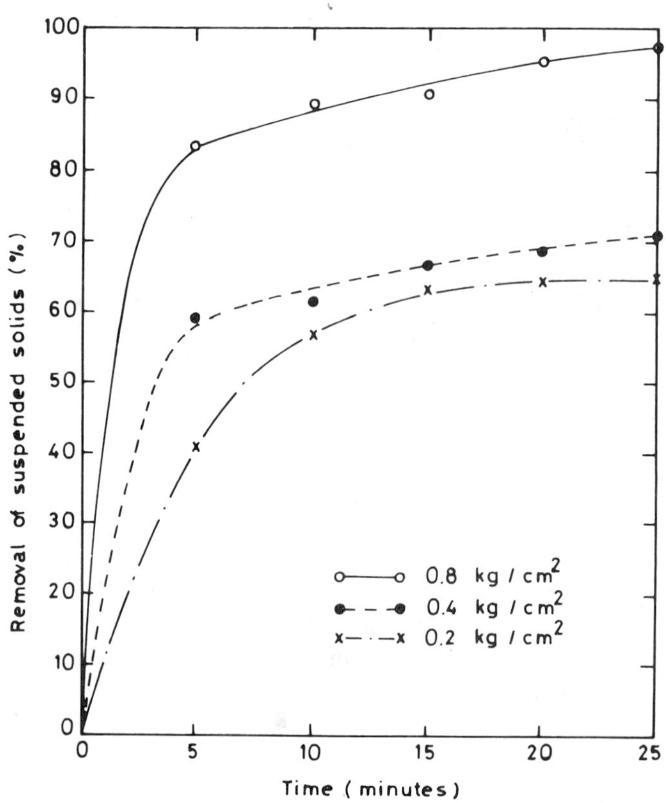

FIG. 4—*Effect of hydrocyclone treatment on suspended solids removal: agglomeration wastes.*

contains small fractions of the raw granulated materials. A trial was performed to evaluate the use of residual iron and CaO in the hydrocyclone effluent to treat the waste without any coagulant added. A series of experiments was carried out to verify the effect of changing the pH of the waste. The sulfate pickling liquor waste was used for this purpose. A pH range varying from 11.25 to 3 was investigated. From the results obtained (Fig. 5), it can be seen that two optimal pH ranges exist. The first falls on the alkaline side, ranging from 10 to 10.5, whereas the second falls within the acidic range, extending from pH 4.5 to 5.0. However, it may be noted that higher clarity levels were attainable in the alkaline rather than the acidic range. Turbidity removals up to 99% were reached in the alkaline pH range, whereas only a 91% level was reached at the lower optimum pH range. Iron and suspended solids removal at a pH value of 10.25

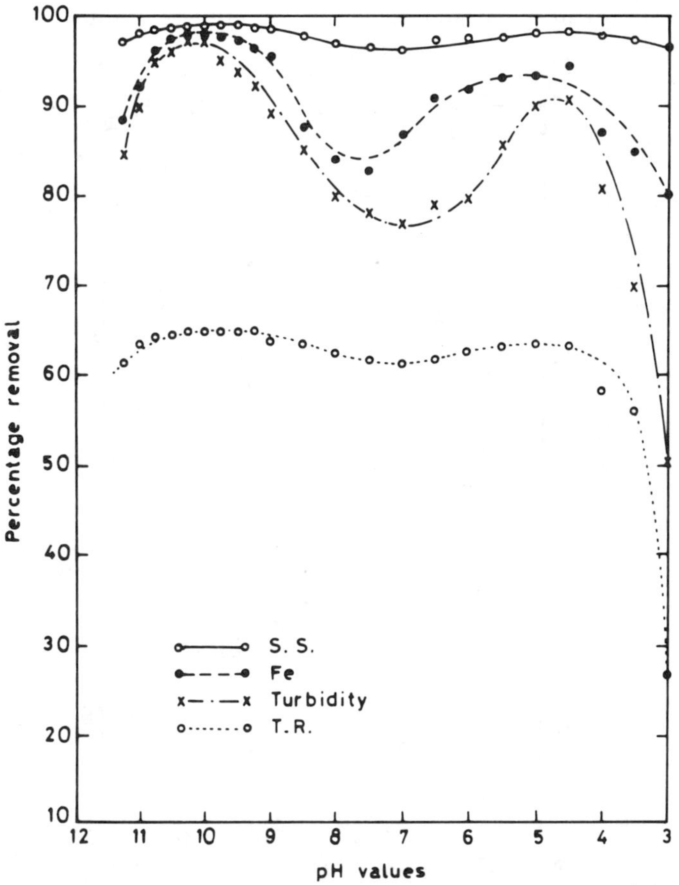

FIG. 5—*Quality of granulation waste at different pH values.*

reached 97.3% and 99.3% levels, respectively. Corresponding residual iron and suspended solids were 2.3 mg/L.

Combined Treatment of Wastes from Blast Furnace, Rolling Mill, and Granulation Processes

Hydrocyclone-treated effluents of rolling mill and granulation wastes were mixed with raw blast furnace gas washing wastes. The experimental combination ratio was selected to represent the workable ratio at the factory.

The iron content in the mixture ranged from 276 to 357 mg/L, with an average value of 311 mg/L. The suspended solids content varied from 945 to 1298 mg/L. The pH of the mixture was around 7.5. The combined mixture was subjected to chemical treatment.

The results obtained showed that the use of 10 mg/L of aluminum in combination with 0.25 mg/L produced the best results. Turbidity removal was 99%. The treated effluent contained less than 1 mg/L of suspended solids.

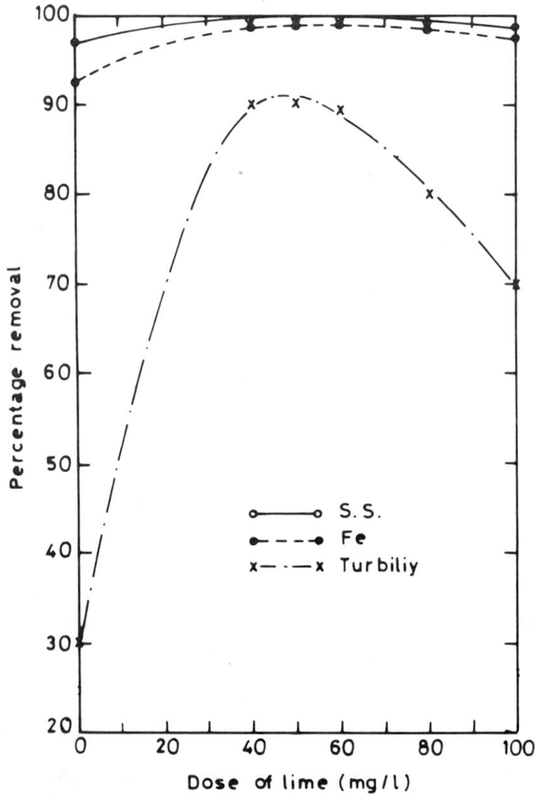

FIG. 6—*Treatment of composite wastes using pickling liquor and lime.*

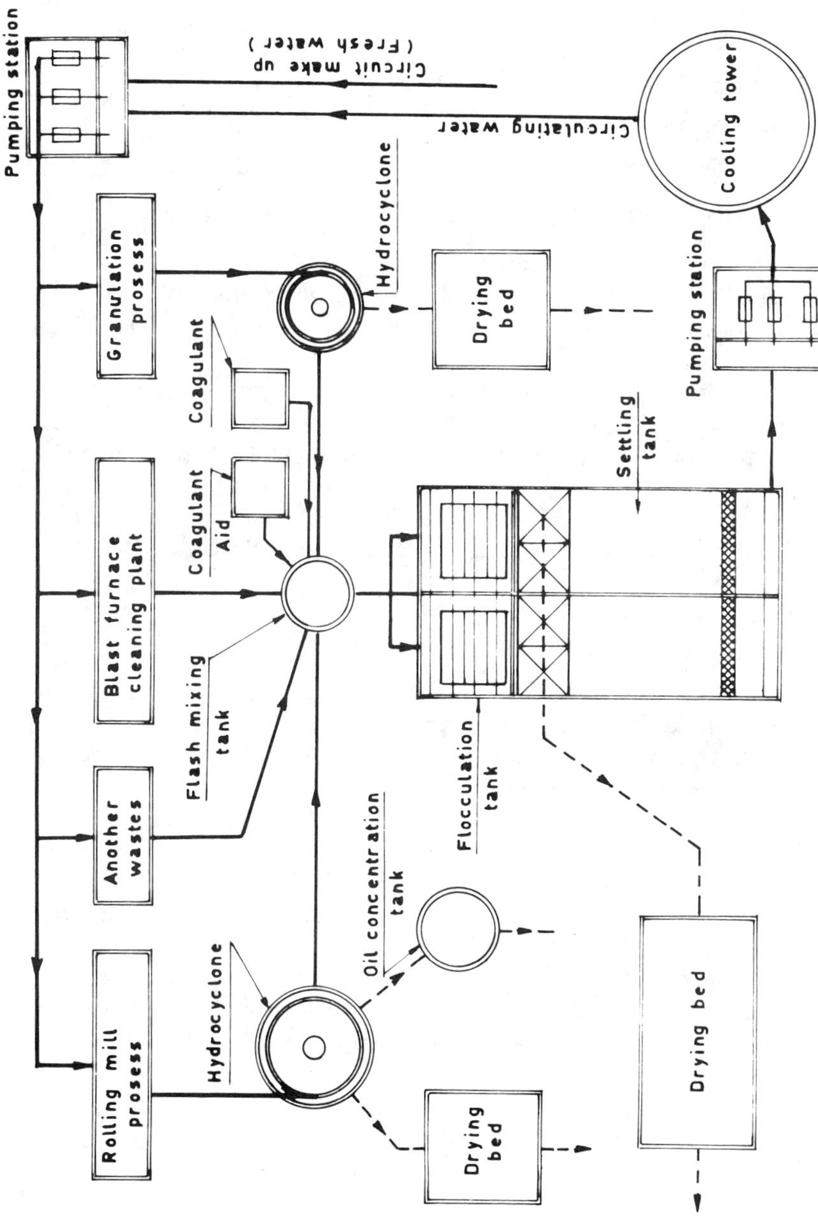

FIG. 7—*Flow diagram of waste management in the iron and steel industry.*

The use of 10 mg/L ferric chloride aided by 0.25 Nalco achieved similar results.

When 0.2 ml/L pickling liquor waste was combined with 50 mg/L CaO, up to 90% turbidity removal was achieved (Fig. 6). Residual suspended solids and iron levels were 3 mg/L and 2 mg/L, respectively.

Analysis of variance for composite waste treatment through coagulation revealed that the three examined coagulants can be used. Economy and handling costs become the crucial factors.

Prospective Design of Waste Treatment in Iron and Steel Industry

In an attempt to utilize the previous findings, a preliminary design based on experimental results has been proposed. The existing treatment facilities are taken into consideration. A schematic layout of the proposed plan is given in Fig. 7.

Wastes from rolling mill and agglomeration processes receive primary treatment with a hydrocyclone under predetermined optimum conditions. The effluent, with other wastes, is directed into a flash mixing tank where the selected chemicals are added. This is followed by flocculation and sedimentation. It must, however, be emphasized that a pilot plant study should be carried out to assess technical as well as economic aspects of the proposed design.

Discussion

The iron and steel industry is a major user of resources such as iron ore, coal, limestone, refractory materials, water, and energy. It is, therefore, important that such resources be used efficiently with as little waste as possible.

The results obtained show that there is generally no difficulty in recovering by sedimentation most of the solid wastes produced in the iron and steel industry. The hydrocyclone installation is more efficient than plain sedimentation when the solid particles are not fine. The hydrocyclone takes a relatively small surface area and can be constructed underground.

In some cases when solids are particularly fine, such as flue dust, or when the treated effluents are recirculated, as in the existing situation, it is not practical to remove these solids without the use of coagulants. The results obtained show that coagulants can greatly speed up the deposition of the bulk of flue dust load, especially the finest particles, which represent 2 to 3% by weight of the total load.

References

[*1*] Emery, J. J., *Canadian Mining and Metallurigcal Bulletin*, December 1975, p. 60.

[*2*] Contract number CPA-22-69-119, April 1973. Section 7.5, U.S. Environmental Protection Agency, Research Triangle Park, N.C.

[*3*] Environmental Protection Technology Series, Industrial Process Profiles for Environmental Use, EPA-600/2-77-023X, U.S. Environmental Protection Agency, Washington, D.C., February, 1977, chapter 24.

Léon Klein[1]

European Community Action on the Management of Waste Products

REFERENCE: Klein, L., **"European Community Action on the Management of Waste Products,"** *Hazardous and Industrial Waste Management and Testing: Third Symposium, ASTM STP 851*, Larry P. Jackson, Alan R. Rohlik, and Richard A. Conway, Eds., American Society for Testing and Materials, Philadelphia, 1984, pp. 373–393.

ABSTRACT: The objectives of the European Economic Community (EEC) in actions regarding waste are prevention, recycling, and elimination. The EEC's environmental action programs set out the priorities and aims of the organization, as well as supply the necessary legal backing for the resulting actions and provide insight into the institutional and legal background of the EEC.

The three main spheres of action, which form the basis of the EEC's environmental policy, are (1) prevention of the production of waste products, (2) elimination, without danger, of nonrecoverable residues, and (3) recycling and reuse of wastes. Statistical data are presented to show the scale of the problem in the ten member states.

KEY WORDS: European Economic Community, waste prevention, waste recycling, clean technology, toxic wastes, waste paper, sewage sludge, hazardous wastes, industrial wastes

The Council of the European Communities on 17 Dec. 1982 approved a third action program for the environment. This action has made the present a good time to consider the future of waste management in the European Economic Community (EEC) within the framework of environmental policy in general. We are indeed at a point of change—a point of reassessment of policies and prospects, both in Europe and in the United States—in the whole environmental protection and waste management field, at such a point it is always a good idea to look back and ask what we have learned, to pause to consider what has changed and what has not, to consider what we can learn from past experience, and to try to discern future initiatives and their implications.

This is a good moment for the European Economic Community to look both backwards and forwards, as it is just over ten years since, at the 1972 meeting of the organization's heads of state and government, the decision was taken to

[1]Deceased; formerly deputy division head, Commission of the European Communities, Brussels, Belgium.

launch the EEC's environmental action program. That was, of course, the time of the Stockholm Conference on the Human Environment—a time of rapidly increasing concern about the impact of human activities on the environment, and especially about pollution problems. It was also a time of relative prosperity, with the result that once action was seen to be necessary there were not too many economic obstacles to finding the necessary resources to take action.

The first major change in the situation in the 1980s, in comparison with that in 1972, is that the whole economic context within which environmental policies are being developed and pursued is much more complicated, and there can sometimes be challenges to environmental policies on the grounds of their cost, which would not have been heard of ten years ago. On the other hand, it is also true that there have been major changes in thinking about the nature and role of environmental policies and the part they must play in the whole field of socio-economic activity. It is this evolution, in particular, that I should like to trace because I think it will help to indicate where the opportunities will arise for "the environmental industry" over the next few years.

Before I trace our progress over these last ten years, I will try to give a brief description of the European Economic Community. To digress briefly, George Bernard Shaw once described England and America as two countries divided by a common language. In this manner, one might describe the EEC as ten countries united by seven different languages and the Treaty of Rome.

The European Economic Community

The European Economic Community is the most important of three European communities collectively known as the European Communities, although reference is increasingly made to "the European Community." The other two are the European Coal and Steel Community and the European Atomic Energy Community. Since the EEC is in many ways unique and different from other regional or world-level organizations, it may be useful to give a brief description of some of its more important characteristics.

The EEC was set up in 1957 by the Treaty of Rome. The original member states were Belgium, the Federal Republic of Germany, France, Italy, Luxembourg, and the Netherlands. The original six were joined in 1973 by Denmark, Ireland, and the United Kingdom, and the nine have become the ten with the accession of Greece in 1981.

The EEC is governed essentially by four institutions: the Council, the Commission, the European Court of Justice, and the European Parliament.

The Council is composed of representatives who are usually members of the governments of the member states. The membership of the Council is not fixed and varies depending on the subject matter being discussed. Thus, when agricultural matters are under review, the Council will be made up of agricultural ministers; when transport matters are being discussed it will be made up of transport ministers. The Council is the final decision-making body of the EEC.

The Commission's main task is to ensure that the provisions of the Treaty of Rome are being applied. Most of the Council's decisions are taken on a proposal from the Commission. The proposal by the Commission is, in most cases, a condition for the validity of a Council decision.

The Commission participates fully in the Council's discussions and is responsible for the implementation of the Council's legislation.

The European Court of Justice ensures that the law is observed in the interpretation and application of the Treaty of Rome. It can annul EEC legislation, either directly on the petition of an institution, a member state, or, in certain cases, an individual, or indirectly by means of a reference from a national court in one of the member states. It ensures the uniformity of interpretation of EEC law by means of requests for interpretation of the Treaty of Rome or of secondary EEC law from national courts.

The European Parliament is, for the moment, essentially a consultative and not a legislative body. The Parliament is composed of directly elected members. Consultation of the Parliament is compulsory in a large number of cases (the legislation is voidable if the Parliament has not been consulted), but it is also consulted on a large variety of matters for which its opinion is not required. It is the Parliament which finally adopts the budget of the EEC and which also has the power to oblige the Commission to resign *en bloc*.

The member states have divested themselves of their sovereignty in certain spheres and have vested that sovereignty in the EEC, where it is exercised by the Council. This is so in the agricultural field, for example, where minimum prices of farm products are set at an EEC, and not at a national, level. Insofar as the EEC has not exercised powers given to it under the Treaty of Rome, the member states may act in those spheres but not in a way which might prejudice any future EEC rules. Likewise, externally (that is, in relations between the EEC and third countries, groups of countries, or organizations) the EEC is able to act in fields where it has or intends to adopt internal rules. Again, the member states may act individually, as long as the EEC has not done so, but may not prejudice future EEC action.

European Economic Community law in the form of regulations and directives is directly applicable in the member states without the intervention of the national parliaments. In the case of directives, the parliaments may select methods for implementing EEC policy, but they are bound by the ends to be achieved by the EEC legislation.

This, then, is a brief overview of some of the general aspects of the EEC.

Environmental Policy

The environmental policy of the EEC stems from the realization that one of the objectives of the organization, continued economic expansion, could not be achieved separately from a concern for the quality of life. Article 2 of the Treaty of Rome—which is fundamental to all action undertaken by the EEC—speaks of the promotion throughout the EEC of "a harmonious development of economic activities and a continued and balanced expansion."

This goal has always been seen as having a qualitative as well as quantitative aspect, and it was thus entirely consistent with the EEC's traditional approach when the heads of state and government at the 1972 summit included in their declaration the following statement:

> Economic expansion is not an end in itself; its first aim should be to enable disparities in living conditions to be reduced. It must take place with the participation of all the social partners. It should result in an improvement in the quality of life as well as in standards of living. Particular attention will be given to intangible values and to protecting the environment so that progress may readily be put at the service of mankind.

Further, this declaration asked the institutions of the EEC to prepare an environmental action program. From this declaration arose the organization's three action programs on the environment, developed in 1973, 1977, and 1982. The work done by the EEC since 1973 has been based on these programs.

First Action Program

The First Community Action Program on the Environment, the first EEC program in the field of the environment, was approved in 1973 [*1*] and was updated in 1977. It still remains entirely valid. Then and since, the EEC has interpreted the term *environment* fairly broadly to include all natural resources (such as the natural environment, land, energy, and raw materials) which may be damaged or over exploited as a result of social or economic development and also to include the "quality of life." The objective has thus been to ensure the sound management of natural resources and to bring qualitative aspects into the planning and organization of economic and social development. One other fundamental point has been critical to the role of the EEC in the field of the environment. It was recognized from the outset that if, in response to the concern over environmental issues that was felt in 1972, the member states of the EEC were to adopt separate measures of a widely differing character, this would inevitably lead to distortions of competition which would interfere with the proper functioning of the Common Market.

This, then, is the background of the EEC's action program on the environment. Before going on to look briefly at what has been achieved under it and considering how it ought to evolve in the 1980s, it is worthwhile to say a few words about some of the features of the first action program.

From the outset there was a strong political commitment on the part of the member states to a wide range of measures to protect the environment.

I would like to underline in particular one or two points from the first action program. In setting out the objectives and principles of EEC's environmental policy, there was a clear commitment to a number of basic approaches which remain important:

1. Already at this early stage it had been recognized that prevention of pollution and nuisances was better than cure.

2. The need to take account of environmental needs at the earliest possible stage in decision-making processes was stressed.

3. The essential features of the "polluter pays principle" were outlined.

4. There was a clear recognition that the environment has a limited capacity to absorb pollution and that this asset, although it can be used, must not be abused.

5. It was always necessary to determine the appropriate geographical level for any suitable actions (whether local, regional, national, EEC, or international); proposals for harmonization at EEC level should thus not be automatic but made only when action at that level is needed in light of the facts of the case.

The first program went on to spell out in some detail the activities that were thought to be needed to secure an adequate degree of environmental protection. A large part of the program was devoted to the reduction of pollution and nuisances. Priority pollutants were identified; overall approaches were defined; and details were given of needed remedial measures related to certain products, to certain industrial sectors, to certain areas of common interest (such as the sea and the river Rhine), and also to wastes and waste management. In the field of pollution control it was clear that considerable priority was going to be given to tackling water pollution problems, and also, more specifically, "dangerous substances," both in water and elsewhere. In the field of waste, the broad lines of the EEC's approach were already set out; the program had this to say:

> The protection of the environment against pollution, sound economic management of resources, the effort to reduce the Community's dependence on imported raw materials, the rational long-term management of natural resources which are either non-renewable or can be renewed only at a certain rate—all these considerations together argue in favor of an immediate and hard-hitting campaign against waste.

So much, then, for the beginnings of the EEC's environmental action program, which dates back more than ten years. What has followed? It would be fair to say that what has followed has been a period of fairly intense activity on the environmental front—particularly in pollution control and prevention—and a continuing high degree of public concern and political commitment. The result has been that, to date, some 60 legislative texts have been adopted by the EEC—most of them directives—and a great deal of work has been done in studying particular environmental problems for the purpose of providing either the basis for proposals for directives or guidance to the Commission and to the member states on the evolution of environmental policy.

Second Action Program

The Second Community Action Program on the Environment was adopted in 1977 [*2*]. It restated the principles and objectives of the first program, continued action initiated under it, and emphasized two aspects of environmental policy:

(*a*) the prevention of damage to the environment, and
(*b*) better management of resources.

To reduce pollution, priority was still to be given to measures for protecting the water and air. To manage resources, this second program emphasized avoidance of the generation of waste and reiterated the three basic concepts of waste management:

(*a*) prevention of the production of waste products,
(*b*) elimination without danger of nonrecoverable residues, and
(*c*) recycling and new uses for wastes.

These basic principles are repeated throughout the EEC's actions on wastes.

The program included the following actions:

1. It recorded the adoption in 1975 of the general directive on waste [*3*], which required member states to designate competent authorities which would draw up plans for the disposal of waste without endangering human health or the environment.

2. It noted also the adoption of the Council directive on the discharge of dangerous substances into the aquatic environment of the EEC [*4*].

3. It recorded the submission to the Council in 1976 of the proposal for the directive now known as the "sixth amendment," under which provision is made for a system of premarket testing and notification of new chemical substances.

The intention of studying environmental impact assessment was also referred to. Clearly, preventive action was already well under way.

The second program made the following statement on waste management:

> While many industrialists and private persons are indeed already aware of the need to combat waste and conserve resources, the structure and unwieldiness of investment, product design and consumer habits often combine to hinder a greater awareness of this need and of the action to be taken as a result.

The aim of waste management policy must be, with due allowance for economic and technical constraints, to avoid the generation of waste and to ensure that materials are used as efficiently as possible at every stage of manufacture and use. This will mean studying the possibility of using alternative materials instead of materials which are very rare or could eventually become so, and encouraging industry to design and manufacture products with an eye both to avoiding wastage and to facilitating the recovery of raw materials (planning of specifications, standardization, lengthening the useful life of products, and so on).

The Commission will study these possibilities with the help of the Committee on Waste Management [5]. Action on improved waste management is thus also under way.

Third Action Program

Let us move on and turn now to the Third Community Action Program on the Environment [*6*]. "Preventive" has become the key word in this program, which is less detailed than its two predecessors and more concerned with the overall approach to be adopted in relation to environmental policy than with listing specific future actions.

It is our conviction in the Commission that arguments in favor of a preventive environmental policy are receiving an increasingly favorable reception from the general public. Public concern on environmental issues remains high. And it seems to be more and more agreed, in all EEC countries—and even by many industrial enterprises—that "cutting corners" where the environment is concerned makes no sense. Prevention is now seen to be cheaper than cure.

Another concept that is increasingly important (and will be of great importance in the 1980s) is that of the interrelationship between economic and environmental factors. One could perhaps characterize the 1960s as a "defensive" period, during which environmental pollution was generally accepted as the price of development. This period was followed by a period of "compromise" in the early 1970s, in which an attempt was made to reconcile economic and environmental requirements—often unsuccessfully and to the detriment of the environment. We have now reached the stage at which "integration" is seen as essential—for both environmental and economic reasons.

There has thus been a real change of perception in environmental thinking. In the Commission's proposals for environmental action for 1982 through 1986, environmental policy has been declared to be a structural policy—part and parcel of the total socioeconomic framework of the EEC and its member states—and a policy to be pursued steadily, irrespective of the short-term ups and downs of the economic situation. Moreover, the Commission has recognized that the environment is the framework and base on which all other human activities depend.

Against this background, the commitment to a preventive approach requires the integration of environmental considerations into planning and the implementation of actions in virtually all other important fields—agriculture, industry, transport, energy, regional policy, tourism, and overseas development, to name only those areas that are specifically mentioned in the program proposals. Considerable emphasis is also placed on the wise management of resources, such as land (especially ecologically sensitive zones), flora and fauna, and water resources. Action to secure better management of waste—reducing wastes, increasing reuse and recycling, and the safe disposal of unavoidable remaining wastes—is seen to be necessary, and to this end, the EEC supports the development of clean or low-waste technologies.

The proposals also stress that, in the socioeconomic context of the 1980s, environmental action, on its part, must contribute to finding solutions to the major problems of the day: unemployment, inflation, the energy supply, and regional disparities. To this end the EEC's environmental policy must *inter alia* seek the following goals:

(*a*) to help in creating new jobs by promoting industries with products, equipment, and processes that are less polluting or which use fewer nonrenewable resources;
(*b*) to reduce pollution, nuisance, or actions creating waste or unacceptable costs for the EEC;
(*c*) to economize in the use of nonrenewable raw materials and to encourage waste recycling, as well as less polluting alternatives; and
(*d*) to counter the negative effects of increased reliance for energy on coal or nuclear power by promoting energy saving and the use of less polluting energy resources.

These are all fairly broad and general concepts. They define an overall approach which will condition thinking in the future. It is being increasingly recognized that to think in terms of economic development or environmental protection is to think in terms of a quite false antithesis. If it is true—and there are many specific examples which demonstrate it—that in the environmental field prevention is cheaper than cure, then environmental protection becomes simply an aspect of good management. All sound economic developments will ensure that environmental needs are built into the decision-making process from the outset so as to avoid environmental problems arising rather than having to tackle them—at greater cost—after the event.

EEC Legislation

It is impossible, in a paper such as this, to cover all aspects of EEC legislation in the environmental field. We will concentrate, therefore, only on those aspects which would seem to be of greatest interest to industry.

Any industrial activity entails risk for man and the environment. Some of these risks may be generated at the processing stage, and in that case, they can be divided into two categories:

(*a*) routine risks in normal operating conditions (such as a certain level of emission of a dangerous substance into water), and
(*b*) exceptional risks in abnormal operating conditions (such as risks of accidents, explosions, fires, or massive emissions of dangerous substances when an activity gets out of control).

But risks may also originate after the marketing of products because of their inherent effects, immediate or delayed, on man and the environment.

This outline of risks from industry is presented to explain the reasons for three major EEC environmental actions:

(*a*) control of the discharge of dangerous substances into the EEC aquatic environment, (*b*) prevention of major accidents which might result from certain industrial activities and the limitation of their consequences for man and the environment, and

(*c*) premarketing notification of new chemicals.

Water Protection

The Council adopted a most important framework directive in May 1976 obliging the member states to limit discharges of dangerous substances into the aquatic environment of the EEC by enforcing emission standards which are themselves fixed in prior authorizations [*4*]. The bulk of the work required to implement this directive relates to the protection of surface water. This framework directive is now being applied.

As this directive explicitly calls for specific EEC rules for the protection of groundwater, a directive to that effect has been adopted by the Council [*7*]. The directive of May 1976 also obliges member states to enforce discharge standards or quality objectives set at the EEC level for a whole range of substances listed in a so-called "black list" annexed to the directive.

First of all, the black list of dangerous substances, for which limit values and quality objectives must be prepared, is not defined. The Council has restricted itself to the selection of families and groups from which individual substances are to be drawn. The process establishing the limit values and quality objectives is already well under way. The Commission has made proposals, currently under discussion in the Council, on mercury and three organohalogens—the "drins" group. Texts on cadmium and on several organic substances (DDT, BHC) will be ready during the course of 1983 [*8*], while a new set of substances is to be defined. A directive on mercury has been adopted [*9*] and will soon be completed for sectors other than the chlor-electrolysis industry.

It may be worthwhile commenting on those directives already adopted by the Council. For mercury as for the drins, both of which are still under discussion in the Council, it has proved possible to define scientifically founded, quality objectives compatible with reliable monitoring procedures. The EEC has therefore been able to adopt two directives, in each case, one on limit values and the other on quality objectives.

All discharges of mercury and drins will be subject to a system of prior authorization. The directives on limit values also lay down, beside those limits, the time limits for their enforcement as well as the sampling and analysis requirements for the discharges. The proposal on quality objectives defines, where appropriate, the maximum permissible levels in fish flesh and in fresh water and seawater and provides deadlines for compliance with these objectives, as well as procedures for checking compliance throughout the area likely to be affected by the discharges.

The present directive on mercury is restricted to the chloralkali electrolysis industry; this is because this sector is the main discharger of mercury inside the

EEC and therefore requires speedy regulatory action. Other mercury discharges will be covered by proposals which the Commission has started to prepare.

Seveso Directive

The EEC has also adopted a directive on major accident hazards of certain industrial activities, the so-called "Seveso directive" [*10*]. The object of this directive is to prevent major accidents which could be caused by certain industrial activities and to limit their effects on human beings (workers and the adjacent population) and the environment. This directive will also avoid the distortion of competition created by large discrepancies between national legislation.

Because it is unrealistic to assume no risk, the first objective of the directive is to reduce, as early as the design stage, and throughout the operation of plant facilities, the probability of major accidents, by studying possible causes, monitoring critical points, anticipating the combinations of events which might lead to an accident, and introducing correspondingly more stringent safety measures to decrease the consequences of accidents for man and the environment.

The next objective is to prevent such an accident, should one occur, from turning into a disaster, and hence to limit the consequences as much as possible. Surveillance and safety machinery must be set up and an emergency plan prepared.

The mechanism itself of the directive can be divided into two parts. The first part provides for obligations of a general nature regarding activities which involve or may involve dangerous substances in general. It provides for taking measures needed to prevent accidents and also to limit the consequences of accidents which do occur. It requires that a safety report be made available to the competent authorities and that workers and the public in the surrounding area be informed.

The second part of the directive's mechanism applies more particularly to those industrial activities which involve or may involve particularly dangerous substances, clearly defined, which are present or potentially present in the industrial activity in excess of a stipulated quantity. In this case, the procedure chosen to supervise the safety of industrial activities is the notification procedure, under which the manufacturer sends the competent authority a detailed safety report concerning the substances and the characteristics of the plant and the emergency plant. This report will also contain an analysis of the reliability of the facilities and will seek to show the hazards they could entail for man and the environment as well as the safety measures needed.

The scope of the notification procedure is restricted to ensure that the competent authorities are not overwhelmed by a huge volume of notifications, but at the same time, it has been tailored to cover major accidents which have occurred in recent years.

This directive applies to new industrial activities, to modifications with safety implications of existing plants, and, within an appropriate time limit, to those existing plants which are identified solely by the list of chemicals indicated in the directive.

The directive requires that the competent authorities be notified of major accidents and that the Commission establish an exchange of information on accident hazards and major accidents which have occurred, in order to improve safety.

Of course, risks cannot be eliminated, but they can be foreseen and reduced. In the opinion of the Commission, this directive constitutes an important step forward in this direction.

Notification of New Chemicals

The third environmental action by the EEC concerns the premarketing notification of new chemicals.

When the Council adopted the first environmental action program in November 1973, it specifically charged the Commission "to investigate the measures required to harmonize and strengthen control by public authorities over chemicals before they are marketed." From the start, the objective was a dual one to protect man and the environment from the potential hazards of new chemicals, and at the same time to avoid and eliminate technical barriers to trade.

In September 1976 the Commission sent a proposal to the Council for a sixth modification to the 1967 directive on dangerous substances. This proposal, which we refer to as the sixth amendment, is in essence the EEC's equivalent of the Section 5 premanufacturing notification requirements of the U.S. Toxic Substances Control Act. The Council of Ministers of the Environment adopted this directive in 1979 [*11*]. The directive specifies that anyone manufacturing or importing into the EEC a new chemical substance has to notify the competent authority no later than 45 days prior to marketing.

The primary duty of the notifier (that is, the manufacturer or importer) is to satisfy himself of the characteristics of the substances he is putting on the market. There cannot be any transfer of responsibility. The role of the competent authority, as it is seen by the directive, is not to license a substance but to supervise its introduction and continued use. The manufacturer or importer, who will have notified the competent authority in accordance with the requirements of the directive, will be entitled—by the sole virtue of the notification—to put this substance on the EEC market.

The directive sets out detailed rules on the contents of the notification, what substances must be notified, and the texts which must be made prior to notification.

Waste Management

This report will now go through some of the most important areas in the field of waste management and show in more concrete terms what action the EEC has taken, starting with what we call "clean technologies," that is, the preventive side of environmental policy.

Clean Technologies

In this field the Commission has already taken practical steps to stimulate the development and growth of clean technologies, by which is meant—according to the Council's 1979 definition—any new production process or modification to an existing process which aims to do any of the following: (*a*) reduce pollution at the source, (*b*) reduce the amount of waste produced at source, or (*c*) achieve more rational use of natural resources.

In 1979 the Commission was asked to take the following preparatory steps toward the formulation of a structured policy on clean technologies:

(*a*) to analyze and compare the member states' policies, to encourage the development of clean technologies, and to formulate recommendations for the most effective ways for the authorities to support these technologies;

(*b*) to survey the sectors most concerned on the basis of work undertaken by the United Nations Economic Commission for Europe and the best technologies developed in the member states;

(*c*) to encourage the dissemination of information on these technologies among the member states;

(*d*) to make proposals as part of the EEC's environmental research program to put more effort behind clean technologies.

The Commission first carried out a number of fact-finding studies in order to discover the prime technologies being used by industries for protection of the environment and management of resources. It also investigated the problems these technologies can cause, the possibilities of technological improvements, and what was being done to effect them.

Particular attention was paid to the following industries: agriculture/food, surface treatment of metals, cement, fertilizers, textiles, paper, and certain branches of the chemical industry.

A number of pilot projects were selected for financial support from the 1982 budget,[2] and more will follow. The criteria applied in these selections have been the following:

(*a*) their innovative character,

(*b*) their demonstrated impact and interest to the EEC,

(*c*) their relationship with other EEC policies,

(*d*) their economic effectiveness (in comparison with conventional technology and pollution-control equipment), and

(*e*) the situation of the industrial sector concerned.

[2] 1.5 million European Units of Account (UA) for 1982 and 1.5 million UA for 1983.

The Commission also organized a European Seminar on Clean Technologies at The Hague from 5 to 7 Nov. 1980. Much information was exchanged, and there were discussions on how to promote research into clean technologies.

The Commission's aim in doing all this is to determine the role of clean technologies in EEC environmental policy and how the institutions of the EEC can be of help.

For the future, the Commission hopes that the financial provision for 1984 and later years will enable the organization to give support to more of the promising clean technologies proposed.

Toxic and Dangerous Wastes

One area in which we hope these new technologies will have an important impact is the area of toxic and dangerous wastes. This problem is a major one, which has occupied us, virtually since the conception of the EEC first environmental action program. This particular kind of waste enjoys a high priority rating in our environmental and waste management policies, not because of the quantities involved (the EEC produces about 20 million metric tonnes per year of toxic and dangerous wastes), but rather because of the special risks these wastes pose in their handling and safe disposal, even in small quantities. The particular properties of the substances involved are well known: high toxicity, health hazards, danger to water, infection, explosion, and so on.

In recent years, the amounts and complexity of toxic and dangerous wastes have increased considerably. This is a direct result of, among other things, the growing awareness of environmental problems; of measures banning the uncontrolled disposal, dumping, or storage of such waste; and of advances made in the physical and chemical understanding of these wastes and in the determination of their specific properties and dangers to the environment.

This has led to the "determination" and "discovery" of a growing number of new kinds of dangerous waste.

Only gradually and in the long term can we expect substantial reduction in the amounts of toxic and dangerous waste produced, as industry adjusts to the legal, administrative, and technical provisions of environmental directives. Another factor is that considerable costs are involved in waste treatment and disposal. This has led, in its turn, to major research and development work.

Industry has already started to take up this challenge in the following ways:

(*a*) by replacing substances, methods, and processes in order to reduce the amounts of dangerous waste produced on potential production sites;

(*b*) by introducing or using processes which produce little waste (for instance, the dangers arising from hardening salts have been largely removed);

(*c*) by increasing the recycling of dangerous wastes to generate energy or as secondary raw materials (for example, with waste solvents and waste galvanization).

Where dangerous waste is concerned, efforts must also be made to move on from disposal to processing. Three treatment and disposal methods are available: chemical and physical treatment, biological treatment, and landfilling. Moreover, it is extremely important that there should be efficient collection and transportation systems, legislation on control and monitoring, and an improvement in methods of chemical analysis.

In its efforts to minimize the risks and dangers from these wastes the EEC passed, in 1978, a directive on the disposal of toxic and dangerous wastes [*12*]. Therein, the EEC has defined, albeit by means of a list of 27 main toxic and dangerous substances or materials, what is meant legally by toxic and dangerous waste in the ten member states.

We are only too aware that the definition given in the directive is not unassailable scientifically. Had we nursed the ambition of achieving a scientifically unassailable definition, there would still be no EEC directive on toxic and dangerous waste, for we should then have had to define the problem of toxicity in its entirety.

The EEC therefore endeavored to arrive at a pragmatic solution by drawing up an agreed upon list of substances to resolve the problems facing all the member states.

Although for the moment the directive does not specify any concentrations, the European Commission is currently studying this question in detail. The need for progress on this question of concentrations becomes more acute as some member states have already adopted differing national provisions.

Briefly, the main provisions of the directive cover the following:

(*a*) prohibition of uncontrolled dumping and tipping;
(*b*) labeling to give the nature, composition, and quantity of the waste;
(*c*) notification of storage sites, to include the types of waste;
(*d*) licencing of storage, treatment, and disposal establishments;
(*e*) control of the transport of these wastes; and
(*f*) planning, by the competent authorities of the member states, of a disposal program.

Although the directive goes into some detail, it is basically designed to be a framework of obligations to be fulfilled by the member states, while at the same time allowing individual member states the latitude to legislate on specific problems as and when they arise.

Finally, aspects relating to disposal sites are among the most urgent problems. Even after the introduction of strict rules and regulations and compulsory monitoring machinery to ensure appropriate waste disposal, all the industrialized countries are finding it difficult to provide a sufficient number of suitable plants and tipping sites. In many countries there is a considerable gap between the legal and administrative requirements and the technical, economic, and practical means at present available to meet them.

Consequently, the priority tasks in the years ahead are (*a*) to set up new centers for treatment and tipping and to replace technically obsolete plants, and (*b*) to have a better regional balance, reliable means of transport, and transfrontier cooperation within the EEC in order to enforce the existing regulations.

Transfrontier Shipment of Hazardous Wastes—As a follow-up to the directive on toxic wastes, the Commission recently presented to the Council a proposal for a directive to supervise and control the transfrontier shipment of hazardous wastes [*13*] between member states of the EEC.

Apart from the toxic and dangerous waste directive just mentioned, the Council has already adopted directives on the disposal of waste oils in 1975 [*3*] and on the disposal of polychlorinated biphenyls (PCBs) in 1976 [*14*].

All these directives lay down common principles and a common system of controls and safety regulations designed to provide man, animals, and the environment with effective protection against dangers which could arise when hazardous wastes are collected, transported, treated, stored, and disposed of.

For several years now, transfrontier shipments of hazardous wastes have been increasing steadily, primarly for the following reasons:

(*a*) because of the lack of suitable firms, plants, and dumps of sufficient capacity in the country in which the wastes originated;

(*b*) because plants or dumps in the country of destination are closer at hand than those in the country in which the wastes originated;

(*c*) because disposal costs, including transport costs, are lower in the country of destination, since more advanced technologies are available or less stringent environmental legislation applies there; and

(*d*) to evade more stringent national controls and requirements.

At present, supervision of these wastes ends at the national frontier. Once the wastes cross these frontiers, they leave the jurisdiction of the competent national authorities, without any guarantee of adequate supervision in the country of destination.

This proposal for a directive on the transfrontier shipment of hazardous wastes is to supplement the three directives referred to previously by ensuring uninterrupted supervision and monitoring of hazardous wastes from the source right through to their final nonpolluting disposal, even if this is to take place on the other side of the national frontier, in the interests of protecting human health and the environment.

Accordingly, under this proposal for a directive it would be mandatory for the consignor to notify the competent authorities in the country of dispatch, in the country of destination, and in the countries of transit, if any, whenever hazardous wastes are to be shipped to another EEC country.

Waste oils are also included in this directive, since recent research into the carcinogenic effects of used engine lubricants and coolants has confirmed that they are toxic and dangerous.

Recycling

Under the generic heading of "recycling" we will deal with two specific kinds of waste, both of which, in their own way, highlight the main problems facing any environmental policy: pollution, especially of groundwaters, and economics. Also, from the point of view of management of resources, both of these wastes can be turned from a nuisance into a profit.

Waste Oils—The first area is waste oils, covered by an EEC directive adopted by the Council in 1975 [*15*]. Since these oils are responsible for a large part of the EEC's water pollution problem, it was necessary to introduce a vigorous system of controls to cover the collection, storage, and elimination or regeneration, or both, of these oils.

The total consumption of lubricants in the EEC is in the region of 4 million metric tonnes per year, giving rise to approximately 2 million metric tonnes per year of waste oil. Of these 2 million tonnes, we are reasonably sure that 1 million are recycled, being collected for regeneration, leaving over 1 million tonnes of potential water pollution unaccounted for.

So, although from a quantitative point of view waste oils represent a small part of the EEC's waste flow, it was deemed necessary to take comprehensive action to protect our environment. The importance of waste oils lies in (*a*) the scale of pollution they can cause and the high cost of remedial action, (*b*) their high economic value coupled with the need to import raw materials to manufacture virgin lubricants, and (*c*) the savings which can result from the recovery of waste oils. Because of the value—in energy or raw materials savings—of these wastes, the EEC directive stages, "as far as possible, the disposal of waste oils is carried out by recycling (regeneration and/or combustion other than for destruction)."

Regeneration and combustion with heat recovery are the two main ways in which waste oils can be reused (and the member states have generally opted for one or the other of these solutions). In the EEC as a whole, regeneration accounts for 80% and combustion for 20% of reuse.

As a result of the increases in the price of oil since 1974, the combustion of waste oils has increased considerably in comparison with regeneration. Leaving aside the economic aspect of this problem, combustion of waste oils without preliminary treatment to remove considerable quantities of heavy metals leads to increased atmospheric pollution.

The Commission is at present examining the possibility of supplementing the 1975 directive so as to allow these oils to be burnt only in suitable plants or if they satisfy quality criteria akin to those for fuel oil.

Wastepaper—The second area of recycling is wastepaper, covered by an

EEC recommendation adopted by the Council in 1981 [*16*]. At present, in the ten member states of the EEC, wastepaper represents, very roughly, about 45% by volume and 20% by weight of municipal waste. These figures represent large problems of accumulation of waste, disposal costs, and so on. In addition, and most important, imports of paper and board, paper fibers, and wood-derived products amount to a staggering 15 million metric tonnes per year, worth $6.675 million in American dollars—a value second only to that for petroleum products.

Therefore, in its recommendation—a legislative device which is advisory rather than mandatory—the EEC is seeking to encourage the member states to define wastepaper management policies, to encourage the use of recycled paper and board, and, in particular, to do the following:

(*a*) encourage the use of recycled and recyclable paper and board, especially in national administrations public bodies, and those national official services which can set an example;

(*b*) encourage, where feasible, the use of recycled paper and board containing a high percentage of mixed wastepaper;

(*c*) reexamine in the light of recent technological advances the existing specifications for paper products which restrict, for reasons other than adequacy of a product for its purpose, the reuse of wastepaper and the use of recycled paper and board;

(*d*) implement programs of consumer and manufacturer education to promote paper and board products made from recycled paper and board;

(*e*) develop and promote uses for wastepaper other than as a raw material for the manufacture of paper and board;

(*f*) encourage the use of products (inks, glues, and so on) which do not preclude the subsequent recycling of paper and board.

Although this recommendation is advisory, the European Commission has reserved the right to introduce mandatory measures in case the aims of the recommendation remain unfulfilled.

We will now turn to three areas in which the Commission has initiated action which has not yet resulted in Council directives.

Beverage Containers—The first area is that of beverage containers—which now constitute about 10% by weight of all municipal waste and which have steadily increasing disposal costs, borne by the local authorities.

Containers for liquids for human consumption give rise to three types of collective costs: the disposal of waste cost, air and water pollution costs, and litter costs. These costs vary depending on the type of container. The costs arising from the manufacture, use, and disposal of containers for liquids for human consumption are borne directly by the consumer or the manufacturer to only a very small extent. They are passed on to some extent to the community as a whole. An estimate of external costs shows that these are not insignificant.

A proposal for a directive on this subject was presented to the Council on 23

April 1981 [*17*]. Its aim was to reduce the environmental impact of used containers and the consumption of energy and raw materials by reducing the amount of, and making better use of, containers for liquids for human consumption contained in household waste.

It is quite clear that an operation of this kind must be carried out in harmonized fashion throughout the EEC, since uncoordinated individual measures by member states would probably have an adverse effect on the free movement of goods.

The proposal calls on the member states to set objectives, in relation to their own particular situation, for reducing the amount of such waste being produced, to attain these objectives progressively, and also to inform the Commission of the objectives adopted, the means used, and the results obtained.

The choice of means to be used to attain the goal set is left to the member states. They may opt for voluntary agreements or for legislation. They may encourage the recycling of used containers or refilling for further use, or both. It is therefore more a question of defining a framework within which the member states are to act than of laying down specific and detailed provisions.

On the basis of data and statistics supplied by the member states, the Commission, with the assistance of the Committee on Waste Management, will assess the progress made and, where appropriate, consider any additional measures.

As new containers may raise certain problems in relation to refilling or recycling, the member states are called upon to take steps to ensure that new containers brought on to the market comply with the objectives of the directive.

A certain amount of standardization of containers for liquids for human consumption may also help to attain the objectives of the directive while facilitating the distribution of the products in question.

Energy from Waste—The second area is energy from waste. The Commission has organized a committee of national experts from the member states which is currently studying this problem. Their report indicates possibilities for deriving energy from waste, taking into account technical ability, economic feasibility, and environmental considerations. This is an interesting approach to waste management and utilization as it is applicable to all types of waste.

In 1977, some 22.2 million metric tonnes of municipal waste were incinerated in the EEC, representing approximately 23% of the total on a weight basis. Over 12.5 million tonnes were incinerated with energy recovery: this amount represented 56% of the total.

The incineration of municipal waste makes only a small contribution to energy saving in the EEC. In 1977, the energy value of municipal wastes incinerated with energy recovery represented approximately 22.6 million Gcal, corresponding to a saving of 3.3 million tonnes of oil, or 0.42% of the total consumption of liquid hydrocarbons. However, the situation differs considerably from one member state to another.

Some conclusions can already be drawn from the work carried out concerning the recovery of energy from municipal waste and the resulting environmental problems:

Given that the waste has a significant calorific value (and that in 1977 only 13% of the waste arising in the EEC was used to derive energy), steps must be taken to encourage the development of many more projects using energy derived from waste or involving the production and use of waste-derived fuels. At present only conventional incineration is in use.

Generally speaking, incineration—even with energy recovery—is not economically feasible. Energy recovery, however, reduces incineration costs and makes it possible to save fuel, and costs fall as the volume of waste to be treated increases. However, in view of the increases in the prices of energy raw materials, several incineration plants in the EEC are now just about breaking even and are able to compete economically with tipping as a method of disposing of municipal waste.

Sewage Sludge—Third, sewage sludge represents one of the major problems of waste management both quantitatively and qualitatively.

Sewage treatment produces enormous quantities of sludge. For the EEC as a whole, about 6 million metric tonnes of dry matter (that is, over 230 million m^3 of fresh sludge) are estimated to be produced every year. Each inhabitant therefore produces approximately 800 kg of sludge a year. In addition to this, there is the biodegradable sludge produced by industry (breweries and the food industry).

The outlook for the future suggests that there will be an appreciable increase in the amount of sewage sludge produced within the EEC.

Sludge is disposed of in the following ways (average figures):

(*a*) 45% is landfilled in controlled tips;
(*b*) 10% is dumped at sea;
(*c*) 7% is incinerated, and
(*d*) 29% is used in agriculture.

Clearly the first three methods of sludge disposal are not satisfactory, as dumping at sea or on land causes odor pollution, eutrophication, and bacteriological and chemical water pollution, and even incineration can still result in water and air pollution, in addition to being expensive. The 29% of sludge used in agriculture is, therefore, a relatively small percentage considering that sludge is of real value to the soil as a source of organic matter, nitrogen, and phosphorus.

The Commission has recently adopted a proposal for a directive [*18*] to encourage the use of sewage sludge in agriculture, stressing the agricultural value of sludge but at the same time specifying precautions to ensure proper use.

The spreading of sewage sludge has developed to different degrees in the member states, and the provisions governing it are not always satisfactory for the management of wastes and resources or for the protection of the environment and agriculture. Moreover, differences in the stringency of rules and regulations from one member state to another may lead to distortion of competition for the building of sewage treatment plants. Disposal of sludge from sewage treatment plants presents a complex problem for the managers of these plants and the local authorities.

The directive applies to sludge from treatment plants for domestic wastewater and plants which also handle industrial wastewater.

This directive sets out to do the following, in particular:

(*a*) to ensure that a large proportion of sewage sludge can be spread on agricultural land in proper fashion;

(*b*) to this end, to set a number of minimum requirements to be satisfied (limit values for heavy metal concentrations in sludge, amounts of heavy metals which may be added annually to farmland, and trace element concentrations admissible in the land to which the sludge is applied);

(*c*) to prevent microbiological contamination of crops: to stabilize sludge which is not ploughed into the soil, to prevent sludge from being spread on crops eaten raw by humans, and to require six weeks to elapse between the spreading of sludge on grassland and putting stock out to pasture;

(*d*) to restrict the spreading of sewage sludge on soil with a pH of less than 6, in view of the mobility of heavy metals in acid metals;

(*e*) to ensure that the spreading is carried out in such a way as to avoid runoff of nutrient elements into surface waters and groundwater;

(*f*) to set, alongside the minimum and mandatory values, recommended values to ensure that the proposal is flexible enough to take into account local situations;

(*g*) last, to limit the scope of the proposal to sludge from treatment plants above a certain size (serving a population of 5000 or more), treating urban or similar effluents or mixed effluents, but not industrial effluents.

Conclusion

We have to acknowledge, indeed, that virtually all the materials that are brought into our industrial processes and thus into the service of man sooner or later become "waste," either as spoil from the winning of raw materials, or as discharges or "leftovers" from our manufacturing processes, or as the very products of those processes themselves, when they have reached the end of their useful lives.

It may be of interest to try to face up to this question and to ask what it is that prevents higher rates of recovery from being achieved.

1. Are there technical difficulties in separating out these metals from the rest of the waste stream? Could changes in the process or product design help?
2. Are the economics of recovery against us? If so what could be done to change the picture?

3. Are there uncertainties of demand for these secondary materials (whether as a result of fluctuations in the price of primary materials or for other reasons) that mitigate against recovery? If so, can anything be done to improve the stability of markets and the predictability of demand?

4. Could governments and other public authorities, who are large purchasers, assist through their specifications, for example, in supply or public works contracts?

Perhaps some ideas on these issues may emerge from this paper. I am convinced that they are all issues that will need to be tackled in the near future and that answers to them could provide guidance to the environmental management industry as to the lines to pursue to maximize both economic opportunities and environmental benefits during the 1980s.

References

[*1*] *Official Journal of the European Communities*, No. C 112, 20 December 1973, p. 1.
[*2*] *Official Journal of the European Communities*, No. C 139, 13 June 1977, p. 1.
[*3*] *Official Journal of the European Communities*, No. L 194, 25 July 1975, p. 39.
[*4*] *Official Journal of the European Communities*, No. L 129, 18 May 1976, p. 29.
[*5*] *Official Journal of the European Communities*, No. L 115, 1 May 1976, p. 73.
[*6*] *Official Journal of the European Communities*, No. C 305, 25 November 1981, p. 1.
[*7*] *Official Journal of the European Communities*, No. L 20, 26 January 1980.
[*8*] *Official Journal of the European Communities*, No. C 146, 12 June 1979, p. 5.
[*9*] *Official Journal of the European Communities*, No. L 81, 27 March 1982, p. 29.
[*10*] *Official Journal of the European Communities*, No. L 230, 5 August 1982, p. 1.
[*11*] *Official Journal of the European Communities*, No. L 259, 18 September 1979, p. 10.
[*12*] *Official Journal of the European Communities*, No. L 84, 20 March 1978, p. 43.
[*13*] *Official Journal of the European Communities*, No. 186, 12 July 1983, p. 41.
[*14*] *Official Journal of the European Communities*, No. L 108, 26 April 1976, p. 41.
[*15*] *Official Journal of the European Communities*, No. L 194, 25 July 1975, p. 23.
[*16*] *Official Journal of the European Communities*, No. L 355, 10 December 1981, p. 56.
[*17*] *Official Journal of the European Communities*, No. C 204, 13 August 1981, p. 6.
[*18*] *Official Journal of the European Communities*, No. C 264, 8 October 1982, p. 3.

Summary

Summary

This book is the third ASTM Special Technical Publication (STP), sponsored by ASTM Committee D-34 on Waste Disposal, on the subject of problem solving in waste disposal. The first volume[1] dealt primarily with laboratory waste analysis and land disposal considerations, such as siting and liners. The second volume[2] expanded on these themes and added sections on sampling and biological assessment. This third volume advances our knowledge of waste and disposal site analysis and the fate of wastes after disposal. Two new topics are introduced: waste management for resource recovery and international perspectives on waste management.

The papers included herein have undergone peer review and revision since their presentation. They provide state-of-the-art information on waste-related analysis and innovative options for waste management.

Analysis of Wastes and Waste Disposal Sites

In the last few years we have seen great activity in developing analytical methods to identify and quantify individual constituents in the unusual and complex waste matrix. Our confidence has grown to the point that we believe we can reasonably say what is in a waste. Although many refinements are still necessary, we can see a shift in effort toward understanding how the waste behaves in the disposal environment. Under the topic of waste and disposal site analysis only two papers deal with waste analysis. The remaining seven are directed toward understanding the effects of wastes on the environment.

Janisz and Butterfield examine correlations between waste site leachates and resident aquatic animals. Their experiences show that such studies are valuable when water contamination is well documented and the contaminants have high bioconcentration factors. The results of such studies are most effective when bioconcentration is shown or species diversity declines.

The study by *Young and Parker* on waste site odors offers a systematic analytical approach toward understanding this common problem. The key ele-

[1] *Hazardous Solid Waste Testing: First Conference, ASTM STP 760*, R. A. Conway and B. C. Malloy, Eds., American Society for Testing and Materials, Philadelphia, 1981.

[2] *Hazardous and Industrial Solid Waste Testing: Second Symposium, ASTM STP 805*, R. A. Conway and W. P. Gulledge, Eds., American Society for Testing and Materials, Philadelphia, 1981.

ments are acquiring representative, quantitive samples of the many diverse odorous chemicals. Three techniques are required for obtaining samples of the three types of odorous compounds. A larger number of techniques is required to identify and quantify the compounds of interest adequately. Finally, it appears that the age and decomposition rate of the waste are the most significant factors in odor generation.

The paper by *Belkin and Bishop* describes an approach for analysis of unlabeled containerized wastes found inside the waste generator property. The paper emphasizes the value of determining possible background information about the container. Once the background is established, general observations can lead directly to the most appropriate analyses.

Côté and Isabel analyzed simulated solidified wastes according to an established fixed repeating leachate generation schedule developed by the nuclear industry. The kinetics are appropriate for radioactive wastes but not for other solidified hazardous wastes. Typically, hazardous wastes have a wide variety of constituents, each of which leaches at a different rate. By using diffusion equations and variable leaching times to improve constituent detection, diffusion constants can be established. These derived constants can be used to predict how rapidly solidified wastes leach individual constituents.

Kilau and Shah researched the problem of acid leaching of chromium-bearing slags. They found that a slag calcium oxide/silicon dioxide concentration ratio of less than 2 minimizes leaching of chromium.

Cation mobility in soil has been studied by *Wong et al.* These authors percolated a complex cation mixture through a soil and measured the amount of each cation in the leachate. They developed a simple relationship using the linear distribution coefficient to predict relative migration rates and the distance of travel of each cation at any given time.

Oil contamination of soil is a common phenomenon that often requires environmental action. *Petros et al* evaluate five methods that could be used for on-site analysis of oil in soil. Useful methods should be reasonably accurate, simple enough for field work, and rapid enough to be used during the actual cleanup. Three methods, each useful for different situations, were found suitable for in-house studies.

Fly ash is generated in huge quantities in many nations. Each individual ash contains leachable metals in varying quantities. *H. W. Young et al* tested three chemical treatments to reduce fly ash leaching potential. Overall, portland cement treatment produced the best quality of leachate. Samples treated with lime and phosphoric acid showed mixed results. Ash from refuse-derived fuel had a more concentrated leachate. Leachates from wastes of different particle sizes showed surprisingly little variation in composition. This work provides a variety of ways to alter leachate quality and can be the basis for new studies.

Baker et al studied the relationship between cadmium loading in soil and passive plant uptake. These authors developed a logarithmic equation to relate leachable cadmium in soil to the cadmium content of plants. This work provides a basis for further study of metal uptake by plants, which should, in turn, provide the means to prevent overloading of agricultural soil with waste.

Amelioration of Wastes in the Disposal Environment

Wastes are often regarded as unmitigated evils, which, by definition, can do no good. The first paper in this section shows that a properly managed waste can become an asset. The next three papers show that well-considered codisposal can actually reduce the deleterious effects of either or both wastes. Finally, an improvement in the common formulation of liquid waste fixation can result in greatly reduced leaching potential.

Disposal of oily sludge on farmland is shown by *Mucsy et al* to be of actual benefit to sandy soil. The oil provides nutrients and improves water retention. If fertilizer is also applied, larger and healthier crops result. However, it is still prudent, on such soil, to raise crops not destined for direct human consumption.

Barber et al looked at disposal of oil emulsion waste with municipal waste. Their experiments show that the oil is retained by the waste but the water flows through. This work provides a set of conditions for environmentally sound codisposal.

Soda ash wastes can provide a suitable base for later disposal of steel manufacturing wastes. *Rinaldo-Lee et al* demonstrate that the characteristics of soda ash waste are well suited for treating and immobilizing leachable chromium from steel manufacturing waste.

Young, Baldwin, and Wilson present another example of one waste ameliorating the effects of another. Cations in industrial waste were attenuated by underlying municipal waste. Furthermore, fatty acid production and the concurrent pH drop in municipal waste was halted by the industrial waste.

Fixation of wastes with cement or fly ash is well known. *Falcone et al* studied sodium silicate addition to these formulations and found less matrix breakdown and less cation leaching.

Waste as Resource

This section naturally follows the first two, which discuss knowledge of waste and ways of putting some of the wastes characteristics to use. The five papers in this section illustrate how some wastes are really resources in disguise.

Food processing wastes characteristically have nutrient loadings so high that they can actually harm natural waterways. *M. H. Wong et al* used five such wastes as algae feed to produce food while reducing pollution.

Sanna and Camilli studied anaerobic digestion using seasonal agroindustrial wastes. They found that fixed-bed anaerobic digestion is more efficient than aerobic digestion, even in seasonal operations. However, methane production steadily increased during operation, indicating that optimization is slow.

Ferraiolo et al have developed optimization criteria for energy recovery from anaerobic digesters. Their equations use easily measured parameters to establish and monitor the process. Sewage is treated by activated sludge. Waste sludge is anaerobically digested, thus providing some energy and less waste for disposal.

Spent chromium plating solutions can become resources rather than wastes if processed using well-known ore processing methods. *Bolto et al* replaced sludge for ore in the process, used high-temperature oxidation and produced hexavalent chromium in useful concentrations.

National Perspectives in Waste Management

This section describes some waste management practices from around the world.

Rapid industrialization causes a wide variety of changes wherever it occurs. One of the changes often overlooked is the generation of many pollutants never seen before in the area. *Sundaresan et al* present in striking detail the magnitude of the problem in India. A similar, but more hopeful, theme is presented in the paper by *Hamza and Gallup* with examples from Alexandria, Egypt. Both papers amply illustrate the need to establish coordinated solid waste programs.

Schomaker describes the U.S. Environment Protection Agency (EPA) land disposal research program. This program requires development of reliable waste analyses, compatibility of the waste with the disposal method, and knowledge of the fate of the waste over time.

Gillen and Tusa examine one EPA approach for determining the fate of wastes over time. These authors categorized and prioritized cases of suspected environmental damage from land waste disposal or storage sites. Such information can be used to determine what wastes and what disposal methods are most dangerous.

Thailand's eastern seaboard has recently seen rapid industrial growth, and more is expected in the near future. *Ludwig et al* explain the Thai government's new role in systematic planning for present and anticipated pollution loads. The initial program calls for solids to be used for coastal land reclamation and liquids to be piped into the Gulf of Thailand. Monitoring of all disposal sites will determine when improved treatment is required.

El-Gohary and El-Khouly discuss waste management in the Egyptian iron and steel industry. Wastes are not merely washed away, as commonly practiced in many places. Solids are removed from waste streams, resulting in some solids and clarified water than can be reused. Wastes in Egypt are beginning to look like resources just as they are in Europe.

The European Economic Community (EEC) consists of ten member states which consider some national problems easier to solve with cooperative action. Pollution is one of the problems addressed by the EEC. As described in the paper by *Klein*, the EEC philosophy is to reduce waste volume and transform waste into a resource whenever possible while being cost-effective at the same time. As part of this approach, effort is made to discover processes and materials that produce less waste and consume fewer scarce resources.

Alan R. Rohlik

Standard Oil Co. (Ohio), Cleveland, Ohio 44128; editor.

Richard A. Conway

Union Carbide Corp., South Charleston, W.Va. 25303; editor.

Larry P. Jackson

Western Research Institute, Laramie, Wyo. 82071; editor and symposium chairman.

Index

Cumulative Index

The Roman numerals printed in boldface type refer to the particular symposium in this series of solid waste testing symposia. The first symposium (**I**) is reported in *ASTM STP 760*,[1] the second (**II**) in *STP 805*,[2] and the third (**III**) in *STP 851*.[3]

Subject Index

[1] *Hazardous Solid Waste Testing: First Conference, ASTM STP 760*, R. A. Conway and B. C. Malloy, Eds., American Society for Testing and Materials, Philadelphia, 1981.
[2] *Hazardous and Industrial Solid Waste Testing: Second Symposium, ASTM STP 805*, R. A. Conway and W. P. Gulledge, Eds., American Society for Testing and Materials, Philadelphia, 1983.
[3] *Hazardous and Industrial Waste Management and Testing: Third Symposium*, L. P. Jackson, A. R. Rohlik, and R. A. Conway, Eds., American Society for Testing and Materials, Philadelphia, 1984.

G

H

I

L

M

N

Author Index